£37.50

Aluminum alloy structures

Surveys in Structural Engineering
and Structural Mechanics

Main editor

W. F. Chen, Purdue University

Advisory Editors

J. H. Argyris, University of Stuttgart
Z. P. Bazant, Northwestern University, Evanston
T. B. Belytschko, Northwestern University, Evanston
P. G. Bergan, University of Trondheim
D. C. Drucker, University of Florida, Gainesville
T. V. Galambos, University of Minnesota
M. P. Gaus, National Science Foundation, Washington, DC
K. H. Gerstle, University of Colorado
C. Massonnet, University of Liège
G. Mehlhorn, University of Kassel
Z. Mróz, Institute of Fundamental Technological Research, Warsaw
F. Nishino, University of Tokyo
E. P. Popov, University of California, Berkeley
E. C. Ting, Purdue University
N. S. Trahair, University of Sydney
R. Wang, Peking University
J. Witteveen, Institute of Building Materials and Building Structures, Delft
J. T. P. Yao, Purdue University
O. C. Zienkiewicz, University College of Swansea

Aluminum alloy structures

Federico M. Mazzolani
Institute of Construction Technology
University of Naples

Pitman Advanced Publishing Program
Boston · London · Melbourne

PITMAN PUBLISHING INC
1020 Plain Street, Marshfield, Massachusetts 02050

PITMAN PUBLISHING LIMITED
128 Long Acre, London WC2E 9AN

Associated Companies
Pitman Publishing Pty Ltd, Melbourne
Pitman Publishing New Zealand Ltd, Wellington
Copp Clark Pitman, Toronto

© F. M. Mazzolani 1985

First published 1985

Library of Congress Cataloging in Publication Data

Mazzolani, Federico M.
  Aluminum Alloy Structures
  (Surveys in Structural Engineering
  and Structural Mechanics: 3)
Bibliography: P.
  Includes index.
  1. Aluminum construction. 2. Aluminum alloys.
I. Title. II. Series.
TA690.M38   1985   620.1′86   84-26599
ISBN 0-273-08653-7

British Library Cataloguing in Publication Data

Mazzolani, Federico M.
  Aluminum alloy structures.—(Surveys in
  structural engineering and structural mechanics; 3)
  1. Aluminium construction 2. Aluminium alloys
  I. Title   II. Series
  624.1′826   TA690

ISBN 0-273-08653-7

All rights reserved. No part of this publication may be reproduced, stored in a retrieval system, or transmitted, in any form or by any means, electronic, mechanical, photocopying, recording and/or otherwise, without the prior written permission of the publishers. This book may not be lent, resold, hired out or otherwise disposed of by way of trade in any form of binding or cover other than that in which it is published, without the prior consent of the publishers. This book is sold subject to the Standard Conditions of Sale of Net Books and may not be resold in the UK below the net price.

Printed in Northern Ireland by The Universities Press (Belfast) Ltd.

To my children
Stefano, Laura Chiara and Gianfredi

To my children,
Nathan, Lora, Hugh, and Geoffrey

Contents

Preface ix
List of symbols xi

1 Aluminum and its alloys 1
   1.1 Introduction 1
   1.2 Birth of aluminum 3
   1.3 Fabrication processes 4
   1.4 Classification of alloys 8
   1.5 Alloy families 20
   1.6 Mechanical and physical properties 25
   1.7 General criteria for design 31
       References 36

2 The structural material 38
   2.1 Types and shapes 38
   2.2 Characterization of σ–ε law 41
   2.3 Definition of industrial bar 64
   2.4 Geometrical imperfections 65
   2.5 Mechanical imperfections 68
       References 125

3 Safety principles 129
   3.1 Trends 129
   3.2 Loading conditions 135

3.3 Allowable stress method 137
3.4 Semi-probabilistic method 139
3.5 Strength of the base metal 149
References 151

4 Welded connections 153

4.1 Technology of welded connections 153
4.2 Strength of welded joints 161
References 175

5 Mechanical joints 176

5.1 Technology of joints 176
5.2 Strength of connections 184
5.3 Bolted shear connections 188
5.4 Friction joints with high-strength steel bolts 197
5.5 Riveted shear connections 202
References 204

6 Strength of structural elements 205

6.1 Serviceability limit state 205
6.2 Ultimate limit state 209
6.3 Bending behavior of the cross-section 219
6.4 Behavior of the cross-section beyond the elastic limit 261
References 281

7 Stability of structural elements 283

7.1 General principles 283
7.2 Members under compression 301
7.3 Members under bending 372
7.4 Members under compression and bending 380
7.5 Plates 395
References 424

Index 429

Preface

In 1979 I realized that it was time to organize all the material acquired through my work with the ECCS Committee on Aluminum Alloy Structures in the form of a book. The text of the book is addressed to engineers working in civil and mechanical design who probably already have an acquaintance with problems relating to steel structures.

This book has grown out of ten years of research, during which time I have subjected aspects of aluminum alloy structures to theoretical problems, experimental tests, design and specifications.

Back in 1970, when I was appointed Chairman of the ECCS (European Convention for Constructional Steelwork) Committee on Aluminum Alloy Structures, my knowledge of aluminum was limited to nonstructural applications. I had been greatly interested in theoretical and experimental research and design of metal structures for some time.

The main purpose of the Committee was to promote studies and research in the field of aluminum alloy structures, to gather all the existing literature and to develop Recommendations based on the same principles that had previously inspired the ECCS in studying steel structures. Problems were systematically studied through research programs which were carried out and successfully completed thanks to cooperation between the members of the Committee, representing several European countries, some of which provided financial support. An active and fruitful collaboration was initiated between the universities of Naples, Liège, Karlsruhe, Bilbao, Cambridge, The Testing Institute ISML in Novara, the Aluminum Pechiney in Paris, the TNO in Delft, the Aluminum Zentrale in Dusseldorf, the Alusuisse in Zurich, and other international bodies. This collaboration was a fine example of the balanced and complementary interaction between university and industry.

Continuing and expanding research initiated by the Committee enabled the first European Recommendations for Aluminum Alloy Structures (ERAAS) to be produced, which were published in 1978.

My friend and colleague Augusto Carpena, whose untimely death occurred a few months ago, was the first Chairman of the Committee. I cannot miss the opportunity to recall him to the people who knew him

Preface

and, as I did, held him in high estimation, especially for his professional qualities and his great humanity.

My work over the last ten years has been helped by Ciro Faella and Attilio De Martino, who carried out several research projects which I extensively refer to. As in every good Preface I will conclude with my acknowledgements, which I express, first of all, to them.

A very special acknowledgement of gratitude is also extended to Antonello De Luca who, after having completed his dissertation on this subject, critically cooperated in the production of this book.

I am also sincerely grateful to the following people: Guido Capria, who produced the graphics; Giuseppe Stasio, who typed the manuscript; and Ian Robertson who improved the text, thus making it readable even to a Cambridge literate.

Federico M. Mazzolani

List of symbols

a	distance; geometrical dimension; depth of the throat section of a fillet weld
b	width
b_{eff}	effective width
b_f	width of flange
b_h	half width of the heat-affected zone
b_r	half width of the reduced-strength zone
b_w	width of web
c	geometrical dimension; specific heat
d	diameter; diameter of the shank of a bolt; length of a finite element
e	eccentricity
e^*	initial eccentricity
$e_x(e_y)$	components of the eccentricity in direction $x(y)$
e_0	distance between centroid and shear centre
f	strength
$\bar{f}$	dimensionless value of stress
f_c	compression strength
$f_{c,k}$	characteristic value of compression strength
$f_{c,m}$	mean value of compression strength
f_d	design value of strength
$f_{d,b}$	design strength of a bolt
$f_{d,N}$	design value of one bolt tensile strength
$f_{d,0}$	design strength of the weld metal
$f_{d,r}$	design strength of a rivet
$f_{d,\text{red}}$	design strength of the heat-affected material
$f_{d,V}$	design value of one bolt shear strength
$f_{d,w}$	design strength of a butt weld
f_e	limit of elasticity
f_k	characteristic strength
$f_{\lim}$	limit strength

List of symbols

$f_{\lim}^{-(+)}$	limit strength in compression (tension) for a material prestressed in tension (compression) with a value $f_{\lim}^{+(-)}$
$f_{0,\lim}$	limit strength of virgin material
f_m	mean value of strength
f_p	strength at the limit of proportionality
f_t	ultimate tensile strength
f_u	ultimate strength
f_y	yield strength
$f_{y,b}$	yield strength of a bolt
f_{ε_0}	stress producing a residual strain ε_0
f_0	reference strength
$f_{0.01}$	conventional strength at 0.01 percent
$f_{0.1}$	conventional strength at 0.1 percent
$f_{0.2}$	conventional strength at 0.2 percent
$f_{0.2}^*$	reduced value of $f_{0.2}$
h	depth of the web in a beam
i	radius of gyration
i_m	average value of the radius of gyration
$i_{\min}(i_{\max})$	minimum (maximum) radius of gyration
i_{nom}	nominal value of the radius of gyration
$i_x(i_y)$	radius of gyration around $x(y)$ axis
k	numerical coefficient; reduction factor; thermal conductivity
$k_{\sigma,x}(k_{\sigma,y})$	buckling coefficient for a rectangular plate submitted to normal stress in $x(y)$ direction applied to both sides
k_τ	buckling coefficient for a rectangular plate submitted to shear stresses
l	see L
m	number of ...; numerical exponent
n	number of ...; numerical exponent
n_b	number of bolts in a joint
n_f	number of friction surfaces in a bolted joint
p	probability; spacing of bolts
p_k	probability corresponding to the characteristic value of a set of random variables
q	distributed live load (for length or surface unit)
q_{nom}	nominal value of a distributed load
s	deviation, standard deviation
t	thickness
t_f	flange thickness
$t_{\min}$	minimum thickness
t_w	web thickness
u, v, w	displacement component in the x, y, z directions

List of symbols

v	speed
v_0	initial out-of-straightness; midspan deflection in a simply supported beam
v_{max}	maximum beam deflection
v_M	deflection due to bending moment
v_r	residual deflection
v_v	deflection due to shear
$\left.\begin{array}{l}x\\y\\z\end{array}\right\}$	coordinates
y	distance of a point from the centroid in a section
A	area
A_d	area of a diagonal bar
A_{exp}	'experimental' area
A_f	area of flange
A_h	area of the reduced-strength region
A_m	average value of an area
A_n	net area
A_{nom}	nominal value of an area
A_{red}	reduced area
A_{res}	resistant area
$A_s (A_{s,l}; A_{s,t})$	area of a stiffener (longitudinal; transversal)
A_t	area of a transversal bar
A_w	area of a web; area of weld
B	parameter of the Ramberg–Osgood law; transformation matrix
C	transformation matrix
D	flexural stiffness of a plate; diagonal matrix
E	work done by external load; Young's modulus
E_s	secant modulus
E_t	instantaneous tangent modulus
$E_{t,m}$	mean value of tangent modulus
F	force, external action, load
F_d	design value of a load
F_k	characteristic value of a force
F_m	mean value of a force
F_N	force causing axial stress
F_s	shrinkage force
F_t	tendon force
F_V	force causing shear stress (in bolted joints)
$F_{V,u}$	ultimate value of a force causing shear stress
$F_{V,f}$	value of the slipping force in a bolted joint
$F_{0.2}$	force corresponding to the achievement of the stress $f_{0.2}$

List of symbols

G	shear modulus; permanent load
G_k	characteristic value of the permanent load
G_t	tangent shear modulus
I	moment of inertia of a plane area
I_m	average value of a moment of inertia
I_{nom}	nominal value of the moment of inertia of a section
I_{red}	reduced moment of inertia
$I_s(I_{s,L}; I_{s,t})$	moment of inertia of a stiffener (longitudinal stiffener, transversal stiffener)
I_t	moment of inertia of a transversal bar
I_T	twisting moment of inertia
$I_x(I_y)$	moment of inertia related to the $x(y)$ axis
I_ω	warping moment of inertia
L	length
L_c	effective length of a compression member
$L_{c,h}$	effective length in the horizontal plane
L_d	length of a diagonal bar
L_t	spacing of a chord
L_w	length of a weld
L_0	initial distance between the reference points in a specimen; distance between nodes in trusses; length of a panel
M	moment, bending moment
$M_a(M_b)$	bending moment at the edge $a(b)$
$M_{cr,D}$	Euler's critical moment for flexural–torsional buckling in elastic range
M_D	moment corresponding to the flexural–torsional buckling of a beam
M_e	elastic moment (corresponding to the achievement of the elastic limit in a point of the section)
$M_x(M_y)$	bending moment around the $x(y)$ axis
M_{eq}	equivalent bending moment
M_m	average bending moment
M_{max}	maximum bending moment
M_{pl}	plastic moment (corresponding to the plastic collapse of the section)
$M_{pl,x}(M_{pl,y})$	plastic moment for bending around the $x(y)$ axis
M_u	ultimate value of the bending moment
M_0	bending moment in the middle span
$M_{0.2}$	bending moment corresponding to the achievement of the stress $f_{0.2}$ in a point of the section
N	axial force
$\bar{N}$	dimensionless value of the axial force
N_c	maximum load-bearing capacity of a compression member

List of symbols

$N_{c,k}$	characteristic failure load in compression
$N_{c,m}$	average failure load in compression
N_{cr}	Euler's critical load
N_d	design strength of a bar
N_f	axial force in a flange
N_{ref}	reference axial force
N_s	prestressing due to the tightening of bolts
N_u	ultimate value of the axial force
$N_{0.2}$	axial force corresponding to the achievement at the stress $f_{0.2}$
Q	variable load; live load; heat quantity
Q_e	exceptional loading condition
Q_k	characteristic value of the variable load
$Q_{k,i}$	characteristic value of the variable load number i
R	generic resistance
R_b	resistance of a bolted joint
R_t	ultimate tensile resistance
$R(\ )$	resistance depending on a given stress
S	first moment of a plane area (static moment); internal action; action effect; stress resultant
S_{red}	reduced static moment
$S_x(S_y)$	static moment referred to the $x(y)$ axis
T_s	torque (per bolt)
V	shear force
$V_{d,b}$	design force for one bolt in bearing
$V_{d,t}$	design force for one bolt in tension
V_{eq}	equivalent shear force
V_f	slipping force in a bolted joint
$V_{f,0}$	friction force for one bolt in shear only
$V_{f,red}$	reduced friction force for one bolt in shear and tension
V_u	ultimate shear force
$V_{u,f}$	ultimate shear force for flanges only
$V_{u,w}$	ultimate shear force for web only
W	elastic section modulus (in bending)
W_c	elastic section modulus referred to the edge in compression
W_{red}	reduced value of the elastic modulus
W_t	elastic section modulus referred to the edge in tension
$W_x(W_y)$	elastic section modulus for bending around the $x(y)$ axis
X	force parallel to the x axis
Y	force parallel to the y axis
Z	force parallel to the z axis; plastic section modulus (in bending)

List of symbols

Greek symbols

α	angle; ratio; exponent; coefficient; deflection limit ($= v_{max}/L$); section shape factor; coefficient of thermal expansion
α_p	geometrical shape factor of a section
β	angle; ratio; exponent; coefficient; buckling factor (for evaluating the buckling length); metallurgical efficiency factor
γ	dimensionless coefficient; shear strain; local buckling coefficient; specific weight; relative flexural rigidity of a stiffener
γ^*	optimum relative flexural rigidity of a stiffener
γ^{**}	actual relative flexural rigidity of a stiffener
γ_f	reduction coefficient of the friction strength in bolted joint
γ_F	partial load factor
γ_G	partial load factor for permanent loads
γ_L	relative flexural rigidity of a longitudinal stiffener
γ_L^*	optimum relative flexural rigidity of a longitudinal stiffener
γ_m	reduction coefficient of the material strength
γ_N	reduction coefficient of the axial strength of a bolt
γ_Q	partial load factor for variable loads
γ_t	relative flexural rigidity of a transversal stiffener
γ_t^*	optimum relative flexural rigidity of a transversal stiffener
γ_V	reduction coefficient of the shear strength of a bolt
δ	variation coefficient; relative axial rigidity of a stiffener
δ_L	relative axial rigidity of a longitudinal stiffener
δ_t	relative axial rigidity of a transversal stiffener
ε	strain
$\bar{\varepsilon}$	dimensionless value of strain
ε_e	elastic limit strain
$\varepsilon_{f_{0.2}}$	deformation corresponding to the stress $f_{0.2}$
ε_{max}	maximum strain
ε_p	proportionality limit strain
ε_r	residual strain
ε_t	strain at failure
ε_u	ultimate strain
ε_0	reference deformation, deformation at the centre of gravity of a section
$\varepsilon_{0.2}$	deformation of 0.2 percent
η	dimensionless coefficient; reduction factor
η_W	efficiency coefficient of a butt weld
ϑ	dimensionless coefficient; angle; rotation

List of symbols

λ	slenderness
$\bar{\lambda}$	normalized slenderness
λ_{eq}	equivalent slenderness
λ_M	normalized slenderness for flexural torsional buckling
λ_0	conventional slenderness of proportionality
$\lambda_x(\lambda_y)$	slenderness for bending around the $x(y)$ axis
μ	coefficient, dimensionless ratio; friction coefficient; ratio between axial load and Euler's critical load
ν	Poisson's coefficient; safety factor
$\nu_{0.2}$	safety factor against elastic limit
ν_c	safety factor against elastic/plastic limit
ν_u	safety factor against ultimate strength
ξ	dimensionless coefficient
ρ	coefficient, dimensionless ratio
σ	normal stress
$\bar{\sigma}$	dimensionless value of stress
σ_{adm}	allowable stress
$\sigma_{adm,b}$	allowable stress for one bolt
$\sigma_{adm,w}$	allowable stress of a weld
σ_c	stress corresponding to the force that causes failure of a compression member
$\sigma_{c,glob}$	failure stress due to overall buckling
$\sigma_{c,k}$	characteristic failure stress
$\sigma_{c,loc}$	failure stress due to local buckling
$\sigma_{c,m}$	average failure stress
σ_{cr}	Euler's critical stress (for compression members)
$\sigma_{cr,0}$	Euler's critical stress (for an indefinite slab)
$\sigma_{cr,x}(\sigma_{cr,y})$	Euler's critical stress in $x(y)$ direction for a rectangular plate loaded by axial forces simultaneously acting in x and y direction
$\sigma_{cr,x,0}(\sigma_{cr,y,0})$	Euler's critical stress for a rectangular plate loaded by axial forces acting in $x(y)$ direction only
$\sigma_{cr,id}$	ideal critical stress
$\sigma_{cr,red}$	reduced critical stress
σ_f	stress in the flanges
σ_g	dead load stress
σ_{gt}	temperature stress
σ_{id}	ideal stress
σ_{lim}	limit stress
σ_{max}	maximum value of stress
σ_q	live load stress
σ_{qs}	snow stress
σ_{qse}	exceptional snow stress

List of symbols

σ_{qsr}	reduced snow stress
σ_{qsre}	reduced stress for exceptional snow
σ_{qw}	wind stress
σ_{qwe}	exceptional wind stress
σ_r	residual stress
$\sigma_{r,c}(\sigma_{r,t})$	compression (tension) residual stress
σ_w	stress in the web
$\sigma_x, \sigma_y, \sigma_z$	components of normal stress
$\sigma_1, \sigma_2, \sigma_3$	principal stresses
$\sigma_\perp, \sigma_\parallel$	normal stress in a weld (referred to the throat section)
τ	shear stress
τ_{adm}	allowable shear stress
$\tau_{adm,b}$	allowable shear stress of a bolt
$\tau_{adm,r}$	allowable shear stress of a rivet
$\tau_{adm,w}$	allowable shear stress of a weld
$\tau_{xy}, \tau_{xz}, \tau_{yz}$	components of shear stress
$\tau_\perp, \tau_\parallel$	shear stress in a weld (referred to the throat section)
φ	dimensionless coefficient; rotation
χ	dimensionless coefficient; curvature
χ_{lim}	limit curvature
χ_r	residual curvature
χ_t	curvature corresponding to the ultimate strain ε_t
$\chi_{0.2}$	curvature corresponding to the strain $\varepsilon = f_{0.2}/E$
ψ	dimensionless coefficient; coefficient taking into account partial stress redistribution in bending (plastic adaptation coefficient)
ψ_i	combination loading factor of variable loads
$\psi_x(\psi_y)$	plastic adaptation coefficient for bending around the $x(y)$ axis
ω	coefficient for increasing the axial load
Δ	ductility factor
ΔA	element area
ΔL	variation of length
Δt	variation of thickness
ΔT_s	expected life of the structure
Δz	part of length
Δx	increment of the magnitude x
Λ	multiplier of the load system
ϕ	diameter of a hole or of a bar

1 Aluminum and its alloys

1.1 Introduction

The use of aluminum alloys in civil engineering represents a new trend. Aluminum alloys have been widely and successfully used in the aeronautical industry, from the early Schwarz and Zeppelin airships to the modern Concorde and Tupolev TU144 aircraft. These materials are now also used successfully in other branches of transportation, such as the rail industry (subway coaches, sleeping cars), the auto industry (containers for trucks, moving cranes) and the shipping industry (civil and military hydroplanes, motorboats).

In what ways can this 'new' material – aluminum and its alloys – satisfy the requirements of civil engineering structures? In which applications can aluminum alloys compete with steel, a more widely used material? This book tries to answer these questions by selecting the main aspects of the subject and discussing each in turn.

It should be noted that the book is mainly addressed to structural engineers. These engineers are more familiar with steel structures, which are better known and more widely used in metal fabrication. Therefore steel will be constantly referred to in the text. In this way a continuous comparison between the two different materials will be given, in order to emphasize the specific characteristics and the advantages and disadvantages of aluminum alloys. It is hoped that this comparison will provide structural engineers with sufficient knowledge of the applications of aluminum alloys to enable them to apply these concepts in design.

For the first time, aluminum alloy members are characterized as 'industrial bars', which means that they are affected by unavoidable mechanical and geometrical imperfections which are linked to the fabrication process. The limit state concept is developed for these members. The theoretical and experimental background which European Recommendations and main national specifications are based upon is explained and commented on. This allows both researchers and designers to get involved in the philosophy of Recommendations instead of uncritically making use of their results.

Aluminum alloy structures

Design criteria are presented in order to emphasize, beyond the main differences with steel structures, all the errors that can be made if these criteria are neglected.

A preliminary presentation of this new building material – aluminum alloys – is made from the metallurgical and technological points of view. The first chapter is devoted to this purpose. It comprises some subjects which may appear trivial to the expert metallurgist, but on the other hand they will provide an essential state-of-the-art to a structural engineer who is trying to understand the different designations and fabrication processes of alloys. Sections are presented on metallurgical history (Section 1.2), fabrication processes (Section 1.3), systems of classification, including chemical composition and treatments (Section 1.4), and, finally, families of alloys (Section 1.5). The varying international classifications have been organized so as to show their significance for the designer. This clarifies a confused nomenclature and thus simplifies the applications of different specifications to aluminum alloy structures. In order to provide general criteria of design in structural applications (Section 1.7), it is necessary to illustrate the wide range of physical and mechanical properties of aluminum (Section 1.6).

Aluminum alloys are presented, in the second chapter, as structural materials, the main differences from steel being emphasized both from the mechanical and cross-sectional properties points of view.

The third chapter is devoted to specification aspects, and provides the general lines of different international codes on safety, loads and design criteria.

Chapters 4 and 5 provide methods of analysis of welded connections and bolted connections, respectively, together with a presentation of different joint types.

Strength and stability of members are, respectively, examined in Chapters 6 and 7.

Unfortunately the available space did not allow applications to be developed. Fatigue analysis is also omitted. It has already been introduced in the Italian Code (UNI 8364) and the ECCS is about to publish an appropriate Recommendation. Since this subject does not seem to be sufficiently researched, I have decided to postpone its exposition to the time when consensus is reached at international level.

The subject is treated in an organic manner by making use of SI units, while symbols are in accordance with ISO 3898 Specifications. The methodology which has been followed in the calculation methods is in accordance with the semi-probabilistic approach even though the allowable stress method is always referred to.

Detailed references are provided at the end of each chapter.

1.2 Birth of aluminum

An English chemist, Sir Humphry Davy of the Royal Institute of London, was the first to foresee the possibility of isolating the aluminum element while he was working on alumina salts (1807). These were discovered at the end of the eighteenth century by Guyton De Morveau during his studies on the ancient 'allume'. This word, derived from the Latin *allumen*, was used when referring to a material of dubious composition which was first mentioned in the Egyptian civilization in the sixteenth century BC.

A few years later Oersted, while testing alkaline metals discovered by Davy, isolated some powdery traces of the new metal.

In 1827 the German chemist Whoeler obtained the first aluminum nugget, which is still on display in the Göttingen University museum.

Thus it may be said that with uncertain parentage, and after twenty years of gestation, aluminum has been delivered with difficulty. However, this birth must be considered full of destiny.

The appearance, lightness and ductility of the new metal aroused people's interest. Efforts were therefore made to produce the metal industrially. Henry Sainte-Claire in 1854 developed a process of electrolytic reduction of aluminum from the double chloride of sodium and aluminum, thus pioneering the development of future industrial manufacturing processes.

The first aluminum produced was used to coin a metal in homage to Whoeler, and the first public order, personally requested by Napoleon III, was to fabricate aluminum eagles to be fixed to the flag lances of the imperial regiments. At that time aluminum was known as 'silver clay' because of its lightness.

A Frenchman, Paul Louis Touissant Héroult (1863–1914), became very interested in this new metal. In 1866 he patented the electrolytic process from which industrial production of aluminum began, especially in France and Switzerland.

By a curious coincidence, Charles Martin Hall (1863–1914), without being in contact with Héroult, achieved the same results at the same time in the USA. His process was to be used in the industry now called the Aluminum Company of America (ALCOA).

Héroult and Hall can be considered to be a very impressive case of coincidence of identity: they were born in the same year, studied the same subject, achieved equivalent results, promoted equivalent technological innovations, and died in the same year at the beginning of the First World War, during which aluminum alloys were first employed on a large scale.

Aluminum alloy structures

1.3 Fabrication processes [1, 2]

1.3.1 From Mineral to Structural Elements

The first step in the production of aluminum is to obtain it in fused form (first fusion) from the mineral bauxite, whose components are alumina, iron oxide, silicon and water of hydration. To be suitable for fabrication processes, the aluminum must then be cast into an ingot of a shape and size suitable for whatever working process is to be employed.

Considerable electric power is necessary to fabricate aluminum (about 20 000 kWh are needed for each tonne). This is the reason why first-fusion aluminum is usually produced in those regions with plentiful cheap electricity (it should be noted that the first centers for aluminum production were close to hydroelectric power plants, e.g. Nordhausen in Switzerland).

On the other hand, owing to the low melting point of aluminum (660 °C), the fabrication of aluminum alloys does not need a large amount of power. Plates 5–6 mm thick are produced from billets by the hot-rolling process. To obtain thinner plates the cold-rolling process is then necessary. This process induces work hardening proportional to the amount of alloy present in the product, and thus annealing processes are required to increase ductility.

One of the biggest advantages of aluminum alloys is that they can be extruded by press (Fig. 1.1a). The extrusion process allows the manufacture of profiles of any shape that cannot be obtained by hot rolling (Fig.

Fig. 1.1a

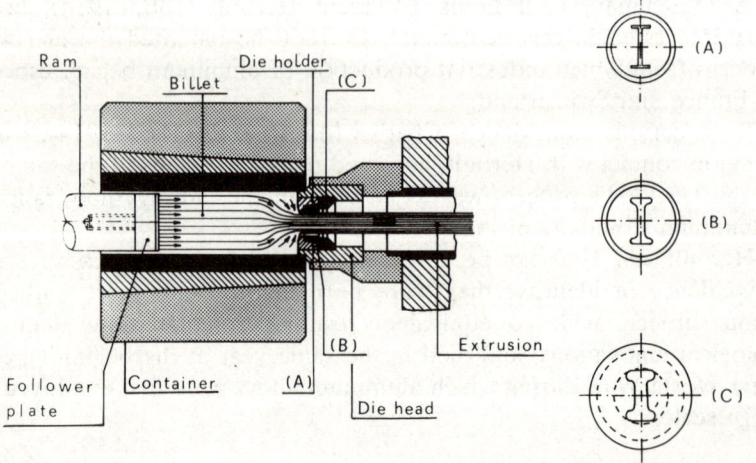

Aluminum and its alloys

Fig. 1.1b

Fig. 1.1c

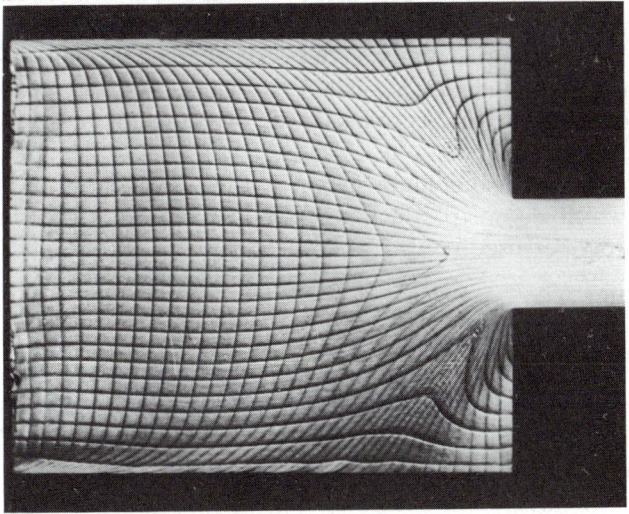

1.1b). This is one of the main advantages of aluminum over steel, and should be taken into account in an economic comparison of aluminum and other metals.

The *extrusion process* consists of the following stages:

(a) The billet is heated up to the extrusion temperature.
(b) The billet is then put in the press container, whose exit is blocked by the drawplate.

Aluminum alloy structures

(c) The shape of the hole in the drawplate, which is usually made of high-strength tempered steels, gives the shape of the profile to be produced.
(d) High pressure is then applied by a pressing bar (the pressure varies from 250 to 1000 N mm^{-2}).
(e) The profile is then expelled from the drawplate (Fig. 1.1c). The length of the profile is affected by the extrusion ratio, which is the ratio between the cross-section of the billet and the cross-section of the extruded profile.
(f) After the extrusion process, a high-tension force is applied to the profile on the straightening table.

The process of *straightening by tractioning* is also peculiar to aluminum alloys. Straightening tables exert tension forces of about 200 tonnes, and can be up to 50 m long.

The possibilities of the extrusion process are limited by the dimensions of the press. Nowadays presses of capacities from 1000 to 12 500 tonnes are available in Europe. The cross-sections of the profiles which can be obtained from these presses can be inscribed either in a circle whose diameter is 500 mm or in a 600×200 mm rectangle. Most hydraulic presses are of the horizontal type, but vertical presses can also be found.

It should be noted that so-called 'strong' alloys are less extrudable because of the higher pressure needed to overcome the greater resistance to plastic deformation.

Another part of aluminum manufacturing technology is the *casting process*, which allows production of monobloc and rigid shapes. The casting processes used nowadays are:

Sand casting
Shell casting
Casting under pressure (Fig. 1.2)

The first process makes use of cheap moulds. Shell casting is used for larger production runs, the moulds being more expensive. The third process is the most recent. It produces components which are extremely accurate and gives an excellent surface finish; it is therefore used when large production runs are requested, so that the sizeable initial investment is justified.

1.3.2 Production of Alloys

Aluminum alloys, which are usually called *light alloys*, have been developed in order to increase the strength of the base metal, aluminum, which is very ductile and corrosion resistant. In contrast to iron–carbon

Aluminum and its alloys

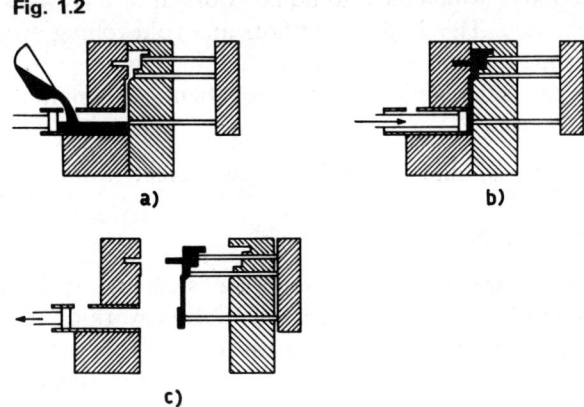

Fig. 1.2

alloys (steel), aluminum alloys need only a small amount of thermic power for their production.

The main elements used in the alloys are magnesium, silicon, zinc, copper and manganese. Nickel, cobalt, chromium, iron, titanium and zirconium are used as additional elements.

The main properties of these elements are as follows:

Magnesium reduces the melting point to 451 °C (this is the reason why it is used as weld metal); it increases work-hardening ability and corrosion resistance to salted water.
Silicon increases strength and ductility. If combined with magnesium it allows precipitation hardening and reduces the melting point.
Zinc drastically increases strength and allows hot or cold precipitation hardening. In some cases it increases susceptibility to stress corrosion.
Copper gives an even greater strength increase and allows cold precipitation hardening, but reduces corrosion resistance, weldability and ductility.
Nickel can increase strength under high-temperature conditions.
Titanium gives decreased grain size.
Zirconium is a stabilizer element.
Chromium increases resistance to stress corrosion.
Iron is usually an impurity which can increase the strength of pure aluminum (Al 99.0, 99.5, 99.7, 99.8 percent) if a low percentage is used.

It should be noted that the increase of strength due to Si, Zn and Cu is caused by the precipitation-hardening process of their compounds (Mg_2Si, $CuAl_2$ etc.) after the solution process.

With respect to the fabrication processes, aluminum alloys can be classified as follows:

(a) *Casting alloys*, which have to be remelted and then casted.

(b) *Wrought alloys*, which have to be hot worked or cold worked without being remelted. This happens in hot- and cold-rolling processes such as extrusion, forging, and drawing.

With respect to their reaction to heat treatment, aluminum alloys can be classified as follows:

(a) *Non-heat-treated alloys*, which are also called work-hardenable alloys.
(b) *Heat-treated alloys*.

Among non-heat-treated alloys are those such as AlMn, AlMg and AlMgMn in which strength is increased by cold working. Cold working can be obtained during the rolling process for plates, or via tractioning for structural shapes. After cold working, strength is increased and ductility is decreased. If the material is heated to 350 °C the alloys return to their initial state. The welding process, therefore, decreases the strength in the so-called *heat-affected zone*.

Heat-treated alloys are AlMgSi, AlZnMgCu and AlCuMg. The heat-treating process provides higher strength and can be separated into the following phases:

Heating up to 450 or 530 °C depending upon the alloy
Tempering with water or air
Aging at room temperature (natural aging)
Aging at a temperature varying between 120 and 180 °C (artificial aging).

The effects of this treatment disappear if the material is heated to between 200 and 350 °C. The heat input from the welding process decreases the strength of the alloy to a value which is bounded by the properties of the material in the annealed stage and in the heat-treated stage. AlZnMg alloys recover most of their strength because of a natural aging process subsequent to welding.

1.4 Classification of alloys [3, 4]

Aluminum alloys are usually classified either with respect to the fabrication process (casting alloys and wrought alloys) and to the heat treatment (heat-treated alloys and non-heat-treated alloys), or with respect to the chemical composition. This classification allows aluminum alloys to be grouped in different families with similar mechanical and technological behavior (see also Section 1.5).

Unfortunately, although the ISO has attempted to unify the nomenclature for aluminum alloys, this is still different in each country. However, aluminum alloys can be classified in two different ways: according to a numerical designation (see Section 1.4.1) or according to an alphanumerical designation (see Section 1.4.2). The symbol which gives the chemical composition is usually followed by another symbol indicating the manufacturing process (see Section 1.4.3).

1.4.1 Numerical Designation

This designation is derived from the American nomenclature (Aluminum Association). It is now also being used in Europe as an alternative to the national nomenclatures.

This type of nomenclature, which is used for wrought aluminum alloys, is represented by a number of four figures. A first figure of 1 denotes those alloys in which the aluminum is greater than 99 percent. A first figure from 2 to 8 denotes the other alloys: the figure identifies the main alloying element. More precisely the following correspondence is established:

1xxx	pure aluminum (greater than 99 percent)
2xxx	copper
3xxx	manganese
4xxx	silicon
5xxx	magnesium
6xxx	magnesium and silicon
7xxx	zinc
8xxx	other elements
9xxx	unused series

In the first group (called pure aluminum) the third and fourth figures indicate the percentage of aluminum above 99 percent, e.g. 1070 alloy has an aluminum content of 99.70 percent. The second figure represents the level of impurities. It is equal to zero if the impurities are uncontrolled, and it can vary between 1 and 9 depending upon the level which the impurity content is not to exceed.

In the other groups, the second figure is equal to zero for the main alloy and varies between 1 and 9 for its modifications. The third and fourth figures identify the specific alloy within each group.

This type of numerical designation was first proposed by the Aluminum Association for wrought alloys. Because it is easy to understand, it is now also being used in European countries.

The Aluminum Association also utilizes the numerical designation for casting alloys, with the following meanings:

1xxx	pure aluminum (greater than 99 percent)
2xxx	copper
3xxx	silicon plus copper and/or magnesium

Aluminum alloy structures

4xxx	silicon
5xxx	magnesium
6xxx	unused series
7xxx	zinc
8xxx	tin
9xxx	other elements

In England a special numerical designation is used:

(a) One or two figures are used for wrought alloys. The letter N is placed before the two figures for a cold-worked alloy, whereas the letter H is placed before the two figures for heat-treated alloys.

(b) One or two figures preceded by the letters LM are used to represent casting alloys.

In Spain and Canada, too, numerical designations different from the American are used.

1.4.2 Alphanumerical Designation

This type of designation is used in most of the European countries. It is also used in the ISO normalization.

Each alloy is identified by a group of letters and figures divided into two sets. The first set is to identify the base metal, i.e. aluminum is referred to as Al (Italy, Germany, Switzerland, Austria, Sweden, Netherlands, Yugoslavia), ALP (pure aluminum; Italy), A (France). In some cases this letter is preceded by another letter (or letters) with the following meanings:

G	casting alloys in the rough stage (Italy, Switzerland, Netherlands): sand casted (G_s), permanent mold casted (G_c), pressure die casted (G_p) (Italy)
GD	pressure die casted (Germany)
GK	permanent mold casted (Germany)
K	casting alloys (Yugoslavia)
P	wrought alloys (Italy)

The second group of letters represents the alloying elements. Sometimes these letters are followed by the percentage of the element. Symbols representing these elements are formed by one letter (France) or two letters (other countries). For example:

Copper	U or Cu
Magnesium	G or Mg
Manganese	M or Mn
Silicon	S or Si
Zinc	Z or Zn

In some countries, such as Switzerland and Italy, some alloys are represented with a name followed by a number. For example:

Peraluman 460   (corresponding to 5083)
Peraluman 500   (corresponding to 5086)
Estrudal 050   (corresponding to 6060)
Anticorodal 100   (corresponding to 6351)
Unidur 100   (corresponding to 7020)

It should be noted that these designations are only used in trade exchanges, and are not recognized by unifying organizations.

1.4.3 *Fabrication Stage Designation*

The metallurgical stage of an alloy is usually identified by a symbol which follows the chemical composition symbol. This symbol can be formed by letters and numbers, which usually differ according to the national designations.

The following symbols are provided by the Aluminum Association:

F   rough stage of fabrication
This symbol is applied to work-hardened products, without control of the cold-working rate and the heat treatment. Mechanical properties cannot be defined in these work-hardenable products.
O   annealed stage
(Only valid in the case of work-hardenable products.) This is the most ductile stage. Conversely, strength is very low.
H   work-hardened stage
(Only valid in the case of work-hardenable products.) This stage represents those alloys whose strength is increased by cold working, and in some cases where a soft-annealing process follows. The symbol H is usually followed by two or three letters.
W   tempered nonstabilized stage
(Not very often used.) This is an unstable metallurgical stage of those alloys which are tempered and aged at room temperature after solution. The symbol is followed by a number indicating the natural aging period (e.g. W $\frac{1}{2}$ hour).
T   heat-treated stage
This represents those products which have been heat treated by a combination of the following processes:

Solution
Tempering
Natural aging
Quenching

Aluminum alloy structures

Sometimes the heat treatment is followed by a cold-working process in order to stabilize the alloy and to eliminate the stresses induced by the tempering process. The symbol is always followed by one or two figures.

Stage H is subdivided as follows:

H1   cold-worked stage
H2   cold-worked and partially annealed stage
H3   cold-worked and stabilized stage

Stage T is subdivided as follows:

T1   cooled after hot working and naturally aged. After being rolled or extruded the material is rapidly cooled to obtain further hardening due to aging at room temperature.

T3   in separate solution, work hardened and naturally aged. After solution followed by a cooling process, the material is work hardened. Specifications take into account the effect due to this process.

T4   in separate solution and naturally aged. After solution, followed by cooling, the material is not work hardened. If work hardening is applied its effects are not taken into account in specifications.

T5   cooled after hot working and artificially aged. After the rolling process or the extrusion process the material is rapidly cooled to obtain further hardening due to artificial aging.

T6   in solution and artificially aged. After solution followed by cooling, the material is artificially aged without any plastic deformation. If work hardening is applied its effects are not taken into account in specifications.

T7   in separate solution and overaged. This process provides specific properties such as high resistance to stress corrosion.

T8   in separate solution, work hardened and artificially aged. Work hardening is used to increase strength, which is taken into account in specifications.

T9   in separate solution, artificially aged and work hardened. After aging the material is cold worked to increase strength.

T10   cooled after hot working, artificially aged and work hardened. After a rapid cooling the material is artificially aged and cold worked to increase strength.

T11   cooled after hot working, work hardened and naturally aged. After cooling the material is cold worked before natural aging to increase strength.

T12   cooled after hot working, work hardened and artificially aged. After cooling and cold working the material is artificially aged.

Aluminum and its alloys

Fig. 1.3 Aluminum Association symbols for fabrication stage designation

Basic treatments				Symbol
Heat treatment with separate solution	without work hardening	naturally aged		T4
		artificially aged		T6
	with work hardening	work hardened	naturally aged	T3
			artificially aged	T8
		artificially aged, work hardened		T9
Heat treatment without separate solution	without work hardening	naturally aged		T1
		artificially aged		T5
	with work hardening	work hardened	naturally aged	T11
			artificially aged	T12
		artificially aged, work hardened		T10

The meaning of the Aluminum Association symbols is summarized in Fig. 1.3.

Special and supplementary treatments are indicated by the following symbols:

T61	'soft' artificial aging
T66	'hard' artificial aging
TX51	relaxation by tractioning
TX52	relaxation by compression
TX53	relaxation by heat treatment

X is the number representing the treatment: 4, 6, 7 etc.

In many countries nonunified systems of representation are used. In England, for example, particular symbols (M, WP) are used. In Sweden numbers are used (00, 02, 03 etc.).

As an alternative to fabrication stage designation it is possible to identify an alloy by a parameter indicating strength. This is the minimum value of ultimate stress ($N\,mm^{-2}$) preceded by the letter:

| R | in France |
| W, F, or G | in Germany (for annealed, cold worked and tempered or cold worked and partially annealed) |

Aluminum alloy structures

Fig. 1.4 European fabrication stage designation for casting alloys

	Stage	France	Germany	UK
Sand casting	Non-heat-treated	Y20	G( )	M
	Tempered and artificially aged	Y23	G( )wa	TF
	Tempered and naturally aged	Y24	G( )ka	TB
	Tempered and stabilized	Y25		TS
	With special treatments	Y29		
Permanent mold casting	Non-heat-treated	Y30	GK( )	M
	Tempered and artificially aged	Y33	GK( )wa	TF
	Tempered and naturally aged	Y34	GK( )ka	TB
	Tempered and stabilized	Y35		TS
	With special treatments	Y39		
Pressure die casting		Y40	GD( )	M

This system of representation is very simple and efficient since it gives the final result of the fabrication process – strength – without explaining how this result has been achieved.

To indicate the metallurgical stage and fabrication process of casting alloys conventional designations can be used. Some examples are given in Fig. 1.4 (under the column for Germany the item in parentheses stands for the complete designation of the alloy).

1.4.4 Comparison of Different Designations

Figures 1.5 and 1.6 represent an attempt to relate the symbols used for the main alloys all over the world. Figure 1.5 refers to wrought alloys, and Fig. 1.6 refers to casting alloys.

In these figures are given the symbols used in Canada, the USA and in various European countries – France, Germany, Spain, England, Italy, Netherlands, Sweden, Switzerland and Yugoslavia. In Austria the German designation applies. The minimum mechanical properties requested by DIN specifications are also shown.

Some of the main properties of the alloys with respect to structural applications in some European countries are given in more detail in Figs. 1.7 to 1.14.

The following data are given in each figure:

International designation (Aluminum Association numerical designation is given for wrought alloys, ISO alphanumerical designation is given for casting alloys)
National designation

Fig. 1.5 National designations for wrought alloys

France	W. Germany	UK	Italy	Netherlands	Spain	Sweden	Switzerland	Yugoslavia	Canada CSA	Canada ALCAN	USA (numerical designation)
A-U2G	AlCuMg0.5	L-86	P-AlCu2.5MgSi	—	—	—	Al4Cu0.3Mg	—	CG30	16S	2117
A-U4G	AlCuMg1	H-14	P-AlCu4MgMn	AlCuMg1	—	—	Al3.5Cu0.5Mg	—	CM41	17S	2017
A-U4SG	AlCuSiMn	H-15	P-AlCu4.5SiMgMn	AlCu4MgSi	—	AlCu4MgSi	—	—	CS41N	B26S	2014
A-U2GN	—	H-18	—	—	—	—	—	—	—	42S	2618–2618A
A-M1	AlMn	—	P-AlMn1.2	AlMn1	—	AlMn1	AlMn	—	—	D3S	3003
—	AlMnCu	N-3	—	—	—	—	—	—	—	—	3103
A-MG	—	—	—	—	—	—	—	—	—	—	3005
A-MIG	—	—	P-AlMn1.2Mg	—	—	—	Al1Mg	—	—	4S	3004
A-G0.6	AlMg1	N-41	P-AlMg0.8	AlMg1	—	AlMg1	Al1Mg	—	—	B57S	5005
A-G2	AlMg2	N-4	P-AlMg2.5	AlMg2	—	AlMg2.5	Al2Mg	—	GR20	M57S	5052–5251
A-G3	AlMg3	N-5	P-AlMg3.5	AlMg3	L332–L341	—	Al3Mg	—	GR40	53S–54S	5154–5254–5154A
A-G4MC	—	N-8	—	AlMg4.5Mn	—	AlMg4.5Mn	Al4Mg	AlMg4	GM40	B54S–D54S	5083–5086
A-G5	AlMg5	N-6	P-AlMg5	AlMg5	—	—	Al5Mg	—	GM50R	A56S	5056–5056A
A-GS	AlMgSi0.5	H-9	P-AlSi0.5Mg	AlMgSi	—	AlMgSi	AlMgSi	AlMgSi0.2	GS10	50S	6060–6063
A-SG	—	H-19	—	AlSi1Mg	L337	—	—	—	—	51S	6005A–6051 6081
A-SGM	AlMgSi1	H-30	P-AlSi1MgMn	AlMg1SiCu	—	AlSi1Mg	AlSiMg	AlMg1SiCu	GS11R	B51S	6181
		H-20	P-AlMg1SiCu						GS11N		6061
A-Z3G2	AlZnMg3	—	—	—	—	—	—	—	—	—	7051
A-Z5G	AlZnMg1	H-17	—	—	—	AlZn5Mg1	AlZnMg1	AlZn5Mg1	—	D74S	7020

Aluminum alloy structures

Fig. 1.6 National designations for casting alloys

France	W. Germany	UK	Italy	Netherlands	Spain	Sweden	Switzerland	Yugoslavia	Canada (CSA)	USA (international designation)
A-U5GT	G-AlCu4TiMg	—	—	—	—	—	G-AlCu5MgTi	K.AlCu4MgTi	CG50	AlCu4MgTi
—	G-AlCu4Ti	—	—	G-AlCu4Ti	—	—	G-AlCu5TiMn	K.AlCu4Ti	—	AlCu4Ti
A-STG	G-AlSi7Mg	LM25	G-AlSi7Mg	G-AlSi7Mg	—	4244	G-AlSi7MgTi	K.AlSi7Mg	SG71	AlSi7Mg
A-S10G	G-AlSi10Mg	LM9	G-AlSi9MnMg	G-AlSi10Mg	L256	4253	G-AlSi9Mg	K.AlSi10Mg	—	AlSi10Mg
A-S13	GD-AlSi12	LM6	G-AlSi13	G-AlSi12	L252	4260	G-AlSi13	K.AlSi12	S12N	AlSi12
A-Z5G	—	—	G-AlZn5MgFe	—	L271	4438	G-AlZn5MgCr	K.AlZn5MgCr	ZG61N	AlZn5Mg

Fig. 1.7 France: alloy properties for structural applications

<table>
<tr><td rowspan="2"></td><td colspan="2">Designation</td><td rowspan="2">Stage</td><td rowspan="2">Products</td><td colspan="3">Mechanical properties</td></tr>
<tr><td>International
USA</td><td>National
NF</td><td>$f_{0.2}$</td><td>f_t</td><td>ε_t</td></tr>
<tr><td rowspan="15">Wrought alloys</td><td>5083</td><td></td><td>0</td><td>profiles</td><td>110</td><td>265</td><td>18</td></tr>
<tr><td></td><td></td><td>0</td><td>plates</td><td>120</td><td>265</td><td>16</td></tr>
<tr><td>5086</td><td>A-G4MC</td><td>0</td><td>profiles</td><td>95</td><td>235</td><td>18</td></tr>
<tr><td></td><td></td><td>H111</td><td>plates</td><td>100</td><td>235</td><td>17</td></tr>
<tr><td>6060</td><td>A-GS</td><td>R20</td><td>profiles</td><td>145</td><td>195</td><td>12</td></tr>
<tr><td>6061</td><td></td><td>R26</td><td>profiles</td><td>235</td><td>255</td><td>8</td></tr>
<tr><td></td><td></td><td>R29</td><td>plates</td><td>235</td><td>285</td><td>9</td></tr>
<tr><td>6081</td><td>A-SG</td><td>R28</td><td>plates</td><td>235</td><td>275</td><td>9</td></tr>
<tr><td>6082</td><td>A-SGM0.7</td><td>R28</td><td>profiles</td><td>235</td><td>275</td><td>8</td></tr>
<tr><td></td><td></td><td>R31</td><td>profiles</td><td>265</td><td>305</td><td>8</td></tr>
<tr><td>6181</td><td>A-SGM0.3</td><td>R28</td><td>plates</td><td>235</td><td>275</td><td>9</td></tr>
<tr><td>7020</td><td>A-Z5G</td><td>R34</td><td>profiles</td><td>270</td><td>335</td><td>10</td></tr>
<tr><td></td><td></td><td>R35</td><td>plates</td><td>275</td><td>345</td><td>10</td></tr>
</table>

<table>
<tr><td rowspan="2"></td><td colspan="2">Designation</td><td rowspan="2">Stage</td><td rowspan="2">Casting</td><td colspan="3">Mechanical properties</td></tr>
<tr><td>International
ISO</td><td>National
NF</td><td>$f_{0.2}$</td><td>f_t</td><td>ε_t</td></tr>
<tr><td rowspan="8">Casting alloys</td><td>Al-Cu4MgTi</td><td>A-U5GT</td><td>Y23</td><td>sand</td><td>195</td><td>290</td><td>5</td></tr>
<tr><td></td><td></td><td>Y33</td><td>permanent mold</td><td>195</td><td>320</td><td>8</td></tr>
<tr><td>Al-Si7Mg</td><td>A-S7G</td><td>Y23</td><td>sand</td><td>175</td><td>225</td><td>1.5</td></tr>
<tr><td></td><td></td><td>Y33</td><td>permanent mold</td><td>175</td><td>245</td><td>3</td></tr>
<tr><td>Al-Si12</td><td>A-S13</td><td>Y20</td><td>sand</td><td>70</td><td>155</td><td>4</td></tr>
<tr><td></td><td></td><td>Y30</td><td>permanent mold</td><td>80</td><td>165</td><td>5</td></tr>
<tr><td>Al-Zn5Mg</td><td>A-Z5G</td><td>Y29</td><td>sand</td><td>130</td><td>185</td><td>4</td></tr>
<tr><td></td><td></td><td>Y39</td><td>permanent mold</td><td>130</td><td>200</td><td>4</td></tr>
</table>

Fabrication stage (according to national standards)
Wrought products (profiles or plates) or the type of casting (sand or shell = permanent mold)
Main mechanical properties (units $N\,mm^{-2}$):
  $f_{0.2}$ conventional yield limit corresponding to a permanent deformation of 0.2 percent
  f_t ultimate strength
  ε_t elongation at rupture

Aluminum alloy structures

Fig. 1.8 West Germany: alloy properties for structural applications

	Designation				Mechanical Properties		
	International	National					
	USA	DIN	Stage	Products	$f_{0.2}$	f_t	ε_t
Wrought alloys	5083	AlMg4.5Mn	F28	profiles	155	275	12
			F30	plates (5–40 mm)	205	295	12
	5086	AlMg4Mn	F28	plates (4–6 mm)	195	275	7
	6060	AlMgSi0.5	F22	profiles	155	215	12
	6061	(AlMgSiCu)	.7	profiles	245	265	9
			.7	plates (0.5–6 mm)	245	295	10
	6082	AlMgSi1	F32	profiles	255	315	10
			F32	plates (0.2–10 mm)	255	315	10
	7020	AlZnMg1	F36	profiles	275	355	10
			F36	plates (0.2–12 mm)	275	355	10

	Designation				Mechanical properties		
	International	National					
	ISO	DIN	Stage	Casting	$f_{0.2}$	f_t	ε_t
Casting alloys	Al-Cu4MgTi	G-AlCu4TiMg	w.a.	sand	235–340	340–410	3–10
		GK-AlCu4TiMg	w.a.	permanent mold	255–370	340–430	3–12
	Al-Si7Mg	G-AlSi7Mg	w.a.	sand	185–235	225–300	2–5
		GK-AlSi7Mg	w.a.	permanent mold	195–275	245–330	5–9
	Al-Si12	G-AlSi12		sand	70–100	155–205	5–10
		GK-AlSi12		permanent mold	80–110	175–235	6–12
		GD-AlSi12		under pressure	135–175	215–275	1–3
	Al-Si10Mg	G-AlSi10Mg	w.a.	sand	175–255	215–315	1–4
		GK-AlSi10Mg	w.a.	permanent mold	205–275	235–315	1–4
		GD-AlSi10Mg			135–195	215–295	1–3
	Al-Cu4Ti	G-AlCu4Ti	t.a.	sand	175–225	275–370	5–10
		GK-AlCu4Ti	t.a.	permanent mold	175–225	310–390	8–18

Unless otherwise stated, the values given in the figures are the minimum required values.

The symbols $f_{0.2}, f_t, \varepsilon_t$ correspond to ISO 3898 specification 'Bases for design of structures – notations – general symbols' (1976), and the units are in the Système Internationale (SI).

Fig. 1.9 United Kingdom: alloy properties for structural applications

	Designation				Mechanical properties		
	International	National					
	USA	BSS	Stage	Products	$f_{0.2}$	f_t	ε_t
Wrought alloys	5083	N8	M	plates (6–25 mm)	130	270	10
	6060	HE9	WP	profiles	160	180	10
	6061	HE20	WP	profiles	230	275	8
		HS20	WP	plates	220	275	5
	6082	HE30	WP	profiles	265	300	8
		HE30	WP	plates	230	290	6
	7020	HE17	WP	profiles	275	390	10
		HS17	WP	plates	265	315	10

	Designation				Mechanical properties		
	International	National					
	ISO	BSS	Stage	Casting	$f_{0.2}$	f_t	ε_t
Casting alloys	Al-Si12	LM6	M	sand		155	5
			M	permanent mold		185	7
	Al-Si7Mg	LM25	M	sand		125	2
			TF	permanent mold		225	
			TF			275	2

It should be noted that other symbols are used in various countries to indicate mechanical properties. The following symbols are used to refer to elastic limit (yield stress) and ultimate strength:

France	$R_{0.2}$	R
West Germany	$\beta_{0.2}$	β_z
Italy	$R_{p0.2}$	R_m
ECCS	$\sigma_{0.2}$	σ_{ult}
USA	F_{ty}	F_{tu}
ISO	$f_{0.2}$	f_t

The ISO designation and SI units will generally be used in this book. However, in some cases national symbols will also be referred to in order not to differ excessively from the original source.

Aluminum alloy structures

Fig. 1.10 Italy: alloy properties for structural applications

	Designation				Mechanical properties		
	International	National					
	USA	UNI	Stage	Products	$f_{0.2}$	f_t	ε_t
Wrought alloys	5083	P-AlMg4.5	R	all	110	245	20
	6063	P-AlSi0.5Mg	TA16	profiles	150	195	12
			TaA	profiles	150	185	11
	6061	P-AlMg1SiCu	TA16	profiles	235	255	8
			TA16	plates	235	285	9
	6082	P-AlSi1MgMn	TA16	profiles	255	305	10
			TA16	plates	235	285	12
	7020	P-AlZn4.5Mg	TA	profiles plates	265	345	10

	Designation				Mechanical properties		
	International	National					
	ISO	UNI	Stage	Casting	$f_{0.2}$	f_t	ε_t
Casting alloys	Al-Cu4Si	G-AlCu4.5	TA	sand	195	245	2
	Al-Si7Mg	G-AlSi7MgMn	TA	sand	175	225	2
			TA	permanent mold	175	245	6
	Al-Si12	G-AlSi13	raw	sand	80	165	4
				permanent mold	90	175	5
	Al-Zn5Mg	G-AlZn5MgFe	TsN	sand	155	215	3
			TcN	permanent mold	165	295	9
	Al-Mg6	G-AlMg5	raw	sand	80	155	3
			raw	permanent mold	90	195	8

1.5 Alloy families

1.5.1 *Pure Aluminum and Alloys of Low Alloy Content (1000 Series)*

This series mainly consists of plates of different aluminum percentages (going from 99.0 to 99.8 percent), which can be used in low-stressed structures.

Pure aluminum has low strength ($f_{0.2} \cong 30 \text{ N mm}^{-2}$); conversely, it is a very ductile material ($\varepsilon_t \cong 30\text{–}40$ percent). If the material is cold worked the strength is increased up to $f_{0.2} \cong 100 \text{ N mm}^{-2}$, whereas ductility is drastically decreased ($\varepsilon_t \cong 3\text{–}4$ percent).

Fig. 1.11 Netherlands: alloy properties for structural applications

	Designation				Mechanical properties		
	International	National					
	USA	NEN	Stage	Products	$f_{0.2}$	f_t	ε_t
Wrought alloys	5083	Al-Mg4.5Mn	0	plates profiles	120 (min)	355 (max)	16
	5086	Al-Mg3.5	0	plates	80 (min)	285 (max)	16
			HA	profiles	175 (min)	245–305	7
	6061	Al-Mg1SiCu	R29	plates profiles	235 (min)	285 (min)	8
	6082	Al-Si1Mg	R29	plates profiles	235 (min)	285 (min)	8

	Designation				Mechanical properties		
	International	National					
	ISO	NEN	Stage	Casting	$f_{0.2}$	f_t	ε_t
Casting alloys	Al-Si12	G-AlSi12	raw	sand permanent mold	80 90	165 195	4 3
	Al-Si7Mg	G-AlSi7Mg	raw	sand	100	135	2
			tempered and quenched	sand	135	205	3

Pure aluminum is highly corrosion resistant, and specific applications therefore include panels, ceilings and tanks.

1.5.2 Aluminum-Copper-Magnesium Alloys (2000 Series)

These types of alloy can generally be found as profiles, plates and pipes. With heat treatment, their strength can increase up to $f_{0.2} \cong 300 \text{ N mm}^{-2}$, with a ductility equal to $\varepsilon_t \cong 10$ percent. Because of the presence of copper, the corrosion resistance is not very high. It is therefore necessary to protect these alloys, especially when they are used in a corrosive environment (plating or painting processes are used for this purpose).

Because of their low weldability, these alloys are not very common in civil engineering. However, they are widely employed in the aeronautical industry, in which riveted connections are used.

Aluminum alloy structures

Fig. 1.12 Spain: alloy properties for structural applications

<table>
<thead>
<tr><th colspan="2">Designation</th><th rowspan="2">Stage</th><th rowspan="2">Products</th><th colspan="3">Mechanical properties</th></tr>
<tr><th>International USA</th><th>National</th><th>$f_{0.2}$</th><th>f_t</th><th>ε_t</th></tr>
</thead>
<tbody>
<tr><td rowspan="14">Wrought alloys</td><td></td><td></td><td></td><td></td><td></td><td></td></tr>
<tr><td>5083</td><td>L332</td><td>F</td><td>profiles</td><td>110</td><td>265</td><td>12</td></tr>
<tr><td></td><td></td><td>0</td><td>plates</td><td>95–120</td><td>245–270</td><td>12–16</td></tr>
<tr><td>5086</td><td>L341</td><td>0</td><td>profiles</td><td>100</td><td>245</td><td>12</td></tr>
<tr><td></td><td></td><td>H112</td><td>plates</td><td>100</td><td>245</td><td>12</td></tr>
<tr><td>6005</td><td>L337</td><td>T5</td><td>profiles</td><td>235</td><td>255</td><td>8</td></tr>
<tr><td>6060</td><td></td><td>T5</td><td>profiles</td><td>125</td><td>175</td><td>8</td></tr>
<tr><td>6261</td><td>L342</td><td>T6</td><td>profiles</td><td>245</td><td>290</td><td>8</td></tr>
<tr><td>6351</td><td>L342</td><td>T6</td><td>profiles</td><td>235</td><td>285</td><td>8</td></tr>
<tr><td></td><td></td><td>T6</td><td>plates</td><td>240</td><td>280</td><td>9</td></tr>
<tr><td>7004</td><td></td><td>T4</td><td>plates</td><td>260</td><td>370</td><td>14</td></tr>
<tr><td></td><td></td><td>T6</td><td>plates</td><td>360</td><td>410</td><td>13</td></tr>
<tr><td>7020</td><td></td><td>T6</td><td>profiles</td><td>265</td><td>330</td><td>10</td></tr>
</tbody>
</table>

<table>
<thead>
<tr><th colspan="2">Designation</th><th rowspan="2">Stage</th><th rowspan="2">Casting</th><th colspan="3">Mechanical properties</th></tr>
<tr><th>International ISO</th><th>National</th><th>$f_{0.2}$</th><th>f_t</th><th>ε_t</th></tr>
</thead>
<tbody>
<tr><td rowspan="10">Casting alloys</td><td></td><td></td><td></td><td></td><td></td><td></td></tr>
<tr><td>Al-Si12</td><td>L252</td><td>F</td><td>sand</td><td>70</td><td>155</td><td>4</td></tr>
<tr><td></td><td></td><td>F</td><td>permanent mold</td><td>80</td><td>165</td><td>5</td></tr>
<tr><td>Al-Si7Mg</td><td></td><td>T6</td><td>sand</td><td>175</td><td>225</td><td>1.5</td></tr>
<tr><td></td><td></td><td>T6</td><td>permanent mold</td><td>175</td><td>245</td><td>3</td></tr>
<tr><td>Al-Si10Mg</td><td>L256</td><td>T6</td><td>sand</td><td>165</td><td>215</td><td>2</td></tr>
<tr><td></td><td></td><td>T6</td><td>permanent mold</td><td>185</td><td>245</td><td>3</td></tr>
<tr><td>Al-Cu4MgTi</td><td></td><td>T6</td><td>sand</td><td>195</td><td>300</td><td>4</td></tr>
<tr><td></td><td></td><td>T6</td><td>permanent mold</td><td>225</td><td>330</td><td>5</td></tr>
<tr><td>Al-Zn5Mg</td><td>L271</td><td>T1</td><td>sand</td><td>135</td><td>205</td><td>5</td></tr>
<tr><td></td><td></td><td>T1</td><td>permanent mold</td><td>145</td><td>215</td><td>7</td></tr>
</tbody>
</table>

1.5.3 Aluminum-Manganese Alloys (3000 Series)

These alloys cannot be heat treated; they thus have a slightly higher strength than pure aluminum. On the other hand they are corrosion resistant. Specific applications are panels and roof systems.

1.5.4 Aluminum-Silicon Alloys (4000 Series)

The properties of these alloys are similar to those of the 3000 series. However, they are not often used.

Fig. 1.13 Sweden: alloy properties for structural applications

	Designation				Mechanical properties		
	International	National					
	USA	SIS	Stage	Products	$f_{0.2}$	f_t	ε_t
Wrought alloys	5083	Al-Mg4.5Mn	00	profiles	110	265 (min)	12
			02	plates (4.5 mm, max)	120	265 (min)	12
			02	plates (1–25 mm)	120	275–350	16
	6060	Al-MgSi	06	profiles	175	215	10
	6082	Al-SiMg	06	profiles	245	285 (min)	8
			06	plates (1–20 mm)	235	285 (min)	8
	7020	Al-Zn5Mg1	06	profiles	275	335 (min)	10
			06	plates (0.5–6 mm)	275	335 (min)	10

	Designation				Mechanical properties		
	International	National					
	ISO	SIS	Stage	Casting	$f_{0.2}$	f_t	ε_t
Casting alloys	Al-Si7Mg	4244	04	sand	175	215	2
			07	permanent mold	215	255	1
	Al-Si10Mg	4253	04	sand	175	215	2
			07	permanent mold	215	255	1
	Al-Si12	4260	03	sand	80	135	1
			06	permanent mold	90	145	2
	Al-Zn5Mg	4438	04	sand	135	215	3
			07	permanent mold	145	215	3

1.5.5 Aluminum-Magnesium Alloys (5000 Series)

Even though these alloys cannot be heat treated, their mechanical properties are higher than those of the 1000, 3000 and 4000 series ($f_{0.2} \cong 100 \text{ N mm}^{-2}$). The strength can be increased if they are cold worked ($f_{0.2} \cong 250 \text{ N mm}^{-2}$), the ductility still being high ($\varepsilon_t \cong 10$ percent). Corrosion resistance is also high, especially in a marine environment. These alloys are often used in welded structures since their strength is not drastically decreased in the heat-affected zone.

Aluminum alloy structures

Fig. 1.14 Switzerland: alloy properties for structural applications

<table>
<tr><th colspan="2">Designation</th><th rowspan="2">Stage</th><th rowspan="2">Products</th><th colspan="3">Mechanical properties</th></tr>
<tr><th>International
USA</th><th>National</th><th>$f_{0.2}$</th><th>f_t</th><th>ε_t</th></tr>
<tr><td rowspan="12">Wrought alloys</td><td>5083</td><td>Peraluman 460</td><td>0</td><td>profiles</td><td>110</td><td>265</td><td>12</td></tr>
<tr><td></td><td></td><td>0</td><td>plates</td><td>115</td><td>265</td><td>16</td></tr>
<tr><td>5086</td><td>Peraluman 500</td><td>0</td><td>profiles</td><td>90</td><td>235</td><td>15</td></tr>
<tr><td></td><td></td><td>H111</td><td>plates</td><td>100</td><td>235</td><td>17</td></tr>
<tr><td>6060</td><td>Extrudal 050</td><td>R20</td><td>profiles</td><td>145</td><td>195</td><td>10</td></tr>
<tr><td>6061</td><td></td><td>R26</td><td>profiles</td><td>235</td><td>255</td><td>8</td></tr>
<tr><td>6082</td><td></td><td>R28</td><td>profiles</td><td>235</td><td>275</td><td>8</td></tr>
<tr><td></td><td></td><td>R31</td><td>profiles</td><td>265</td><td>300</td><td>8</td></tr>
<tr><td>6081–6181</td><td></td><td>R28</td><td>plates</td><td>235</td><td>275</td><td>9</td></tr>
<tr><td>6351</td><td>Anticorodal 100</td><td>R29</td><td>plates</td><td>235</td><td>285</td><td>9</td></tr>
<tr><td>7020</td><td>Unidur 100</td><td>R34</td><td>profiles</td><td>270</td><td>330</td><td>10</td></tr>
<tr><td></td><td></td><td>R35</td><td>plates</td><td>275</td><td>340</td><td>10</td></tr>
</table>

<table>
<tr><th colspan="2">Designation</th><th rowspan="2">Stage</th><th rowspan="2">Casting</th><th colspan="3">Mechanical properties</th></tr>
<tr><th>International
ISO</th><th>National
VSM</th><th>$f_{0.2}$</th><th>f_t</th><th>ε_t</th></tr>
<tr><td rowspan="10">Casting alloys</td><td>Al-Cu4MgTi</td><td>G-AlCu5MgTi</td><td>tempered and aged</td><td>sand</td><td></td><td>330</td><td>7</td></tr>
<tr><td></td><td></td><td></td><td>permanent mold</td><td></td><td>350</td><td>11</td></tr>
<tr><td>Al-Cu4Ti</td><td>G-AlCu5Ti</td><td>tempered and aged</td><td>sand</td><td></td><td>310</td><td>6</td></tr>
<tr><td></td><td></td><td></td><td>permanent mold</td><td></td><td>320</td><td>9</td></tr>
<tr><td>Al-Si7Mg</td><td>G-AlSi7MgTi</td><td>tempered and quenched</td><td>sand</td><td></td><td>215</td><td>2</td></tr>
<tr><td></td><td></td><td></td><td>permanent mold</td><td></td><td>225</td><td>5</td></tr>
<tr><td>Al-Si12</td><td>G-AlSi13</td><td>raw</td><td>sand</td><td></td><td>135</td><td>3</td></tr>
<tr><td></td><td></td><td></td><td>permanent mold</td><td></td><td>175</td><td>5</td></tr>
<tr><td>Al-Zn5Mg</td><td>G-AlZn5MgCr</td><td>aged</td><td>sand</td><td></td><td>215</td><td>3</td></tr>
<tr><td></td><td></td><td></td><td>permanent mold</td><td></td><td>265</td><td>8</td></tr>
</table>

1.5.6 *Aluminum-Silicon-Magnesium Alloys (6000 Series)*

Magnesium and silicon are the main alloying elements. They form the intergranular compound Mg_2Si which readily passes into solution in the aluminum. Hence by heat treatment the strength of the alloy is increased ($f_{0.2} \cong 250 \, \text{N mm}^{-2}$), with a ductility of $\varepsilon_t = 10$ percent. These alloys are corrosion resistant.

Rolled sections as well as pipes and extruded profiles can be found. These alloys are used either in welded structures or in bolted and riveted connections.

1.5.7 Aluminum-Zinc-Magnesium Alloys (7000 Series)

These alloys consist of extruded or rolled heat-treated alloys. They can be divided into two subfamilies depending upon the percentage of copper as the third alloying element:

AlZnMgCu alloys are the highest-strength alloys after heat treatment ($f_{0.2} \cong 500 \text{ N mm}^{-2}$); conversely, they have low weldability and are not corrosion resistant, therefore requiring protection by plating or painting.

AlZnMg alloys give a strength of $f_{0.2} \cong 200 \text{ N mm}^{-2}$ with a ductility of $\varepsilon_t \cong 10$ percent. They are also corrosion resistant. These alloys are generally used in structural applications; they are particularly useful in welded structures owing to their self-tempering behavior, which allows the same strength in the heat-affected zone.

1.6 Mechanical and physical properties

1.6.1 Physical Properties

Figure 1.15 shows the main physical properties of aluminum at room temperature compared with those of steel and stainless steel.

With regard to the parameters most important for structural behavior,

Fig. 1.15 Room temperature properties of aluminum, steel and stainless steel

	Aluminum	Steel	Stainless steel
Average weight density (kg m^{-3})	2700	7850	7900
Melting point (°C)	658	1450–1530	1450
Linear thermal expansion coefficient (°C^{-1})	24×10^{-6}	12×10^{-6}	17.3×10^{-6}
Specific heat (cal g^{-1})	0.225	0.12	0.12
Thermal conductivity (cal cm s °C)	0.52	0.062	0.035
Electrical resistivity ($\mu\Omega$ cm)	2.84	15.5	70
Young's modulus (N mm^{-2})	68 500	206 000	206 000

Aluminum alloy structures

it can be said that:

The density of aluminum is approximately one-third of the density of steel (in different alloys it varies from 2600 to 2800 kg m^{-3}).

Young's modulus is also approximately one-third that of steel (in different alloys it varies from 68 500 to 74 500 N mm^{-2}.

The thermal expansion coefficient of aluminum is twice that of steel (in different alloys it varies from 19×10^{-6} to $25 \times 10^{-6}\,°C^{-1}$).

1.6.2 Mechanical Properties

The principal mechanical properties can be derived from a tension test. The load–displacement diagram (Fig. 1.16) is usually represented by a continuous curve without yielding.

The diagram can be divided into a linear elastic portion up to the proportional stress f_p (which is usually the stress corresponding to a residual strain of 0.01 percent), a nonlinear portion up to a 'knee', and a strain-hardening portion (the slope of which depends upon the type of alloy).

The stress corresponding to a residual strain of 0.2 percent is assumed to be the stress beyond which the behavior of the material is not elastic (it

Fig. 1.16

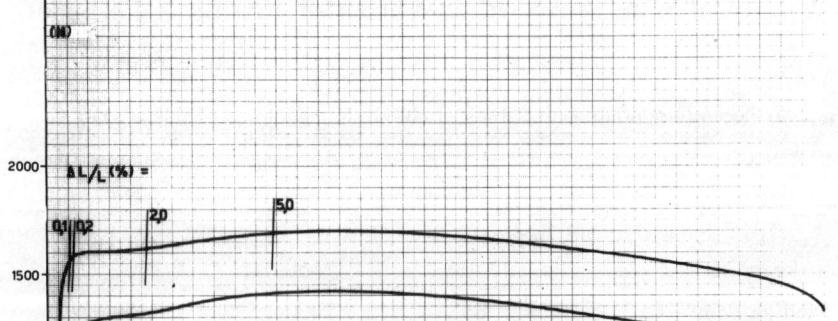

is also conventionally called the elastic limit). This stress $f_{0.2}$ is now internationally used in structural computations, with the same meaning as yield stress in steel.

Ultimate strength f_t corresponds to a strain value which can be defined as the limit of 'uniform elongation' corresponding to necking of the specimen. At this point stress increases at the neck and large deformations of the cross-section can be observed. At the same time the testing machine records a load which is decreasing owing to the failure of the specimen, with a value of elongation equal to ε_t. This value is usually used to evaluate the ductility of the material, that is, the capacity of the material to withstand large plastic deformations without experiencing failure.

The load–displacement diagrams in Figs 1.17 and 1.18 show how the heat-treatment process drastically changes the behavior of the material by affecting either strength or ductility.

It has already been noted (see Section 1.5) that the large family of aluminum alloys covers a wide range of values of strength, ranging from about 10 N mm^{-2} for the elastic limit of pure aluminum to 500 N mm^{-2} for 7000 series alloys. All these alloys are sufficiently ductile to be used in

Fig. 1.17

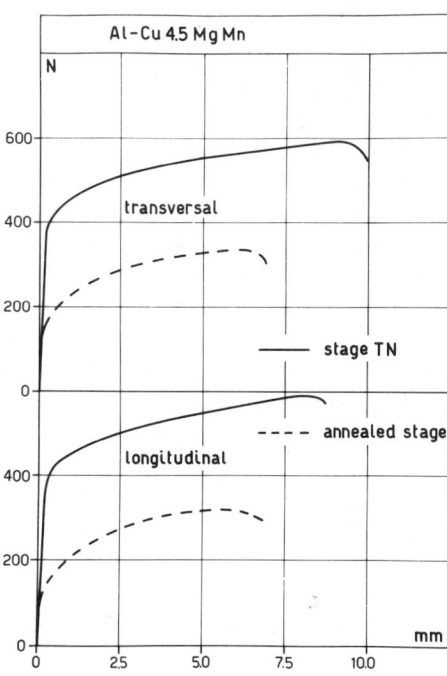

Aluminum alloy structures

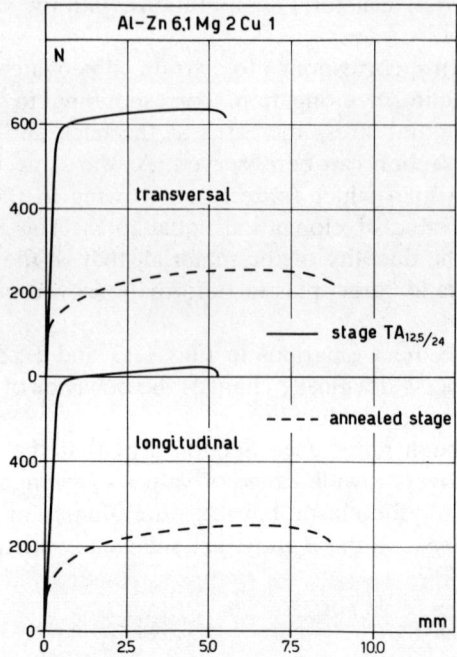

Fig. 1.18

structural applications. It should also be pointed out that these alloys do not lose strength at low temperatures; in fact, aluminum-magnesium alloys increase in strength at low temperatures.

Furthermore, in aluminum alloys a transition temperature cannot be defined. The transition temperature is defined in steel as the temperature below which brittle fracture is most likely to occur. It can be said that aluminum alloys are less susceptible than steel to brittle fracture either at low temperatures or at room temperature. Tests to measure fracture toughness are not usually required in specifications.

Starting from a temperature of about 80–100 °C, the mechanical properties of aluminum alloys decrease in the way described in Fig. 1.19, which refers to alloys fabricated in Switzerland.

1.6.3 Corrosion resistance

Usually aluminum needs no protection against atmospheric or chemical corrosive agents. In contrast to steel, in which the corrosion process is not naturally inhibited (apart from self-protecting steels), corrosion processes in aluminum tend to cease naturally.

Fig. 1.19

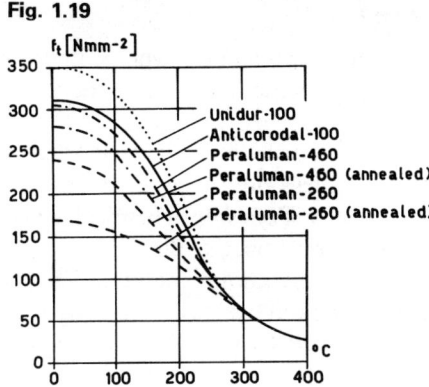

Corrosion can be divided into two classes: that which damages the esthetics of the structure, and that which damages the safety of the structure. Corrosion in aluminum can be considered to belong to the first type.

This fact can be explained from an electrochemical point of view. Aluminum oxide, being more stable than pure aluminum, closely adheres to the surface and thus protects the body of the metal. This natural film is usually a few microns thick; however, the thickness can be artificially increased to between 20 and 100 microns.

The more pure the aluminum, the higher is the corrosion resistance. It has excellent properties in an industrial environment, in proximity to the sea and when in contact with water with a neutral pH value.

The corrosion resistance of aluminum alloys can be estimated with respect to the alloying elements. The following can be said about additions of the various elements to pure aluminum:

Manganese in small percentages does not significantly vary corrosion resistance.
Magnesium in small percentages (5–6 percent) increases corrosion resistance with respect to some corrosive agents.
Silicon decreases corrosion resistance in some cases, although when combined with magnesium it provides corrosion resistance almost equal to that of pure aluminum.
Zinc together with magnesium provides good corrosion resistance if given proper heat treatment.
Nickel, iron and copper, even if combined with the previous elements, definitely reduce corrosion resistance.

The effects of corrosion depend on chemical composition, the fabrication process, the heat treatment and the stress field.

With respect to common atmospheric corrosion, the following types of

Aluminum alloy structures

corrosion can be defined:

(a) *Surface corrosion* is the most general type of uniform corrosion on all surfaces. It usually stops immediately after the formation of a protective oxide film.
(b) *Concentrated corrosion* is usually irregularly distributed on different parts of the surface. Unless due to a persistent chemical action, this corrosion ceases naturally. Persistent chemical action can be observed in the case of smoke acting on a ceiling.
(c) *Intergranular corrosion* is a complex metallurgical process usually only present in the 2000 series alloys, which are used in the aeronautical industry. It can be avoided by specific heat treatments.
(d) *Lamellar corrosion* produces layers very similar to rust, and can be considered as a particular case of the previous type of corrosion. It occurs in those products in which the fabrication processes produce a stratified microcrystal structure (especially rolled sections) without a soft-annealing process.
(e) *Stress corrosion* is the most dangerous type of corrosion. It can be observed when microcracks grow in the tension zones of the stress field owing to the presence of a corrosive agent. Tension can be caused either by external loads or by residual stresses. Alloys susceptible to intergranular corrosion also exhibit this type of corrosion. Alloys generally used in civil engineering can be considered not susceptible to this type of corrosion, provided they are properly fabricated and erected.
(f) *Corrosion by contact* (electromechanical). When aluminum is in contact with another metal with a liquid between the two metals, a voltaic pile is established. The material with the lower voltage flows in solution and covers the other metal, unless this has been neutralized.

The contact voltage of steel with respect to aluminum is 850 mV. It is therefore advisable to avoid direct contact between the two metals, and insulating material can be used to separate them. When steel bolts are used to make a connection, the bolts should be galvanized or electroplated.

In a humid and corrosive environment, contact between aluminum and its alloys and nonmetallic materials such as concrete, mortar, timber and bricks should be avoided. In these cases the use of bituminous or epoxy coatings is sufficient to prevent corrosion. When the environmental conditions are such that protection of the surfaces of the metal is required, the following systems can be chosen:

Painting
Anodization
Metal spraying
Plastic coating

1.7 General criteria for design [5–9]

1.7.1 Comparing Aluminum with Steel [10]

In order to clearly and efficiently compare the physical and mechanical properties of steel and aluminum alloys, it seems logical to summarize the concepts of the previous sections and to outline the main peculiarities of different alloys.

The main properties of aluminum are:

Lightness The specific weight is $\gamma = 2700$ kg m^{-3}, equal to one-third that of steel.

Corrosion resistance When exposed to atmospheric agents, the metal is naturally covered by a protective oxide film.

Aluminum is also very ductile ($\varepsilon_t \cong 40$ percent), but on the other hand its strength is very low for structural applications ($f_{0.2} \cong 20$ N mm^{-2}). In order to increase strength, a cold-working process can be used; however, this process does not greatly increase strength ($f_{0.2} \cong 100$ N mm^{-2}), and ductility is drastically decreased (up to one-tenth of the initial value).

Another way of increasing the strength of the material is to alloy aluminum with other elements (AlMn, AlMg alloys). After this process, the strength can be higher than 100 N mm^{-2}, with a ductility equal to $\varepsilon_t = 10$ percent; however, the corrosion resistance is decreased.

Even higher strengths can be obtained if heat treatment is applied. The strength goes up to $f_{0.2} = 250$ N mm^{-2} in AlSiMg alloys, and can reach $f_{0.2} = 350$–400 N mm^{-2} in AlZn and AlCu alloys.

Figure 1.20 is a summary comparison of some aluminum alloys and the

Fig. 1.20 Comparison of properties of aluminum alloys and two common mild steels

	Aluminum alloys		Steel	
$f_{0.2}$ (f_y) (N mm^{-2})	AlMg4.5Mn AlMgSi1 AlZnMg1	~140 ~260 ~360	Fe360 Fe510	~235 ~350
f_t (N mm^{-2})	AlMg4.5Mn AlMgSi1 AlZnMg1	~280 ~320 ~410	Fe360 Fe510	~360 ~510
E	70 000 N mm^{-2}		200 000 N mm^{-2}	
ε_t	10–25%		25–30%	
γ	26 500 N m^{-3}		77 000 N m^{-3}	
α	0.000 02		0.000 01	

Aluminum alloy structures

most commonly used mild steels Fe360 and Fe510. In particular, the following mechanical properties are given:

Elastic limit ($f_{0.2}$)
Ultimate strength (f_t)
Young's modulus E

It should be noted that AlMgSi and AlZnMg alloys have an elastic limit equal to that of steels Fe360 and Fe510, respectively. Generally speaking it can be said that aluminum alloys offer a wider range of strength than steel. Therefore the concept of a 'fourth dimension' in metal construction seems to be more appropriate to aluminum alloys than to steel.

On the other hand, Young's modulus is one-third that of steel, thus giving more frequent problems of deformation and of instability.

In a comparison of the stress–strain curves of AlMgSi alloy and Fe360 steel (Fig. 1.21) it can be observed that:

Aluminum alloys have a strain-hardening portion without a horizontal line corresponding to yielding.
Ultimate elongation is lower than that of steel.
The $f_t/f_{0.2}$ ratio is lower than that of steel (1.2 against 1.5).

Both materials behave linear elastically up to the elastic limit, which basically represents the working range of structures. They differ, instead, in inelastic behavior. A concise comparison can be made if the ratio f/γ – termed the 'strength rating' – between strength ($f_{0.2}$ or f_t) and specific weight (γ) is considered. Figure 1.22 shows that this comparison is extremely positive for aluminum alloys. However, it should be pointed out that it is not always possible to take advantage of this benefit offered by aluminum alloys, especially when the material is under compression. In this case, owing to the small value of Young's modulus, buckling is more likely to occur than in steel structures.

It should also be noted that the value of the coefficient of thermal expansion (α) of aluminum is twice that of steel. This means that the structure is more sensitive to thermal variations, and thus has higher

Fig. 1.21

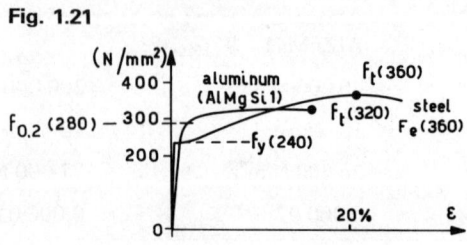

Fig. 1.22

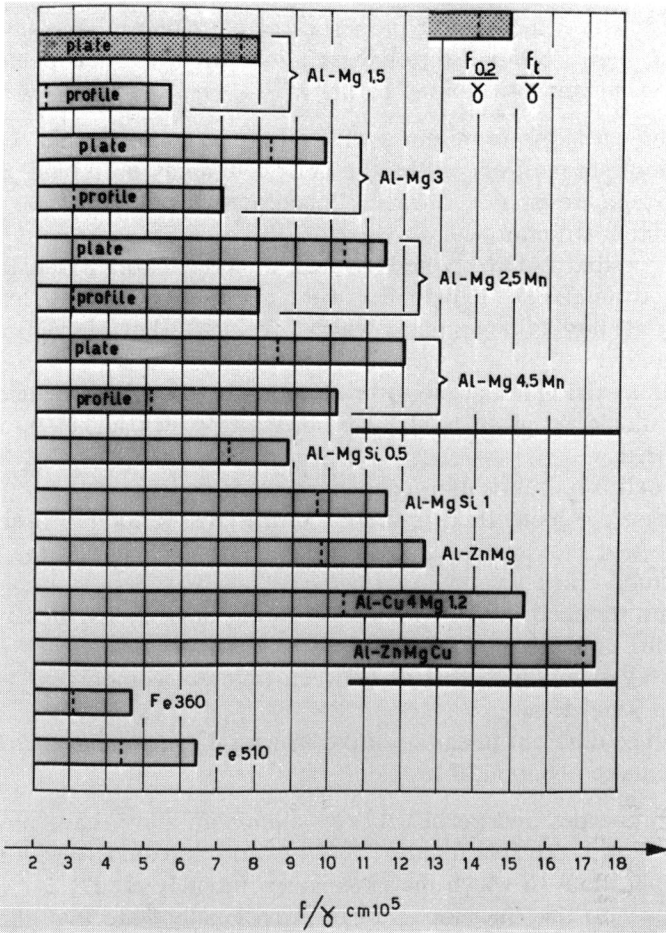

deformation when it is not constrained. This fact has to be taken into account when designing support apparatus.

When the structure is constrained against thermal deformation, residual stresses will be about 30 percent lower than those in steel structures since they are proportional to the product αE. In contrast to steel structures, aluminum alloys do not have a transition temperature.

Other differences between aluminum and steel are due to industrial fabrication processes. Structural components can be fabricated by rolling, extrusion, casting and drawing processes. Of particular interest is the extrusion process, which allows fabrication of profiles of any shape and in particular those not obtainable by a rolling process (see Section 2.1).

1.7.2 Possible Structural Applications [11–20]

The physical and mechanical properties of aluminum alloys have been examined, and a comparison of these alloys with steel has been made. The following conclusions may be drawn:

(a) Aluminum alloys represent a wide family of materials and cover the range of strength offered by the most commonly used mild steels.
(b) Corrosion resistance normally makes it unnecessary to protect aluminum structures.
(c) It is possible to have bolted, riveted, or welded connections.
(d) The advantage of the lightness of the material is offset by the higher deformability of aluminum, which gives a higher susceptibility to instability.
(e) The material is not prone to brittle fracture, though particular attention should be given to those problems in which higher ductility is required.
(f) The extrusion fabrication process makes it possible to:
Increase the geometrical properties of the cross-section by designing a shape that simultaneously gives the minimum weight and the highest structural efficiency.
Obtain stiffened shapes without using built-up sections, thus avoiding welding or bolting.
Simplify assembly processes between different components and improve joint details.
Combine different functions of the structural component, thus achieving a more economical profile.

All the properties and peculiarities of aluminum alloys have now been considered. Is it possible for this 'new' material to compete with steel in those applications in which the latter is traditionally used?

In answer to this question, we can provisionally state that aluminum alloys are economical in those applications in which full advantage is taken of their properties. In particular, lightness makes it possible to have:

Simpler erection
Transport of fully prefabricated components
Reduction of loads transmitted to foundations
Economy of energy either during execution and/or service
Reduction of physical labor

and corrosion resistance makes it possible to have:

Reduction of maintenance expenses
Better performance in corrosive environments

What are the structural applications that best fit these properties? The following cases are typical:

(a) Long-span roof systems in which live loads are small compared with dead loads.
(b) Structures located in inaccessible places far from the fabrication shop. In these cases transport economy and ease of erection are of extreme importance. An example is electrical transmission towers, which can be carried by helicopter completely assembled.
(c) Structures situated in corrosive or humid environments such as swimming pool roofs, river bridges and hydraulic structures.
(d) Structures having moving parts, such as moving bridges, where lightness means economy of power under service.

Over the last thirty years aluminum alloys have been successfully used in different countries around the world in civil engineering.

1.7.3 International Specifications [21–24]

Owing to a greater use of aluminum alloys in construction, several countries have published specifications for the design of aluminum structures.

The USA Aluminum Association has revised its specifications, and in Europe the English (CP 118), Italian (UNI 8634), Swedish (SVR), French (DTU), German (DIN 4113) and Austrian specifications have been published.

The European Convention for Constructional Steelwork (ECCS-CECM) was founded in 1955. The aim of this convention is to study, through about fifteen committees and several working groups, all aspects of steel construction, and to use the results to provide European specifications. In 1967 aluminum was studied for the first time by a specific board of the ECCS (subcommittee 1.1, chairman Carpena). In 1970 the chairmanship passed to the author. In 1973 this subcommittee was upgraded to committee 16. Since 1978 it has been called committee T2, and in 1979 Valtinat replaced the author as chairman. It is due to the efforts of this ECCS aluminum committee and of its working groups that the first edition of the European Recommendations for Aluminum Alloy Structures became available in 1978 [25]. These recommendations represent the first attempt internationally to unify computational methods for the design of aluminum alloy constructions in civil engineering and in other applications. A semi-probabilistic limit state methodology has been used (see Chapter 3), on which the main international specifications are based.

The philosophy of the ECCS recommendations and their implications will be explained in the following chapters. Comparisons with different national specifications will also be provided.

References

1. Panseri, C., *Manuale di Tecnologia delle Leghe Leggere da Lavorazione Plastica* (*Handbook of Technology of Wrought Light Alloys*), Hoepli, Milan, 1957.
2. Varley, P. C., *The Technology of Aluminum and its Alloys*, Newnes-Butterworths, London, 1970.
3. Veschi, D., *Alluminio e sue Leghe* (*Aluminum and its Alloys*), Pàtron, Bologna, 1976.
4. ECCS, *Note general sur l'utilisation des alliages d'aluminum* (*General remarks on the use of aluminum alloys*), Doc. 16-76-1, 1976.
5. Stussi, F., *Tragwerke aus Aluminium* (*Aluminum Constructions*), Springer-Verlag, Berlin, 1955.
6. Aluminium Federation, *Aluminium in Structural Engineering, Proceedings of a Symposium*, London, June 11–12, 1963.
7. Sedlacek, H., Aluminum in civil engineering bearing structures (in German), *Bauingenieur*, 1967, No. 4, pp. 117–23; No. 5 pp. 182–8.
8. Steinhardt, O., Aluminum im Konstruktiven Ingenierbau (Aluminum constructions in civil engineering) *Aluminium*, **47**, 1971, pp. 131–9; 254–61.
9. Massonnet, Ch. and Frey, F., Conception et calcul des structures en aluminium (Design and calculation of aluminum structures), *Metallurgie*, **XV**, 1975, No. 1, pp. 17–33.
10. Mazzolani, F. M., Concezione e calcolo delle strutture in leghe d'alluminio (Design and calculation of aluminum alloy structures), *Alluminio*, 1979, No. 6, pp. 279–97.
11. Alusuisse, Ponts roulants en aluminum (Cranes of aluminum), *Revue Suisse de l'Aluminum*, 1961, No. 5.
12. Dirilgen, N., *Grands profilés et leurs applications* (*Big profiles and their applications*), Aluminum Suisse SA, Zurich.
13. Dirilgen, N., Emploi de l'aluminum dans les structures portantes (Use of aluminum in bearing structures), *Revue Suisse de l'Aluminum*, 1969, No. 2.
14. Rossi, D., L'alluminio nell'industria chimica (Aluminum in the chemical industry), *Alluminio e Nuova Metallurgia*, 1969, No. 6/7, pp. 281–98; 335–51.
15. KOBA, Aluminum, Granges Essem AB, Vasteras, 1971.
16. CEDAL, *Estructuras de Aluminio* (*Aluminum Structures*), Graficas Rey, Madrid.
17. CIDA, *Aluminum Structures*, Aluminum-Verlag GmbH, Dusseldorf, 1972.
18. Koser, J., *Handbuch uber das Konstruieren mit Aluminum* (*Handbook on How to Build in Aluminum*), Fachverband der Metallindustrie Osterreichs, Vienna.
19. Flogl, H., Naderer, R. and Koser, J., Ein Turmbehalter aus Aluminum (A water-tower in aluminum), *Osterreichische Ingenieur-Zeits-Schrift*, 1974, pp. 61–6.
20. Besnard, Delay, Jacomet, Molina, Re (Aluminum Pechiney); Clade; Cuille, Lorin, Martin, Moreau (Electricité de France), L'emploi des alliages d'aluminum dans les supports de lignes de transport d'énergie électrique (The

use of aluminum in power transmission towers), *Institut Technique du Batiment et des Travaus Publiques*, 1976, No. 340, pp. 111–34.
21. Massonnet, Ch., *Comparison des principales normes sur le calcul des barres comprimées en alliage d'aluminum (Comparison of principal codes on the calculation of compression members of aluminum alloys)*. CIDA, Doc. No. 67.77, 1967.
22. Ramirez, J. L., Aplicacion de las Recomendaciones a Casos Practicos (Application of Recommendations to Practical Cases). *Proceedings of Jornadas Tecnicas sobre Estructure en Aluminio*, Bilbao, November 16–17, 1978.
23. Mazzolani, F. M., The bases of the European Recommendations for design of aluminum alloy structures, *Alluminio*, 1980, No. 2, pp. 77–94.
24. Mazzolani, F. M., European Recommendations for Aluminum Alloy Structures and their comparison with National Standards, *Proceedings of the 7th Int. Light Metal Congress*, Vienna, June 26, 1981.
25. ECCS, *European recommendations for aluminum alloy structures*, First edn, 1978.

2 The structural material

2.1 Types and shapes

The industrial fabrication processes described in Section 1.3 allow the manufacture of a wide range of elements suitable for structural engineering construction. Structural shapes can be obtained by:

Extrusion
Welding built-up members
Cold forming

Special components, such as joints in three-dimensional triangulated structures, can be obtained by casting.

Extruded profiles allow the manufacture of any shape, the only limitation being to the capacity of the press.

The examples given in Fig. 2.1 show the various possibilities offered by the extrusion process of obtaining profiles commonly used in metal construction. Standard double T profiles, which are commonly used in steel, can be made even more complex by adding bulbs or lips at the end of the flanges (Fig. 2.1a). Analogous reinforcements can be added to standard channel sections from which Ω sections can be obtained (Fig. 2.1b). Among these sections, in addition to standard shapes, improved profiles with bulbs added to webs and/or flanges can also be obtained. The presence of a double web can facilitate the connection to another element (Fig. 2.1c). Hollow sections, box sections and tubes are made in various shapes such as rectangular, square and circular, with the thickness being constant or varying along the cross-section. Internal or external longitudinal stiffeners can be added to these sections. External stiffeners can also be used as connection plates to join other members (Fig. 2.1d).

The high flexibility, which allows standard shapes to be modified by adding bulbs, stiffeners and reinforcement, can be observed in these examples; the design is more rational and economical because of better mass distribution and rigidity. Instead of standard steel rolled shapes, analogous profiles, even though individually tailored, can be obtained by extrusion.

The structural material

Fig. 2.1

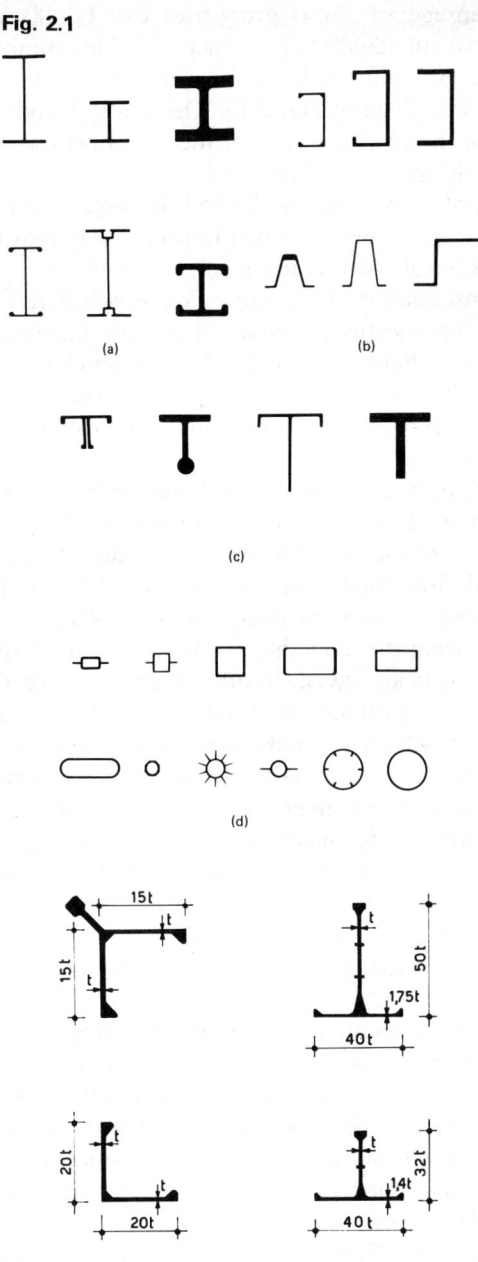

Aluminum alloy structures

Further advantage of these properties can be taken if a profile is designed to have an ideal shape which provides minimum weight and maximum structural efficiency at the same time. An example of this design procedure is given in Fig. 2.1e. The angle, T and Y sections shown in this figure are the cross-sections of the members used in a triangulated structure truss designed in France as a support for power cables. The accurate design of these profiles allowed the weight to be minimized and the assembly system to be simplified in such a way that to use aluminum was more economical than to use steel.

The main consequence of the great flexibility offered by the extrusion process is the impracticality of standardizing the shapes of aluminum alloy profiles and, hence, having a structural shape handbook like that used in steel structures. In aluminum alloy structures, sections can be designed individually with respect to the functional and structural requirements of each component.

A specific drawplate has to be built for each newly designed cross-section. This new drawplate, when added to those already existing, enlarges the range of the available profiles produced by a given press. The complete set of drawplates can be considered to be a catalog of the current production of each extrusion centre. When a new requirement can be satisfied using the available profiles, the cost of the drawplate and its amortization costs are saved. In other cases it is necessary to accurately evaluate how many profiles are needed in order to establish the cost effectiveness of producing a new drawplate. However, when extruded profiles are to be used, it is generally advisable to insure co-operation between the structural engineer and the fabrication shop expert. In this way it will be possible to match structural shape requirements with the technological and economical possibilities offered by the extrusion process.

If the dimensions of the cross-section are larger than those obtainable by extrusion, a welded section can be built up from rolled and/or extruded plates (Fig. 2.2).

Nowadays, welding techniques in aluminum alloys are so improved that there are no difficulties involved. Usually, automatic or semi-automatic processes under inert gas are used (MIG and TIG).

At the same time, in the last few years, theoretical and experimental research on the structural behavior of welded built-up members has also improved. The most important aspects of the resistance and stability of these sections have been studied. Three typical welded cross-sections are given in Fig. 2.3. They have been theoretically and experimentally analyzed in research carried out by the committee for light alloy structures of the European Convention for Constructional Steelwork (ECCS). They are two double T shapes (P and T profiles) and a box section (C

The structural material

Fig. 2.2

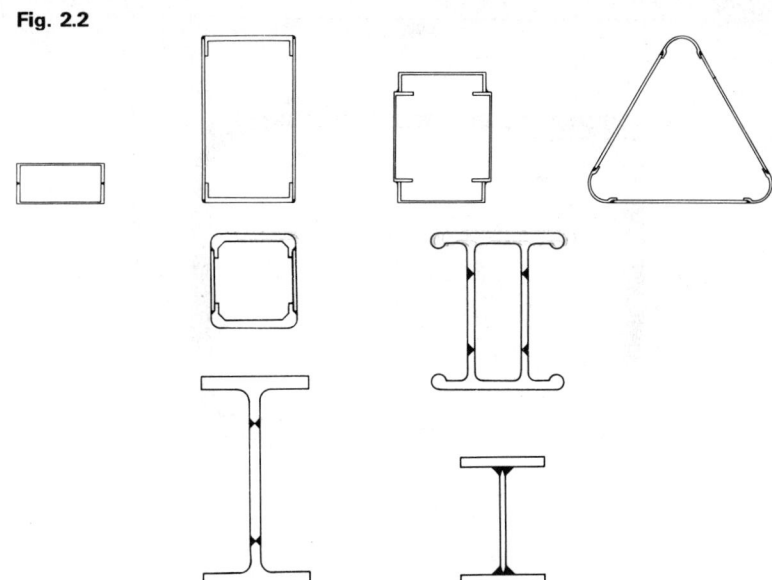

profile). The double T shape can be obtained by adding two extruded Ts, which represent the flanges, to a plate which represents the web (see Fig. 2.3a, P profile), or traditionally, by adding three plates as for steel structures (see Fig. 2.3b, T profile). The first solution offers the advantage of having welds closer to the center of gravity of the section, thus moving the heat-affected zones away from the zones of greatest bending. In particular, moving from solution (b) to solution (a) leads to a less unfavorable residual stress field, as will be explained in Section 2.5.1.

Analogous to steel shapes, cold-formed thin-walled shapes can be obtained using bending machines (see Fig. 2.4a) and sheets thinner than 4 mm are used to fabricate L, C, Z, V, Ω profiles in this way (see Fig. 2.4b). Corrugated and trapezoidal sheets can also be obtained, and are used for ceilings and cladding. These elements' strength is due to their shape, which allows extremely light structures. On the other hand, special care has to be given to local instability problems (local buckling), which are more likely to occur since the sheets used are extremely thin.

2.2 Characterization of σ–ε law

2.2.1 Need for an Analytical Model

Of the several difficulties which are encountered in the theoretical analysis of static and stability problems for aluminum alloy structures, the first is related to the idealization of the tensile material properties.

Aluminum alloy structures

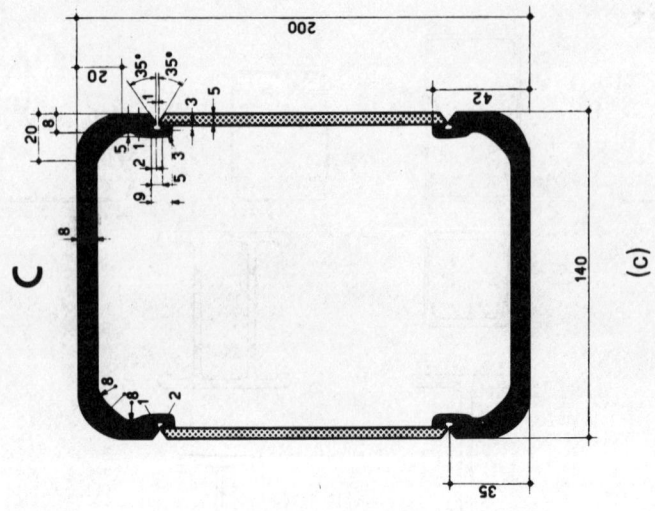

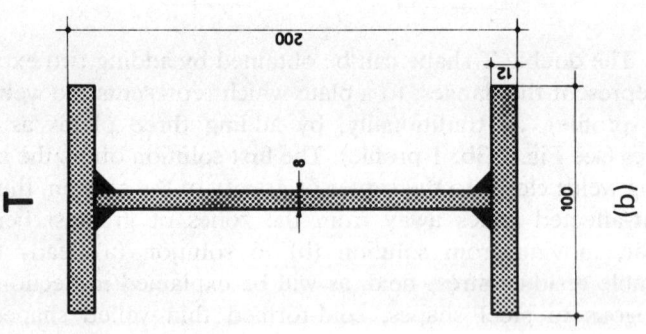

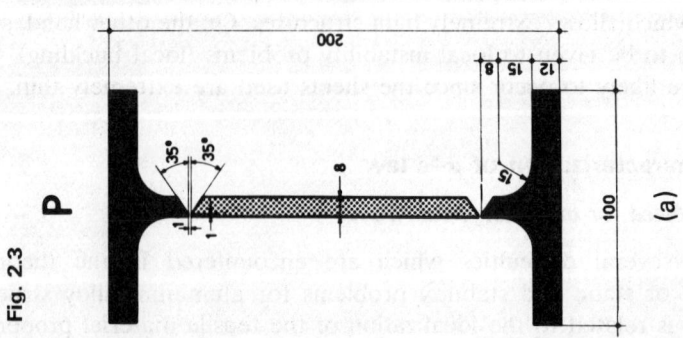

Fig. 2.3

Fig. 2.4

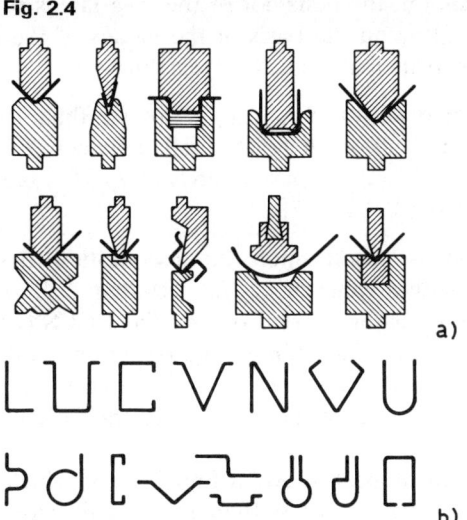

It was shown in Chapter 1 that the aluminum alloys commonly used in structures possess mechanical properties considerably different from each other. Even properties of the same alloy differ when it undergoes different fabrication processes and heat treatments.

Furthermore, the continuous behavior of the σ–ε curve, obtainable in a tension test, cannot be simplified to an elastic/perfectly plastic behavior as is done for mild steel. More complex models have to be used.

The yield stress, which is conventionally identified as the elastic limit $f_{0.2}$ in aluminum alloys, is not sufficient to realistically model the stress-strain law and to classify different alloys on the basis of their mechanical properties. It is also necessary to take into account the different values of Young's modulus, which is not constant as in steels, and the strain-hardening effect related to the fabrication process. These factors lead to σ–ε curves which differ greatly from each other.

These are the reasons why an accurate structural analysis cannot be based upon simplified models, as in steel structures. Instead the analysis has to be based upon a material with a generalized inelastic behavior. This need has been recognized by several authors who have formulated various proposals on this subject. The most significant of these will be explained in the following sections.

2.2.2 Sutter Classes [1]

Sutter [1] proposed a classification for aluminum and magnesium alloys based upon the observation that the σ–ε diagram is greatly affected by the $f_{0.2}/f_{0.1}$ ratio, known as the *strain-hardening parameter*.

Aluminum alloy structures

In this classification, the behavior of the σ–ε law is related to the heat treatment of the alloy on the basis of the values of the strain-hardening parameter. Three principal classes are established:

Class 1   annealed alloys (if $f_{0.2}/f_{0.1}$ is greater than 1.060)
Class 2   tempered alloys (if $f_{0.2}/f_{0.1}$ falls between 1.045 and 1.060)
Class 3   quenched and tempered alloys (if $f_{0.2}/f_{0.1}$ falls between 1.030 and 1.045)

This classification is based upon the assumption that there are no aluminum alloys with a value of $f_{0.2}/f_{0.1}$ less than 1.030. In reality, it has been shown during testing carried out by the ECCS committee on light alloy structures in 1972 that aluminum alloys with a value of $f_{0.2}/f_{0.1}$ less than 1.030 can be found. This can be explained by the fact that zinc and magnesium alloys have been developed since the Sutter classification was proposed.

It therefore seems logical to add a fourth class, which comprises those alloys with a value of $f_{0.2}/f_{0.1}$ less than 1.030. This class takes account of the most recent alloys.

2.2.3 Piecewise Idealization

The simplest way to model the stress–strain relationship of aluminum alloys is with two straight lines representing an elastic-hardening diagram (see Fig. 2.5a). The first part of the diagram, which represents the elastic portion, starts from the origin with a slope equal to the Young's modulus E_0. The second part of the diagram, which represents the strain-hardening portion, has a slope equal to the tangent modulus E_1. The intersection of the two lines defines the conventional value f_p of the elastic limit of proportionality.

This type of approach can be improved if an intermediate line which is a tangent to the actual curve at the 'knee' is introduced (see Fig. 2.5b). The polygon which derives from this idealization is therefore characterized by three moduli E_0, E_1 and E_2, and by two reference stresses f_1 and f_2.

A model similar to this, though on the safe side, has recently been used in the latest German specification (DIN 4113, first calculation method) (Fig. 2.5c).

2.2.4 Continuous Models of Form $\sigma = \sigma(\varepsilon)$

It is very difficult for a law of the form $\sigma = \sigma(\varepsilon)$ to be both very general and very close to the actual inelastic behavior of the material; the law is based upon one expression. In any case, to be utilized in design, whatever

The structural material

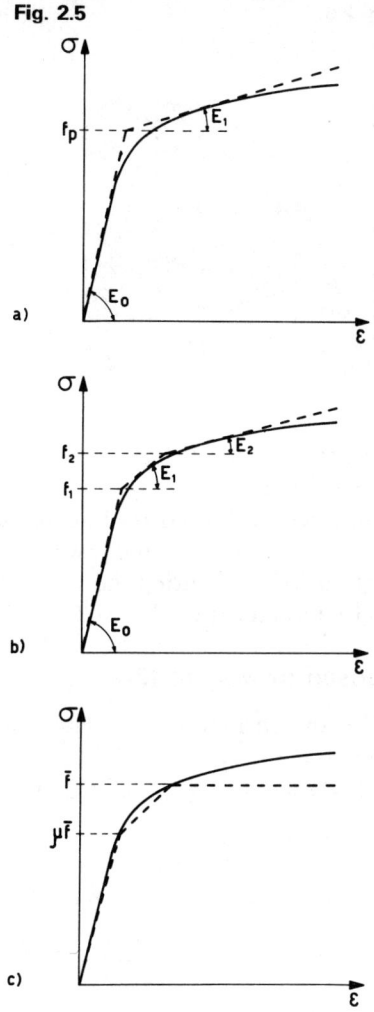

Fig. 2.5

expression is used for the σ–ε law has to be related to the conventional experimental values such as $f_{0.1}$, $f_{0.2}$ and E.

It is usually convenient to identify three separate portions of the function $\sigma = \sigma(\varepsilon)$, in each of which the behavior of the material differs. They can be defined in the following way (Fig. 2.6):

Region 1   elastic behavior
Region 2   inelastic behavior
Region 3   strain-hardening behavior

In each region, the σ–ε relationship which represents the behavior has to

Fig. 2.6

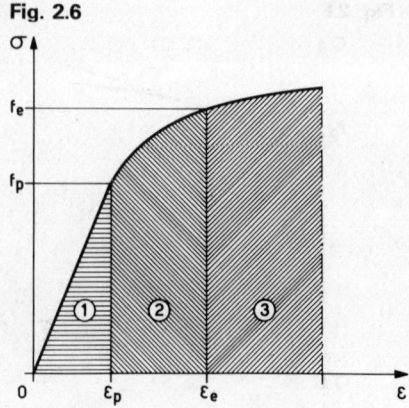

be found. To ensure continuity the three different laws have to produce coincident points at their limits.

This methodology has been followed by Baehre (see Section 2.2.4.1), whose studies have been applied in the Swedish specifications, and by Mazzolani (see Section 2.2.4.2) during the research carried out by the committee on light alloy structures of the ECCS.

2.2.4.1 Model proposed by Baehre [2]

Refer to Fig. 2.7. The dimensionless σ-ε laws in the three different regions can be expressed as follows:

Region 1 (for $0 < \varepsilon/\varepsilon_{0.2} \leq \varepsilon_p/\varepsilon_{0.2}$ or $\varepsilon_{max}/\varepsilon_{0.2} \leq 0.5$):

$$\frac{\sigma}{f_{0.2}} = \frac{\varepsilon}{\varepsilon_{0.2}} \qquad (2.1)$$

Region 2 (for $\varepsilon_p/\varepsilon_{0.2} < \varepsilon/\varepsilon_{0.2} \leq \varepsilon_{f_{0.2}}/\varepsilon_{0.2}$ or $0.5 < \varepsilon_{max}/\varepsilon_{0.2} \leq 1.5$):

$$\frac{\sigma}{f_{0.2}} = -0.2 + 1.85 \frac{\varepsilon}{\varepsilon_{0.2}} - \left(\frac{\varepsilon}{\varepsilon_{0.2}}\right)^2 + 0.2\left(\frac{\varepsilon}{\varepsilon_{0.2}}\right)^3 \qquad (2.2)$$

Region 3 (for $\varepsilon_{f_{0.2}}/\varepsilon_{0.2} < \varepsilon/\varepsilon_{0.2} \leq \varepsilon_u/\varepsilon_{0.2}$ or $1.5 < \varepsilon_{max}/\varepsilon_{0.2} \leq \varepsilon_u/\varepsilon_{0.2}$):

$$\frac{\sigma}{f_{0.2}} = \frac{f_u}{f_{0.2}} - 1.5\left(\frac{f_u}{f_{0.2}} - 1\right)\frac{\varepsilon_{0.2}}{\varepsilon} \qquad (2.3)$$

$\sigma/f_{0.2}$ dimensionless value of the stress referred to the elastic limit $f_{0.2}$
$\varepsilon/\varepsilon_{0.2}$ value of the strain referred to the value $\varepsilon_{0.2} = f_{0.2}/E$
$\varepsilon_{f_{0.2}}$ value of the strain corresponding to the stress $f_{0.2}$
ε_u value of the strain corresponding to the stress f_u
ε_p value of the strain at the proportional limit
ε_{max} maximum value of the strain at the extreme fiber of a cross-section subjected to bending.

Fig. 2.7

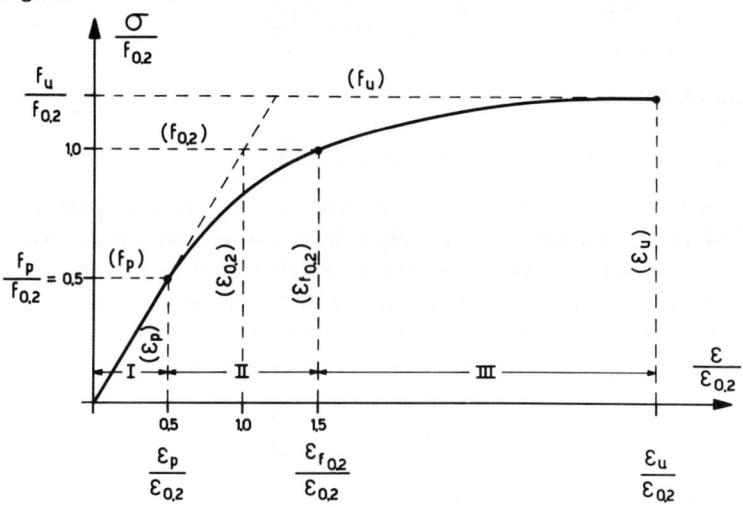

f_u and ε_u values are reference values, bounding the range of validity of the stress–strain law.

If we take

$$\varphi = f_u/f_{0.2} \qquad \vartheta = \varepsilon_u/\varepsilon_{0.2} \qquad \eta = \varepsilon/\varepsilon_u \qquad (2.4)$$

then Eqs 2.1, 2.2 and 2.3 will be rearranged as follows:

Region 1

$$\frac{\sigma}{f_u} = \eta \frac{\vartheta}{\varphi} \qquad (2.5)$$

Region 2

$$\frac{\sigma}{f_u} = \frac{1}{\varphi}[-0.2 + 1.85\vartheta\eta - (\vartheta\eta)^2 + 0.2(\vartheta\eta)^3] \qquad (2.6)$$

Region 3

$$\frac{\sigma}{f_u} = 1 - \left(1 - \frac{1}{\varphi}\right)\frac{1.5}{\vartheta}\eta \qquad (2.7)$$

This approach has been used by the author to formulate moment–curvature laws for aluminum alloy cross-sections. This enables the safety factor, required by specifications with respect to the elastic limit $f_{0.2}$, to be checked. This can be directly compared with the safety corresponding to reaching the yield limit in steel structures.

The same approach has also been used to find instability curves. In this case the equation of the third region has been modified and simplified in

Aluminum alloy structures

the following way:

$$\frac{\sigma}{f_{0.2}} = 1.3 - 0.45 \frac{\varepsilon_{0.2}}{\varepsilon} \qquad (2.8)$$

instead of Eq. 2.3.

2.2.4.2 Model proposed by Mazzolani [3]

With respect to the three regions defined in Fig. 2.6, the upper limit of the first linear portion is f_p, usually referred to as the *proportional limit stress*. This is defined as corresponding to a deformation of approximately 0.005 percent (smaller than the conventional limit of proportionality, defined for a residual deformation equal to 0.01 percent).

The following formula can be used to define f_p:

$$f_p = f_{0.2}\left[1 - \left(1 - \frac{f_{0.1}}{f_{0.2}}\right)^m\right] \qquad (2.9)$$

Consequently ε_p is defined as

$$\varepsilon_p = f_p/E \qquad (2.10)$$

The exponent m in Eq. 2.9, is a function of the strain-hardening parameter $f_{0.2}/f_{0.1}$. The validity of the following relation has been experimentally proved:

$$m = 2.30 - \frac{f_{0.2}}{f_{0.1}} 1.75 \qquad (2.11)$$

Hence this parameter can be related to Sutter classes (see Section 2.2.2). Each of these classes can therefore be represented by its mean value m_{med} (Fig. 2.8).

The second portion of the curve, with its characteristic 'knee' behavior, is valid between the values of f_p and f_e, the latter being coincident with the conventional *elastic limit* of 0.2 percent. Therefore we have

$$\begin{aligned}f_e &= f_{0.2}\\ \varepsilon_e &= 0.002 + f_{0.2}/E\end{aligned} \qquad (2.12)$$

Starting from the point with coordinates (f_e, ε_e), the third portion of the diagram is formed by a curve which is asymptotic to a given value of the stress f_u.

When stresses and strains of the $\sigma-\varepsilon$ law are nondimensionalized with respect to the proportional limit values (f_p, ε_p), the variables of the dimensionless $\bar{\sigma}-\bar{\varepsilon}$ law become

$$\bar{\sigma} = \sigma/f_p \qquad \bar{\varepsilon} = \varepsilon/\varepsilon_p \qquad (2.13)$$

where f_p and ε_p are given by Eqs 2.9 and 2.10.

Fig. 2.8

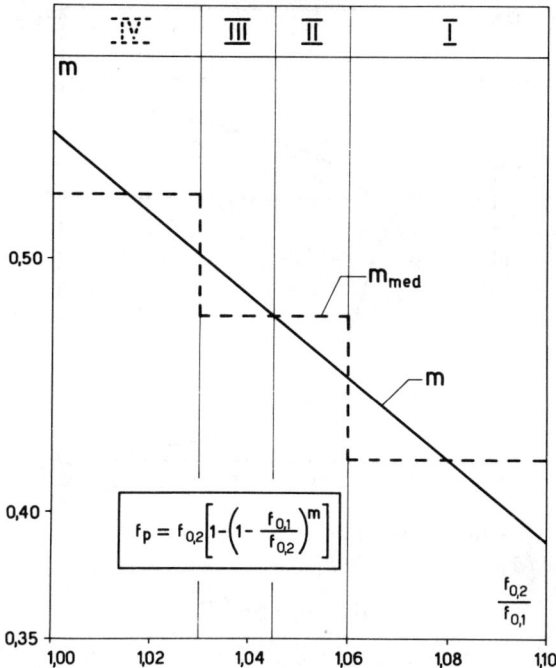

The nondimensionalized relations $\bar{\sigma}-\bar{\varepsilon}$ in the three regions can then be expressed in the following way (Fig. 2.9):

First region $(0<\bar{\sigma}<1;\ 0<\bar{\varepsilon}<1)$:

$$\bar{\sigma} = \bar{\varepsilon} \qquad (2.14)$$

Second region $(1<\bar{\sigma}<\bar{\sigma}_1;\ 1<\bar{\varepsilon}<\bar{\varepsilon}_1)$:

$$\bar{\sigma} = \bar{\varepsilon} - \beta(\bar{\varepsilon}-1)^\alpha \qquad (2.15)$$

where:

$$\bar{\sigma}_1 = \frac{f_e}{f_p} = \left[1-\left(1-\frac{f_{0.2}}{f_{0.1}}\right)^m\right]^{-1}$$

$$\bar{\varepsilon}_1 = \frac{\varepsilon_e}{\varepsilon_p} = \left[1+0.002\frac{E}{f_{0.2}}\right]\left[1-\left(1-\frac{f_{0.2}}{f_{0.1}}\right)^m\right]^{-1} \qquad (2.16)$$

and the semiempirical coefficients α and β are given by:

$$\alpha = \frac{1-\bar{\varepsilon}_1}{\bar{\sigma}_1-\bar{\varepsilon}_1}\left[1-f_{0.2}\left(1-\frac{f_{0.1}}{f_{0.2}}\right)\right]$$

$$\beta = \frac{\bar{\varepsilon}_1-\bar{\sigma}_1}{(\bar{\varepsilon}_1-1)^\alpha} \qquad (2.17)$$

Aluminum alloy structures

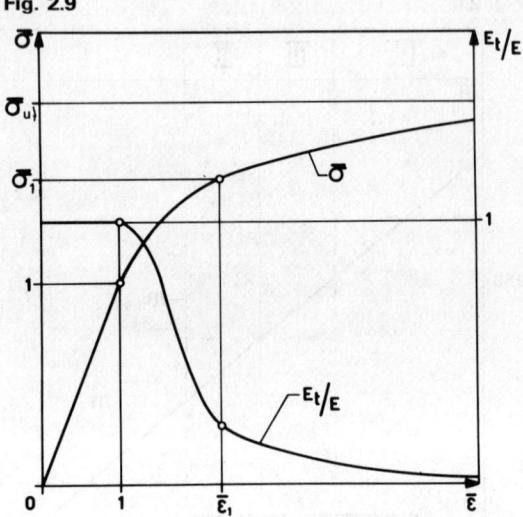

Fig. 2.9

The units of $f_{0.2}$ are $t\,cm^{-2}$.

Third region $(\bar{\sigma}_1 < \bar{\sigma} < \bar{\sigma}_u; \bar{\varepsilon} > \bar{\varepsilon}_1)$:

$$\bar{\sigma} = \bar{\sigma}_u - \psi e^{-\gamma(\varepsilon-\bar{\varepsilon})} \tag{2.18}$$

where

$$\bar{\sigma}_u = \bar{\sigma}_1 \left(\frac{f_{0.2}}{f_{0.1}}\right)^3$$

$$\psi = \bar{\sigma}_u - \bar{\sigma}_1 = \bar{\sigma}_1 \left[\left(\frac{f_{0.2}}{f_{0.1}}\right)^3 - 1\right] \tag{2.19}$$

$$\gamma = \frac{f_{0.2}}{\gamma}\left(1 - \frac{f_{0.1}}{f_{0.2}}\right)$$

The units of $f_{0.2}$ are $t\,cm^{-2}$.

The dimensionless tangent modulus

$$\frac{E_t}{E} = \frac{d\bar{\sigma}}{d\bar{\varepsilon}} \tag{2.20}$$

is equal to:

1	in the first region
$1 - \alpha\beta(\bar{\varepsilon}-1)^{\alpha-1}$	in the second region
$\gamma\psi e^{-\gamma(\bar{\varepsilon}-\bar{\varepsilon}_1)}$	in the third region

The continuity of the curve in Fig. 2.9 at the points $(1, 1)$ and $(\bar{\sigma}_1, \bar{\varepsilon}_1)$ is insured by Eqs 2.15 and 2.18, as well as their first derivatives (Eq. 2.20).

The quantities $\bar{\sigma}_1$, $\bar{\sigma}_u$, ψ and $\gamma/f_{0.2}$ are only affected by the ratio

$f_{0.2}/f_{0.1}$, already defined as the *strain-hardening parameter*, and can therefore be related to the Sutter classes (see Section 2.2.2).

The quantities $\bar{\varepsilon}_1$, α and β are affected by the ratio $f_{0.2}/f_{0.1}$, and additionally by the values of $f_{0.2}$ and E, which are given by experiment and tabulated for each aluminum alloy.

The variations of α, β and of $\bar{\sigma}_1$, $\bar{\sigma}_u$, ψ, $\gamma/f_{0.2}$ with respect to the ratio $f_{0.2}/f_{0.1}$ are given in Figs 2.10 and 2.11, respectively. This graphical presentation allows Eqs 2.15 and 2.18 to be used more easily.

Fig. 2.10

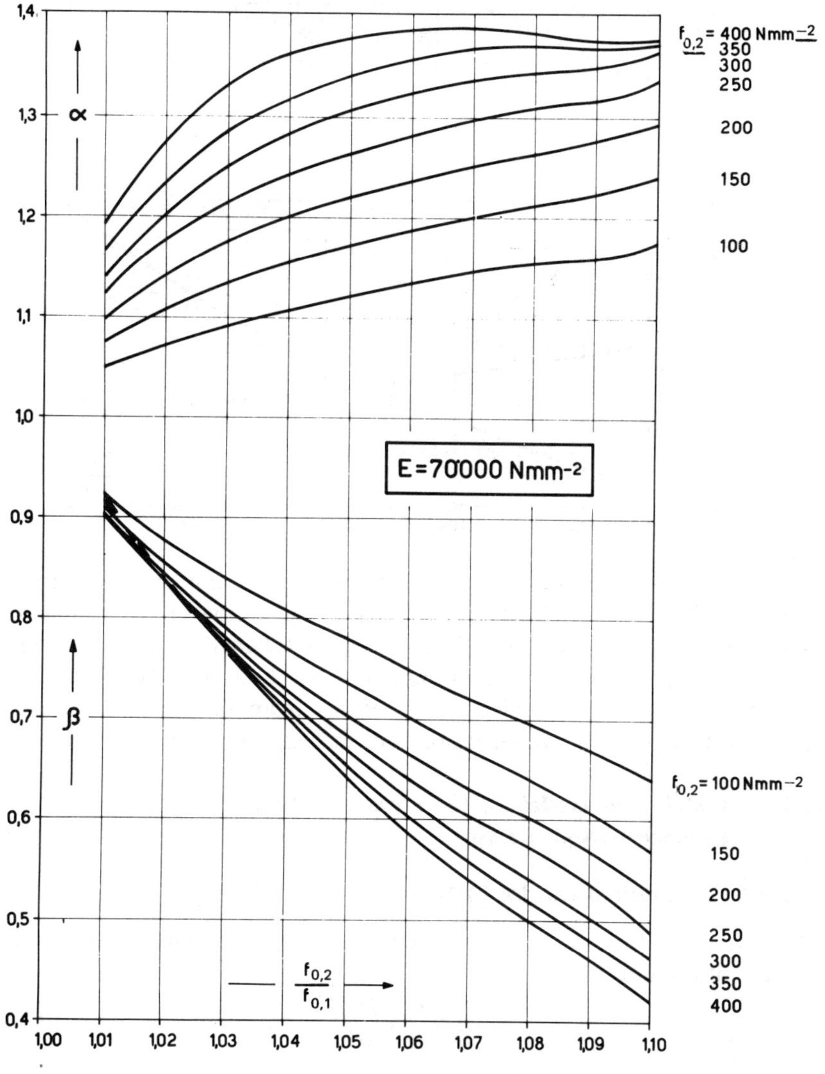

Aluminum alloy structures

Fig. 2.11

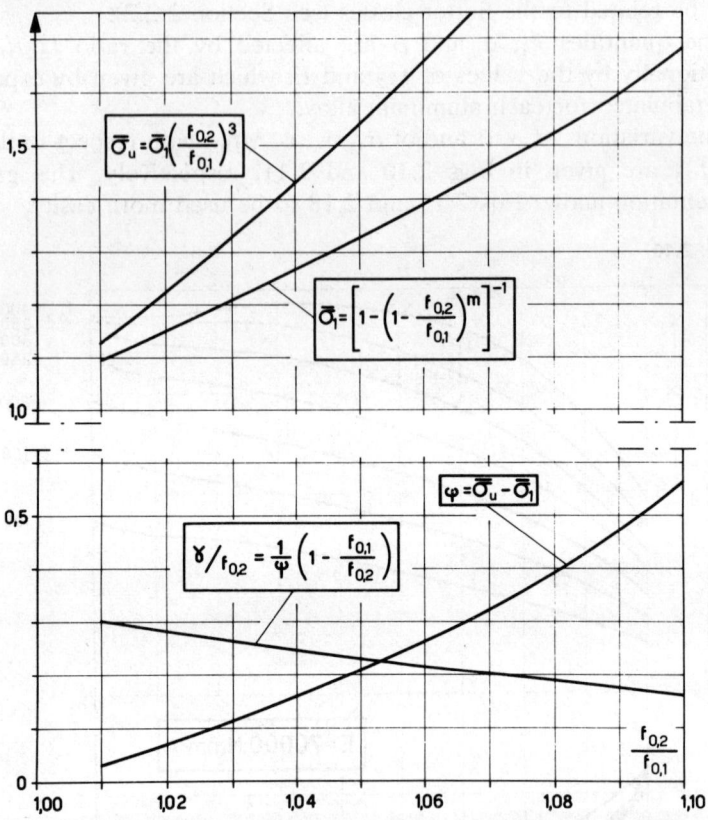

However, the complexity of this formulation disappears if computers are used. This approach has been checked by comparing its results with tests carried out on double T profiles and tubes in various European countries. The curves derived in this way (Fig. 2.12) are always within the statistical range. The maximum deviations from the average value are within a few percent, which represents the experimental approximation.

2.2.5 Continuous models of the Form $\varepsilon = \varepsilon(\sigma)$: Ramberg-Osgood Law

2.2.5.1 Analytical model [4, 5]

A generalized law $\varepsilon = \varepsilon(\sigma)$ has been proposed by Ramberg and Osgood [5] for aluminum alloys:

$$\varepsilon = \frac{\sigma}{E} + \left(\frac{\sigma}{B}\right)^n \tag{2.21}$$

The structural material

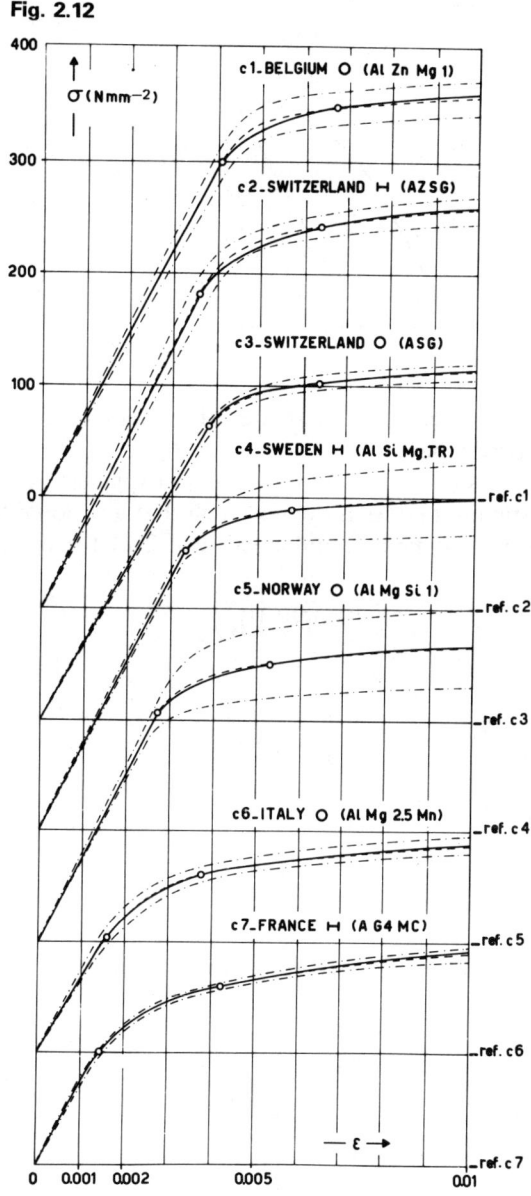

Fig. 2.12

Aluminum alloy structures

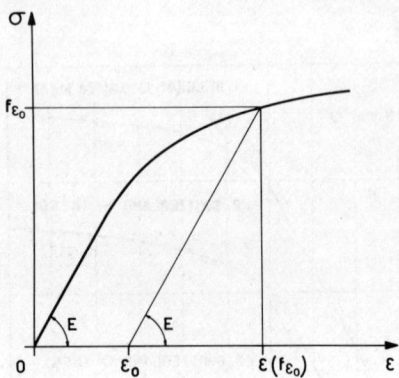

Fig. 2.13

where E is the Young's modulus at the origin, and B and n have to be determined by experiment.

The physical meaning of B and n can be explained as follows (Fig. 2.13). If an elastic limit stress is defined such that a residual strain will be provided when the specimen is unloaded, Eq. 2.21 for $\sigma = f_{\varepsilon_0}$ will be:

$$\varepsilon = \frac{f_{\varepsilon_0}}{E} + \left(\frac{f_{\varepsilon_0}}{B}\right)^n \qquad (2.22)$$

We also have

$$\frac{f_{\varepsilon_0}}{E} = \varepsilon - \varepsilon_0 \qquad (2.23)$$

By substituting Eq. 2.23 in Eq. 2.22 we obtain

$$\varepsilon_0 = \left(\frac{f_{\varepsilon_0}}{B}\right)^n \qquad (2.24)$$

Therefore

$$f_{\varepsilon_0} = B\sqrt[n]{\varepsilon_0} \qquad (2.25)$$

If we assume $\varepsilon_0 = 0.002$, then

$$f_{0.2} = B\sqrt[n]{0.002} \qquad (2.26)$$

Alternatively, if we assume $\varepsilon_0 = 0.001$, then:

$$f_{0.1} = B\sqrt[n]{0.001} \qquad (2.27)$$

For the ratio between Eq. 2.26 and Eq. 2.27 we obtain:

$$\frac{f_{0.2}}{f_{0.1}} = \sqrt[n]{2} \qquad (2.28)$$

The structural material

Equation 2.28 relates the exponent n to the strain-hardening parameter $f_{0.2}/f_{0.1}$ and hence, according to Sutter, to the heat treatment of the material. The exponent n of the Ramberg–Osgood law is therefore characteristic of the strain-hardening rate of the inelastic portion of the σ–ε diagram. It can be expressed as:

$$n = \frac{\ln 2}{\ln\left(\dfrac{f_{0.2}}{f_{0.1}}\right)} \quad (2.29)$$

When the ratio $f_{0.2}/f_{0.1}$ tends to 1, the value of n tends to infinity, and Eq. 2.22 then represents the behavior of mild steels. In this case we have:

$$\varepsilon = \frac{\sigma}{E} + \left(\frac{\sigma}{B}\right)^{\infty} \quad (2.30)$$

which gives:

$$\varepsilon = \frac{\sigma}{E} \text{ for } \frac{\sigma}{B} < 1 \text{ (perfectly elastic portion)}$$

$$\varepsilon = \infty \text{ for } \frac{\sigma}{B} > 1 \text{ (perfectly plastic portion)}$$

The two regions are separated by the value $\sigma/B = 1$, which corresponds to the knee of the elastic/perfectly plastic σ–ε diagram typical of mild steels. The parameter B has the physical meaning of the limit stress of the elastic part of the curve when $n = \infty$. More generally it can be said that, for a finite value of n, the parameter B shows the extent of the curve for which the first term (σ/ε) of the Ramberg–Osgood law is more significant than the second $(\sigma/B)^n$. Furthermore, the ratio $f_{0.2}/f_{0.1}$ tends to 1 in Eq. 2.28 when $n \to \infty$, whereas for $n = 1$ Eq. 2.21 becomes linear and the ratio $f_{0.2}/f_{0.1}$ is equal to 2. It is therefore interesting to compare the values of the exponent n of the Ramberg–Osgood law and the values of the ratio $f_{0.2}/f_{0.1}$ which give the thresholds for the Sutter classes (see Figs 2.14 and 2.15).

It is possible to classify aluminum alloys according to the values of the exponent n of the Ramberg–Osgood law:

$$n < 10\text{–}20 \text{ (non-heat-treated alloys)}$$

$$n > 20\text{–}40 \text{ (heat-treated alloys)}$$

This has also been confirmed by experimental results.

Equation 2.21 can be simplified by substituting B from Eq. 2.26:

$$B = \frac{f_{0.2}}{\sqrt[n]{0.002}} \quad (2.31)$$

Aluminum alloy structures

Fig. 2.14 Comparison of n and $f_{0.2}/f_{0.1}$

		$f_{0.2}/f_{0.1}$	n
Linear law		2	1
Sutter	I class	1.060	11.89
	II class	1.045	15.75
	III class	1.030	23.45
IV class		1	∞
mild steel			

We obtain:

$$\varepsilon = \frac{\sigma}{E} + 0.002\left(\frac{\sigma}{f_{0.2}}\right)^n \tag{2.32}$$

The exponent n is a function of $f_{0.2}$ and $f_{0.1}$, and hence this form of the law is in terms of parameters that can all be determined experimentally from a tension test.

The Ramberg–Osgood law is now widely used because its predicted behavior is very close to the actual behavior of aluminum alloys.

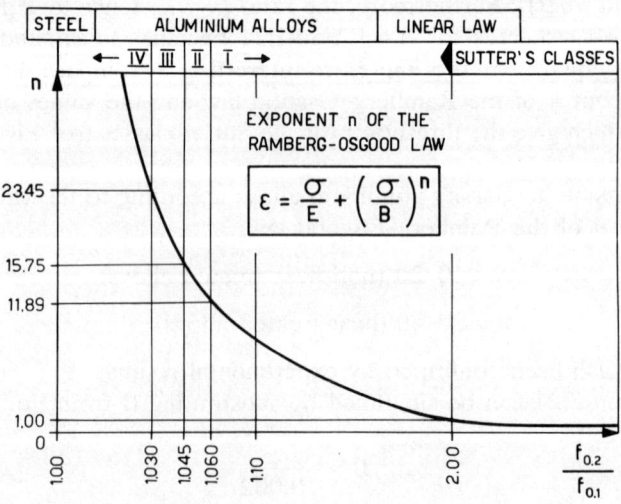

Fig. 2.15

Fig. 2.16

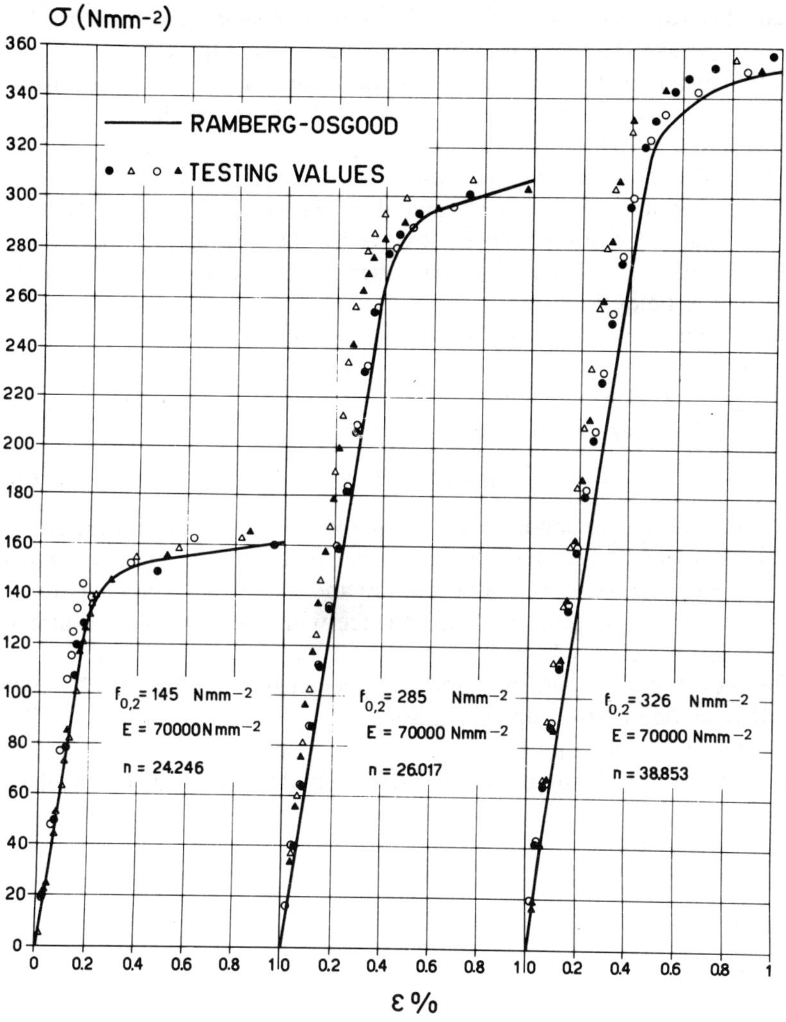

Figure 2.16 shows some comparisons for different non-heat-treated and heat-treated alloys. Ramberg–Osgood curves give a lower bound to the experimental curves owing to the use of the minimum value of Young's modulus ($E = 70\,000$ N mm^{-2}).

For these reasons the Ramberg–Osgood law has been used in the computations carried out by the ECCS committee on light alloy structures when developing the European recommendations for aluminum alloy structures (1978) [42].

2.2.5.2 Proposals for practical applications

When the Ramberg–Osgood law is used in specifications, a practical difficulty arises. Specifications usually do not provide the value of $f_{0.1}$ with the other mechanical properties; only the values of $f_{0.2}$ and E are usually given. Therefore it is not possible analytically to determine the value of the exponent n (Eq. 2.29) by using Eq. 2.32. The exponent could be determined experimentally each time.

Since it is not possible to test each alloy used in design to get the value of $f_{0.1}$, this difficulty has been overcome by means of approximate methods which allow the Ramberg–Osgood law to be used without the value of $f_{0.1}$ being known.

Two proposals of this type are explained below.

Steinhardt proposal [43]

E and $f_{0.2}$ values are assumed to be the minimum required by specifications, and it is also assumed that:

$$10n = f_{0.2} \quad (\text{N mm}^{-2}) \tag{2.33}$$

This proposal is very simple and concise. We observe that in practice the actual values of the elastic limit range between 100 and 150 N mm^{-2} in non-heat-treated alloys and between 200 and 300 N mm^{-2} in heat-treated alloys, and these ranges are in accordance with the classification already given in Section 2.2.5.1.

The latest edition of the German specifications DIN 4113 (1975) [44, 45] has been based upon this assumption.

Mazzolani proposal [6]

Starting from the minimum values required by specifications for mechanical properties:

$f_{0.2}$ elastic limit
f_t ultimate strength
ε_t elongation at rupture

the following approximate expression for the exponent n is proposed:

$$n' = \frac{\ln 2}{\ln(1 + k\chi)} \tag{2.34}$$

where

$$\chi = \frac{f_t - f_{0.2}}{10\varepsilon_t} \frac{f_t}{f_{0.2}} \quad (\text{N mm}^{-2})$$

and k is a dimensional constant.

This expression has been verified using the statistical results of testing carried out at Liège University [46]. On this basis it has been assumed that $k = 0.028 \text{ mm}^2 \text{ N}^{-1}$.

The structural material

Fig. 2.17 Comparison between n' and n

Alloy	Standard values (min)				$\dfrac{f_t - f_{0.2}}{\varepsilon_t} \cdot \dfrac{f_t}{f_{0.2}}$	n'	Testing values (min)		n
	$f_{0.2}$ (N mm^{-2})	f_t (N mm^{-2})	E (N mm^{-2})	ε_t (%)			$f_{0.2}$ (N mm^{-2})	E (N mm^{-2})	
5052	80	200	70 000	12	25	10.25	161.8	68 150	10.93
5083	95	240	71 000	12	30.6	8.40	159.1	67 660	8.82
6060	130	180	70 000	10	6.9	36.00	260.5	72 860	57.03
6061	240	280	70 000	8	5.8	46.00	218.2	72 920	23.79
7020	275	340	72 500	10	8.1	33.90	332.4	69 730	37.98
7020	275	340	72 500	10	8.1	33.90	333.1	71 980	32.60
6051	240	280	70 000	8	5.8	46.00	298.5	66 460	30.47

The comparison between approximate values of n' and experimental values of n is given in Fig. 2.17. Values using Eq. 2.34 have been computed on the basis of the minimum values of the parameters $f_{0.2}$, f_t and ε_t given in French specifications (DTU régles Al, 1971); experimental n values come from the statistical use of test results.

The σ–ε curves derived from the Ramberg–Osgood law, when the

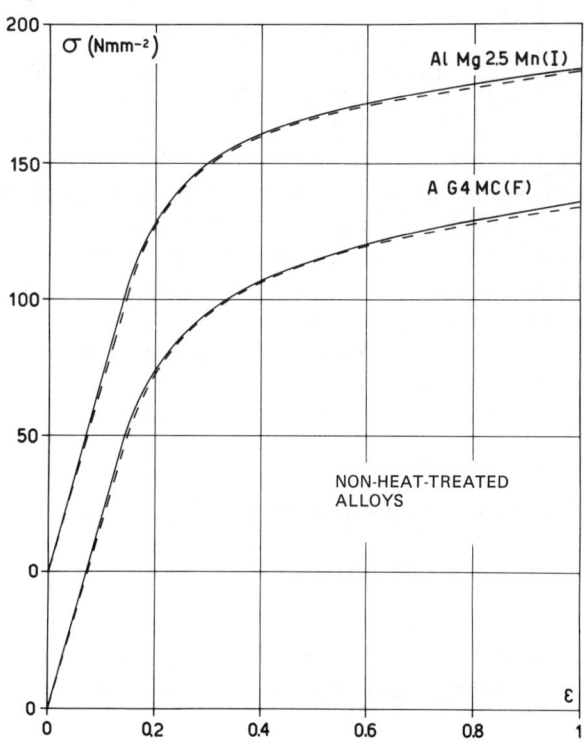

Fig. 2.18

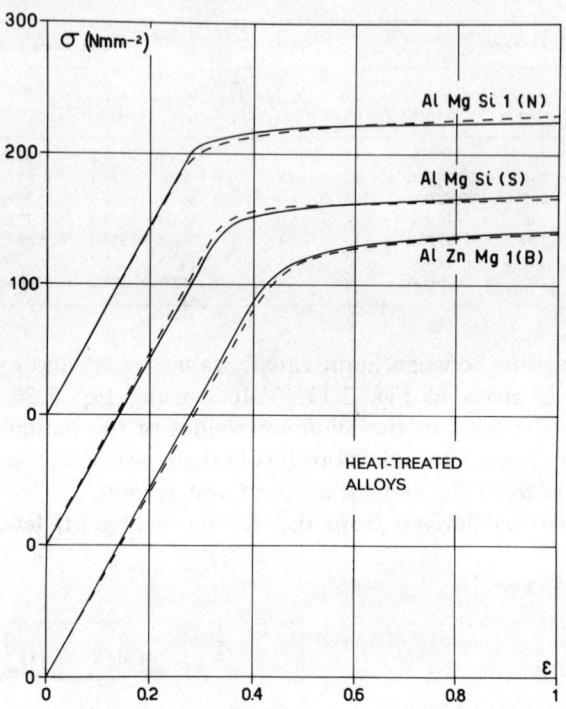

Fig. 2.19

exponent n' and the minimum value required for E are used (continuous line), are compared with experimental values (dashed line) in Figs 2.18 and 2.19. Figure 2.18 is for non-heat-treated alloys (Italy, France) and Fig. 2.19 is for heat-treated alloys (Belgium, Norway and Sweden). The comparison is between alloys having the same elastic limit, in order to emphasize the effect of the exponent n. The approximate curves are very close to the experimental ones.

2.2.5.3 General formulation

If in Eq. 2.21 we substitute for B from Eq. 2.25, we get:

$$\varepsilon = \frac{\sigma}{E} + \varepsilon_0 \left(\frac{\sigma}{f_{\varepsilon_0}}\right)^n \tag{2.35}$$

Equation 2.35 is very general in that it can be successfully applied to other structural metallic and nonmetallic materials as well as to aluminum.

The meanings of the parameters are:

E Young's modulus at the origin
f_{ε_0} conventional elastic limit
ε_0 residual strain corresponding to the elastic limit

$$n = \frac{\ln \alpha}{\ln (f_{\varepsilon_0}/f_{\varepsilon_0/\alpha})} \qquad \alpha > 1 \qquad (2.36)$$

Some σ–ε curves obtained from Eq. 2.35 for various values of n and for $\varepsilon_0 = 0.001$ and 0.002 are given in Fig. 2.20.

It has already been observed that the behavior of mild steel is a particular case of this formulation, and with $n \to \infty$ the σ–ε curve is independent of the value of ε_0. Actually, the elastic/perfectly plastic curve typical of mild steel can be approximated by Eq. 2.35 even with finite values of n, provided that they are greater than 100. In this way the singularity at yielding is overcome.

However, the Ramberg–Osgood law gives its best results when used for aluminum alloys and high-strength steels. In these materials the constitutive law is usually characterized by the conventional values of $f_{0.1}$ and $f_{0.2}$. Therefore we have $\varepsilon_0 = 0.002$ and $\alpha = 2$, which corresponds to the classical formulation already explained in Section 2.2.5.1.

Another particular application of Eq. 2.35 is the perfectly plastic behavior for $\varepsilon_0 = 0$, which corresponds to the stress–strain law of brittle materials such as plastic and composite materials.

2.2.5.4 Proposal for nondimensionalization [7]

In order to nondimensionalize the generalized form of the Ramberg–Osgood law (Eq. 2.35), it is useful to take as the reference stress f_{ε_0} that stress f_0 which corresponds to equal plastic and elastic deformations. This means that the total deformation is equal to $2\varepsilon_0$ (see Fig. 2.21). In this case we have:

$$f_0 = f_{\varepsilon_0} = \varepsilon_0 E \qquad (2.37)$$

If we divide all terms of Eq. 2.35 by this value, we get:

$$\frac{\varepsilon}{\varepsilon_0} = \frac{\sigma}{f_0} + \left(\frac{\sigma}{f_0}\right)^n \qquad (2.38)$$

If we now take

$$\bar{\varepsilon} = \varepsilon/\varepsilon_0 \qquad \bar{\sigma} = \sigma/f_0 \qquad (2.39)$$

the nondimensionalized expression will be

$$\bar{\varepsilon} = \bar{\sigma} + \bar{\sigma}^n \qquad (2.40)$$

Aluminum alloy structures

Fig. 2.20

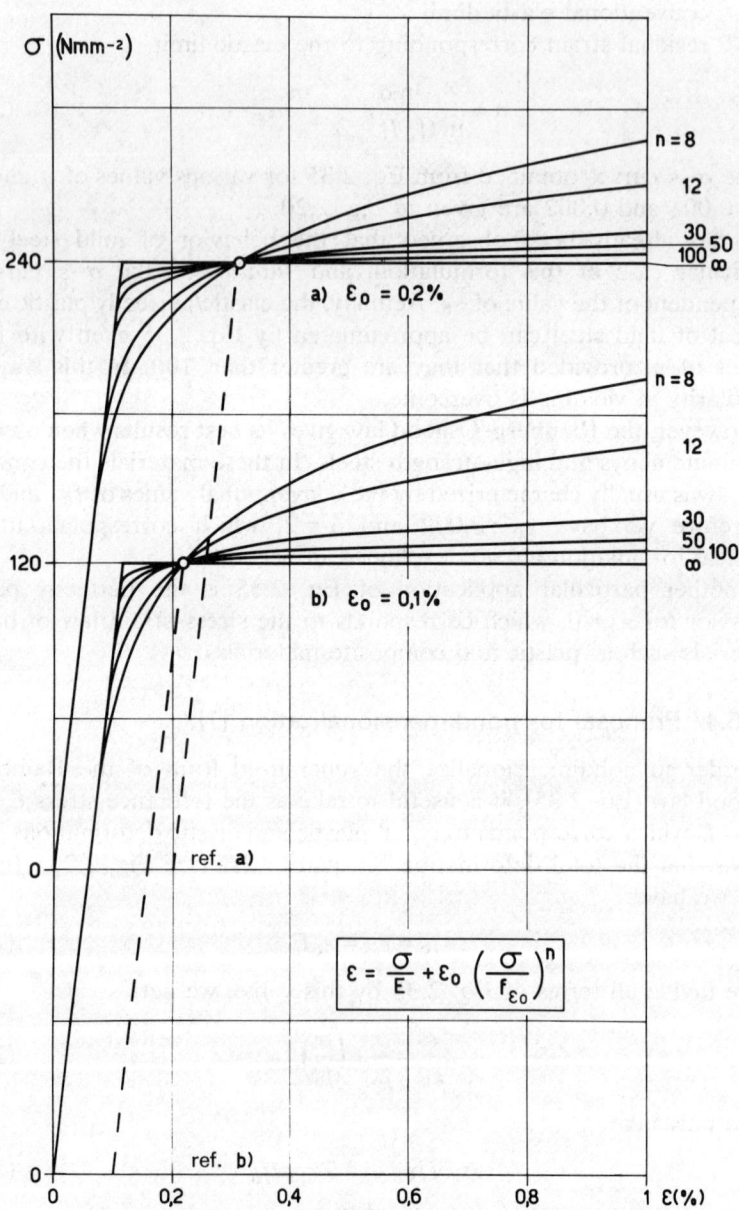

Fig. 2.21

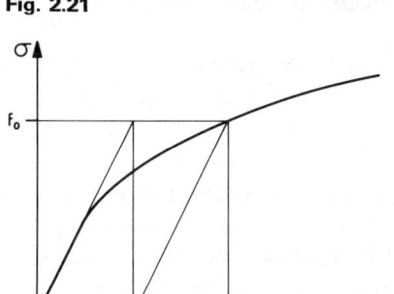

This approach has been followed by Marincek [7], who proposed Eq. 2.40 with the variation ($n = n - 1$)

$$\bar{\varepsilon} = \bar{\sigma}(1 + \bar{\sigma}^n) \qquad (2.41)$$

This approach is more convenient since it is not affected by E and $f_{0.2}$. It also gives σ–ε diagrams which are a function of the exponent n alone and always pass through the point with coordinates (1, 2) (see Fig. 2.22).

The author used this generalized Ramberg–Osgood law to investigate the unstable and stable behavior of different aluminum alloys in order to establish the relationship between ductility and limit states.

The practical application of Eq. 2.41 is somewhat cumbersome because the reference parameters f_0 and ε_0 are not very familiar in laboratories carrying out routine tests. However, when the tension test is undertaken, f_0 and ε_0 values are obtained directly from the complete diagram.

Fig. 2.22

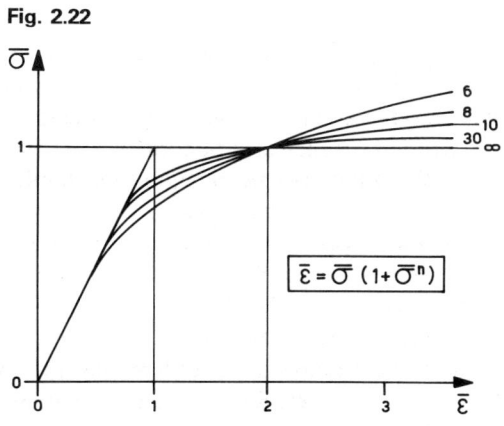

2.3 Definition of industrial bar [8–10]

The most recent trends in international research into the load-bearing capacity of structural metallic components have led to the definition of the 'industrial bar'.

This definition, which was used for the first time in steel structures during the 1960s, details the unavoidable imperfections in structural components produced during the fabrication process. These random imperfections cannot be ignored since they affect, sometimes in a significant way, the load-bearing capacity of the bar.

The industrial bar possesses geometrical and mechanical imperfections closely related to the fabrication process. The industrial bar replaces the 'ideal bar' which is usually assumed as an isotropic, homogeneous, perfectly linear bar with no residual stress. This ideal bar cannot be identified with any structural component, although its use was at one time justified because it was impossible to take account of all the imperfections in an analysis. After computers were introduced on a large scale it was possible to model more closely the actual behavior of the industrial bar.

Geometrical imperfections are usually understood to be the differences between the nominal and the actual geometry of the structural element, either for the longitudinal dimensions or for the transverse dimensions of the cross-section (see Section 2.4).

Mechanical imperfections can be grouped, instead, into two different types:

The residual stress distribution (see Section 2.5.1)
The inhomogeneous distribution of mechanical properties (see Section 2.5.2)

These imperfections depend upon the fabrication process of the structural component, and therefore they have to be examined with respect to the fabrication process involved.

Another mechanical imperfection, also present in steel structures, is the difference in resistance in compression and in tension which is due to the Bauschinger effect, produced by the tractioning process used for straightening aluminum extruded profiles (see Section 2.5.3).

In steel structures the major geometrical and mechanical imperfections and their effect on the load-bearing capacity of structures have already been widely investigated, especially by international organizations. These include the CRC (Column Research Council; now SSRC, the Structural Stability Research Council) in the USA, and the ECCS (European Convention for Constructional Steelwork) in Europe (since 1950).

In contrast, for aluminum alloys this problem was not investigated until 1967, when ECCS decided to establish a specific committee which was

The structural material

devoted to the study of light alloy structures. This committee was called subcommittee 1.1 from 1967 to 1973 and committee 16 from 1973 to 1978, and from 1978 has been known as committee T2. During the period in which the committee was most active, under the chairmanship of the author (1970–1979), it undertook theoretical and experimental research which led to the first definition of the industrial bar in structural elements of aluminum alloys.

On the basis of these studies the ECCS committee developed its European Recommendations for Aluminum Alloy Structures (1978), which is the first available international document concerning specifications of this type [42].

The main results dealing with imperfections in aluminum alloy industrial bars will be given and explained in Sections 2.4 (geometrical imperfections) and 2.5 (mechanical imperfections).

2.4 Geometrical imperfections

2.4.1 Out-of-straightness

Industrial bars are never perfectly straight; they possess an initial out-of-straightness. This deformation can be idealized by a sinusoidal or parabolic expression which is characterized by the parameter v_0 (displacement at midspan). This displacement is commonly expressed as a percentage of the total length of the structural component (Fig. 2.23a). A systematic analysis carried out on extruded profiles from several European countries showed that the difference between an industrial bar and an ideal straight bar is equal to about $L/2000$.

This imperfection is usually higher in steel profiles which have been straightened by rolling. In these profiles a value of $v_0 = L/1000$ is universally assumed in order to compute the load-bearing capacity of the bar under compression.

Extruded aluminum profiles are straighter because of the more severe

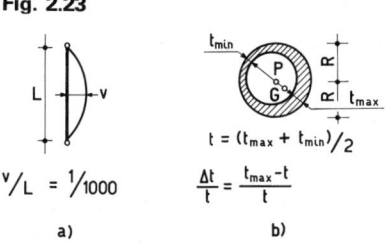

Fig. 2.23

Aluminum alloy structures

Fig. 2.24

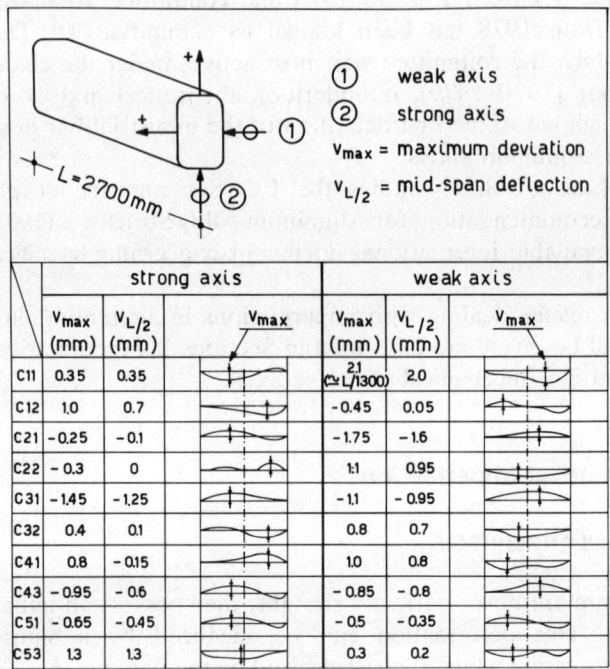

tractioning process. However, national specifications are usually less restrictive since they accept v_0/L ratios equal to $1/500$–$1/1000$.

Further measurements of dimensions carried out on welded double T and box sections (of dimensions shown in Fig. 2.3) always gave v_0/L less than $1/1300$. Some results obtained on box sections (type C), which gave the highest deviations from linearity, are given in Fig. 2.24.

In the light of the results obtained directly and of the specifications which control the fabrication processes in different European countries, the ECCS committee decided to compute the instability curves given in its recommendations on the basis of an initial sinusoidal curvature. This takes $v_0/L = 1/1000$, thus being on the conservative side and in accordance with steel structures.

2.4.2 Variation of the Dimensions of the Cross-section

Measurements taken on some extruded profiles produced in several European countries showed that the dimensions of the cross-section (depth, width and thickness) are very close to the nominal dimensions. Different national specifications allow an average tolerance equal to 1 percent on general dimensions and 5 percent on the thickness of the different parts which form the open section. This value can reach 10 percent in the case of thin profiles less than 5 mm thick.

In the case of hollow sections the dimensions are less precise and national specifications allow higher tolerances. This has also been proved by dimensional measurements carried out on European tubes. This fact can be explained by the extrusion process of hollow sections which produces nonuniform thickness.

This imperfection, called 'eccentricity' by the ECCS committee, can be characterized by the ratio between the highest deviation of the thickness Δt and the average thickness t (see Fig. 2.23b):

$$\frac{\Delta t}{t} = \frac{t_{max} - t}{t} \qquad (2.42)$$

According to specifications, this ratio cannot exceed 10 percent. Measurements gave 'eccentricities' lying between 3 and 9 percent.

The most important consequence of this imperfection is to produce an initial eccentricity of the load applied at the ends of the bar. This is due to the fact that the load acts on the center of the external perimeter, which is not coincident with the center of gravity G (Fig. 2.23b). The negative effect of this type of eccentricity on the load-bearing capacity of columns has to be superimposed on that due to the initial out-of-straightness. This prediction has been confirmed by buckling tests on tubes which gave experimental values, for the failure load of columns, lower than those computed by the numerical analysis for the ideal tube.

Measurements of the dimensions of welded profiles (Fig. 2.3) showed that the deviations from the nominal values were always within the limits of tolerance allowed for extruded profiles (see Fig. 2.25).

An important imperfection in welded double T profiles is the eccentricity of the web in the direction of the weak axis bending (Fig. 2.26). This imperfection, even if within the tolerance limits, causes an eccentricity of the load which has to be added to the initial out-of-straightness.

Fig. 2.25

P	mean (mm)	+%	-%	T	mean (mm)	+%	-%	C	mean (mm)	+%	-%
h	195.09	0.88	-0.66	h	199.74	0.18	-0.22	h	196.61	0.35	-0.46
b	100.17	0.22	-0.27	h_f	198.13	0.59	-0.67	b	140.05	0.25	-0.25
t	11.95	0.38	-0.71	b	100.21	0.58	-0.61	b_m	139.52	0.55	-0.66
e	8.29	1.22	-0.71	t	12.00	2.45	-1.71	t	7.90	0.57	-0.94
				e	8.34	0.72	-1.67	e	5.13	1.24	-2.26

Aluminum alloy structures

Fig. 2.26

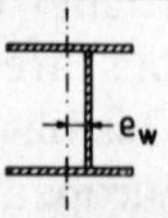

e_w = web eccentricity (mm)
e = load eccentricity (mm)

P	e_w	e	T	e_w	e
P11	−0.40	−0.10	T11	−0.13	−0.05
P12	−0.45	−0.11	T12	1.54	0.60 ≃ L/1800
P14	0.65	0.15	T14	−0.23	−0.09
P21	−0.33	−0.08	T21	0.04	0.02
P22	−0.75	−0.18 ≃ L/6000	T22	0.02	0.01
P24	0.43	0.10	T24	1.73	0.67 ≃ L/1600
P31	−0.03	−0.01	T31	0.	0.
P32	−0.28	−0.07	T32	−0.15	−0.06
P33	0.38	0.09	T33	0.26	0.10
P41	−0.65	−0.15	T41	0.34	0.13
P42	0.35	0.08	T42	−0.68	−0.26 ≃ L/4200

As would be expected, this imperfection is usually larger in those profiles in which the web is directly welded to the flange by fillet welds (type T) than in those profiles whose flanges are extruded and butt welded to the web (type P). However, even the most severe condition, $e = L/1600$, when added to the initial out-of-straightness v_0, is such that it is less than the limit value $L/1000$:

$$(e + v_0) \leq L/1000 \qquad (2.43)$$

This shows that the $L/1000$ value can also be conveniently used to take account of the latter imperfection.

Numerical analysis carried out by the ECCS showed that variations in the dimensions of the cross-section from the nominal values produce a percentage reduction in the load-bearing capacity of columns. This can be assumed to be of the same order of magnitude as the dimensional variation itself.

It is, however, very difficult to take account of this effect from a probabilistic point of view, particularly when considering the interacting effect of other mechanical and geometrical imperfections.

2.5 Mechanical imperfections

2.5.1 Residual Stresses

2.5.1.1 Origin

Under the heading 'residual stresses' are grouped those self-equilibrated internal stresses present in metal members and closely dependent upon the industrial manufacturing process.

They arise in a member when it undergoes nonuniform plastic deformation. In the absence of external forces to counter these stresses they are always elastic.

The inhomogeneous deformation field which generates residual stresses in aluminum alloy profiles is caused by thermal processes such as cooling after extrusion and welding, and mechanical processes such as cold rolling and straightening by tractioning.

2.5.1.2 Measurement methods [11]

To measure residual stresses in aluminum alloy profiles, any of the experimental methods suitable for metallic materials can be used.

Destructive methods (sectioning methods) are based upon the technique of cutting the specimen to be analyzed and measuring the subsequent deformation. When residual stresses are relieved in this way, the specimen finds a new equilibrium condition by changing its deformed shape. Since the residual stress and the stress-relieving process are essentially elastic processes, it is possible to determine the stress field by measuring the corresponding deformation and applying Hooke's law.

In metal profiles, today's universally used method is the sectioning test method, first used in the USA by the Column Research Council (CRC) around 1950 and afterwards adopted in Europe. This method consists of slicing different parts of the profile into longitudinal, parallel strips and measuring deformation through mechanical strain gauges (Fig. 2.27).

Fig. 2.27

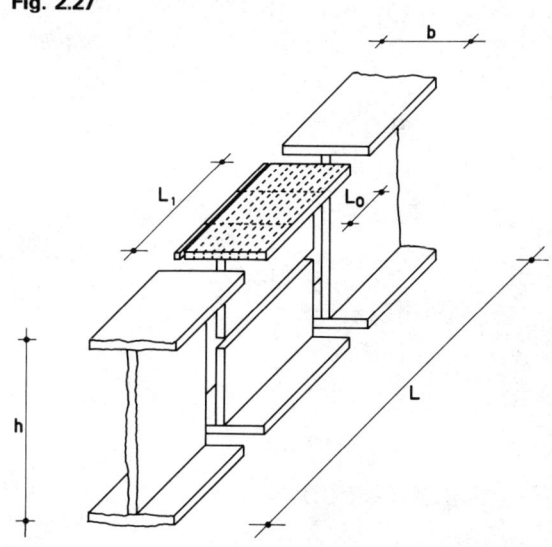

Fig. 2.28

(a)

(b)

The structural material

If L_0 is the basic section length, a specimen of length $L = L_0 + 3h$ (h being the largest transverse dimension) is used to ensure negligible end effects.

Profiles can be cut by a mill, but care should be taken in selecting a cutting velocity which does not cause overheating of the member (Fig. 2.28).

In this method only longitudinal residual stresses are measured. However, as has already been pointed out, these residual stresses are the most interesting from a structural point of view because they must be added to stresses caused by external loads, thus affecting the load-bearing capacity of the member.

2.5.1.3 Extruded profiles [12, 13]

There are several physical reasons for predicting that thermal residual stresses caused by cooling in extruded aluminum alloy profiles are smaller than those in similar hot-rolled steel profiles.

It is known that the intensity of residual stresses is closely related to the rate of inhomogeneity of plastic deformation generated during the cooling process. This is dependent upon the distribution of temperature along the cross-section.

This rate of inhomogeneity is higher if the thermal conductivity k is lower. It is also directly proportional to the specific heat c, the thermal expansion coefficient α and the specific weight of the material γ. As a result, these parameters can be combined in a thermal diffusion factor:

$$\frac{k}{\gamma c} \qquad (2.44)$$

which is proportional to the uniformity of distribution of temperature at different points of the cross-section of a profile. This factor in aluminum alloys is approximately ten times higher than in steel.

In order to experimentally prove these ideas, testing has been carried out on extruded profiles of different alloys using the sectioning method (see Section 2.5.1.2).

Double T $63 \times 63 \times 4$ mm extruded profiles were first examined, the material being the French alloys A-GSM, A-SGM, A-U4G and A-Z5G. It was observed from these test results (Fig. 2.29) that the distribution of residual stresses is very irregular and does not follow any law like that for steel structures. The highest values of residual stresses (end of flanges, center of web) are not very high in compression (lower than 50 N mm^{-2}) and are even lower in tension.

It should also be noted that these values have been measured on the surface of the profiles. At the center of the material the values are

Aluminum alloy structures

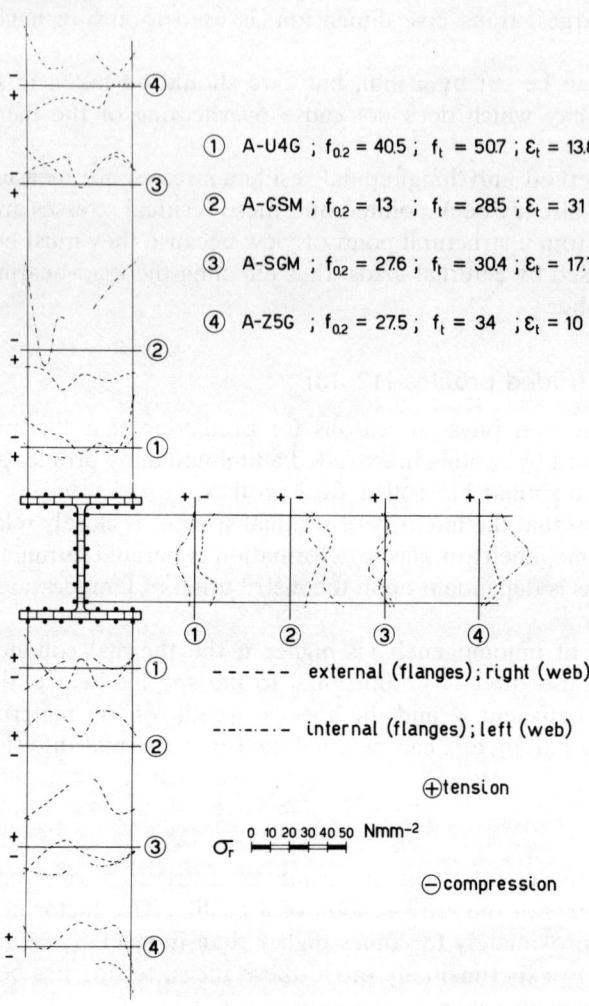

Fig. 2.29

① A-U4G ; $f_{0.2} = 40.5$; $f_t = 50.7$; $\varepsilon_t = 13.8$
② A-GSM ; $f_{0.2} = 13$; $f_t = 28.5$; $\varepsilon_t = 31$
③ A-SGM ; $f_{0.2} = 27.6$; $f_t = 30.4$; $\varepsilon_t = 17.7$
④ A-Z5G ; $f_{0.2} = 27.5$; $f_t = 34$; $\varepsilon_t = 10$

------ external (flanges); right (web)
-·-·-·- internal (flanges); left (web)

⊕ tension

σ_r 0 10 20 30 40 50 Nmm⁻²

⊖ compression

probably lower, especially when, as is usual, residual stresses change sign from one side of the profile to the other.

The mechanical properties of the material did not affect the intensity and distribution of residual stresses.

Austrian profiles were then examined (Fig. 2.30). These consisted of an asymmetrical double T profile (A) and hollow sections with constant thickness with rectangular (B, C) and square (D) cross-section. The materials were AlZnMg1 (Perradur) for profiles A and C, and AlMgSi0.5 (Dekoral) for profiles B and D.

Fig. 2.30

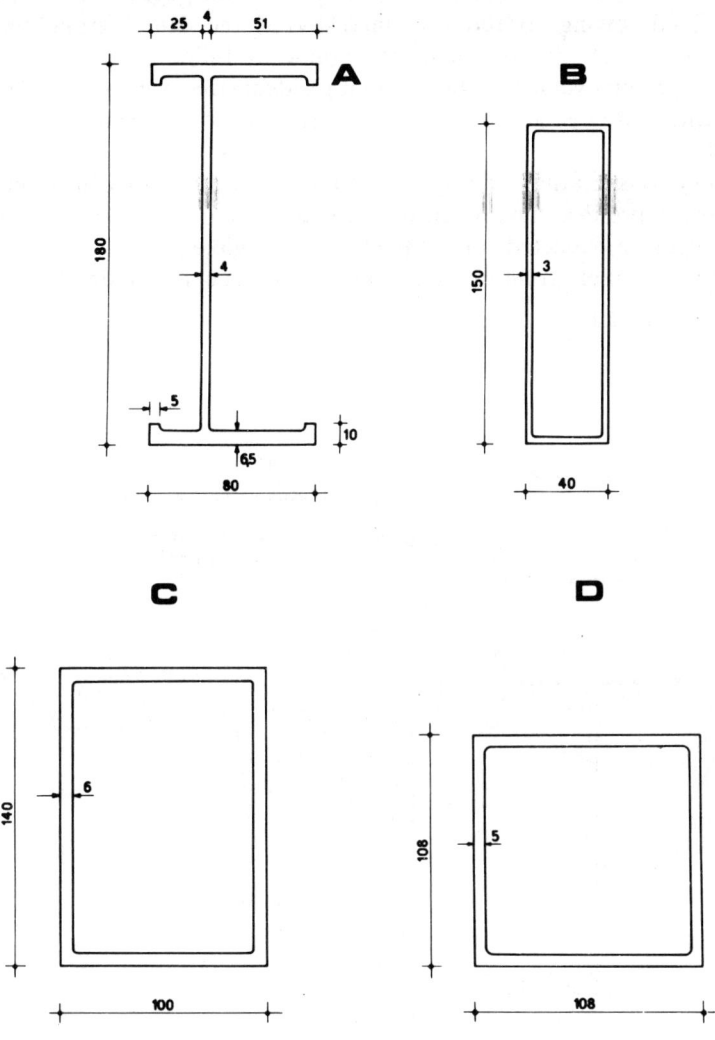

The aim of the testing was to identify the influence of the usual manufacturing phases on the formation of residual stresses.

Measurements were taken on profiles produced using different manufacturing processes:

(1) Extruded and cooled by air
(2) Extruded and straightened (about 1 percent)
(3) Extruded, straightened and artificially tempered (100 °C for 4 hours and 140 °C for 24 hours).

Aluminum alloy structures

The results are given in Figs 2.31–2.34. These results are not easily explained, owing to the low intensity of the initial straightening (1 percent) which did not lead to significant relief of residual stresses. However, final values confirmed that residual stresses produced by manufacturing are generally very low in extruded profiles (less than 20 N mm^{-2}).

Very conservative values of residual stresses, obtained in symmetrical double T profiles, have been used to study their influence on instability. As might be expected, their effect was negligible.

Thus in steel structures residual stresses due to cooling processes in

Fig. 2.31

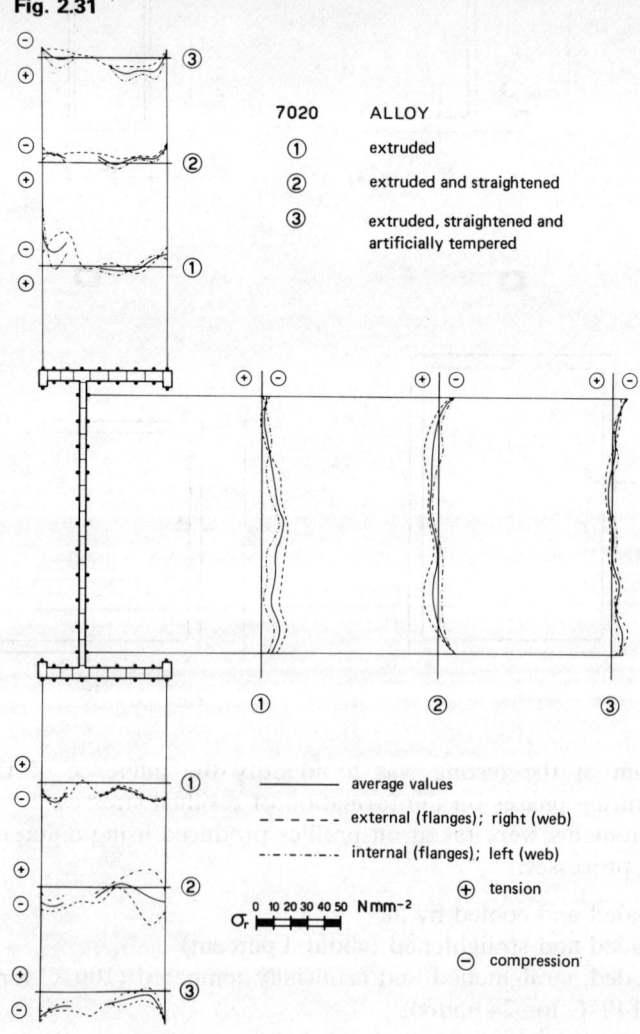

Fig. 2.32

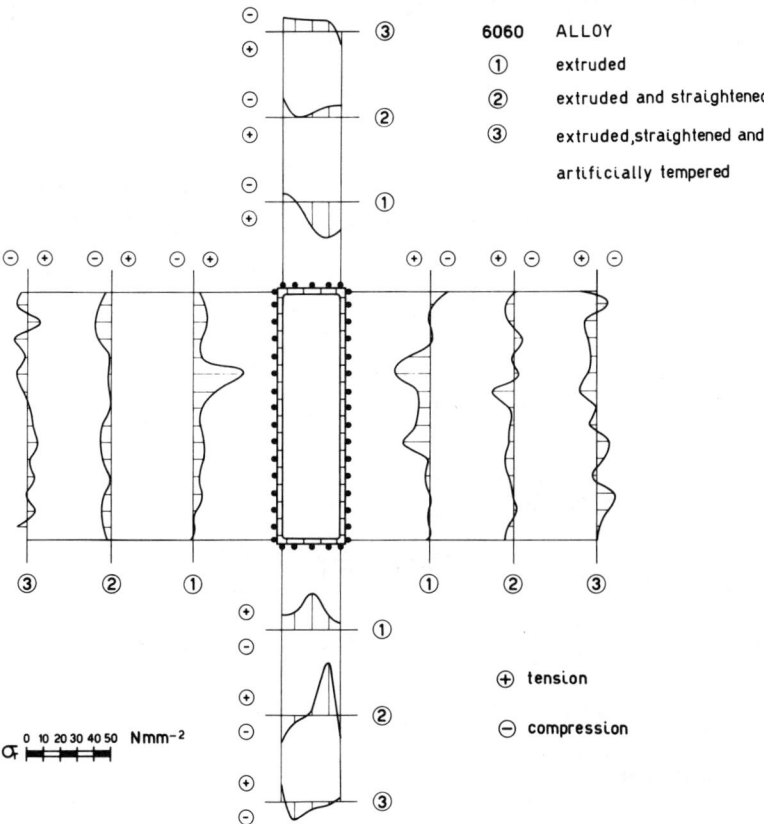

hot-rolled profiles represent an important imperfection because they reach values approximately equal to 0.3–0.5 of the yield limit. By contrast, in extruded aluminum alloy profiles, whatever the heat treatment, residual stresses have very small values; for practical purposes these have a negligible effect on load-bearing capacity.

2.5.1.4 Welded profiles [14–18]

In contrast to what has been said for extruded profiles, residual stresses represent a mechanical imperfection which cannot be neglected in welded profiles. In fact these profiles are subjected to heat treatment which is very inhomogeneous. The welding process produces a concentrated heat input, the intensity of which is closely related to the type of procedure used – in particular to the weld sequence, the pass size and the depth of penetration of the weld.

Aluminum alloy structures

Fig. 2.33

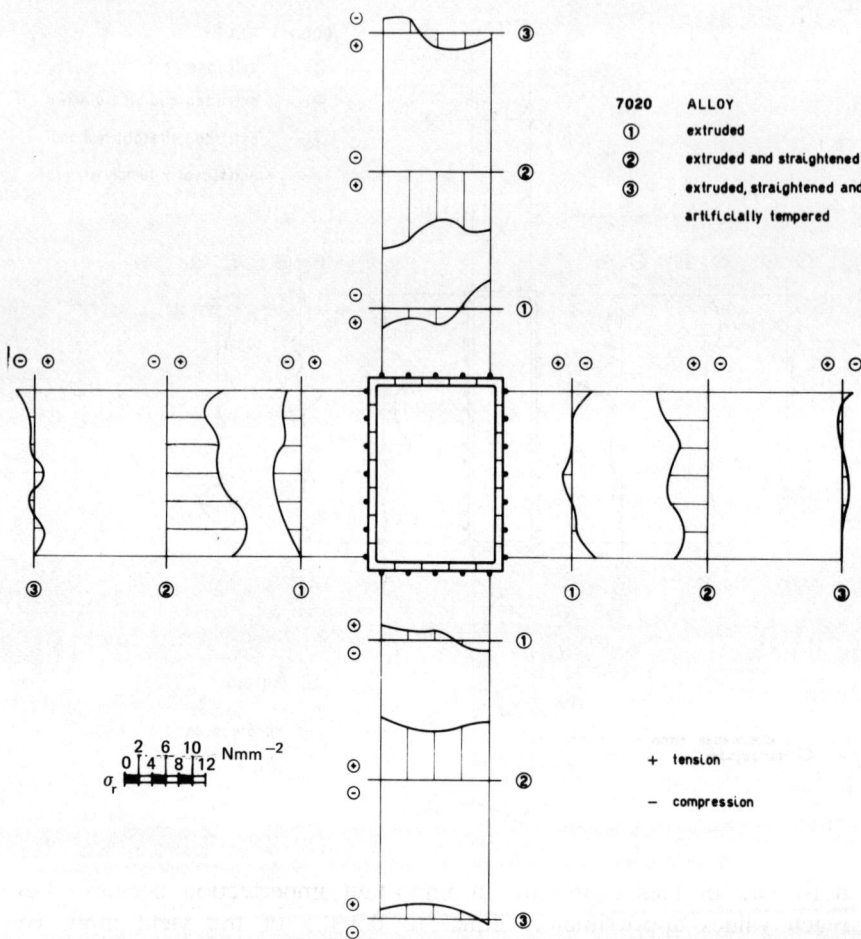

The zones close to the welds are heated to very high temperatures and tend to expand, but this expansion is prevented by the regions further from the weld which are at lower temperatures. As a consequence of this restraint, stresses are generated which cause creep of the fibers. When the member has completely cooled, residual tension stresses close to the weld reach the yield limit, whereas equilibrating compression stresses arise further from the weld.

The intensity and distribution of residual stresses due to welding are always related to the thermal diffusion factor (Eq. 2.44) of the material, and are therefore lower than in steel structures.

Preliminary research has been carried out by Mazzolani [15] on simple

Fig. 2.34

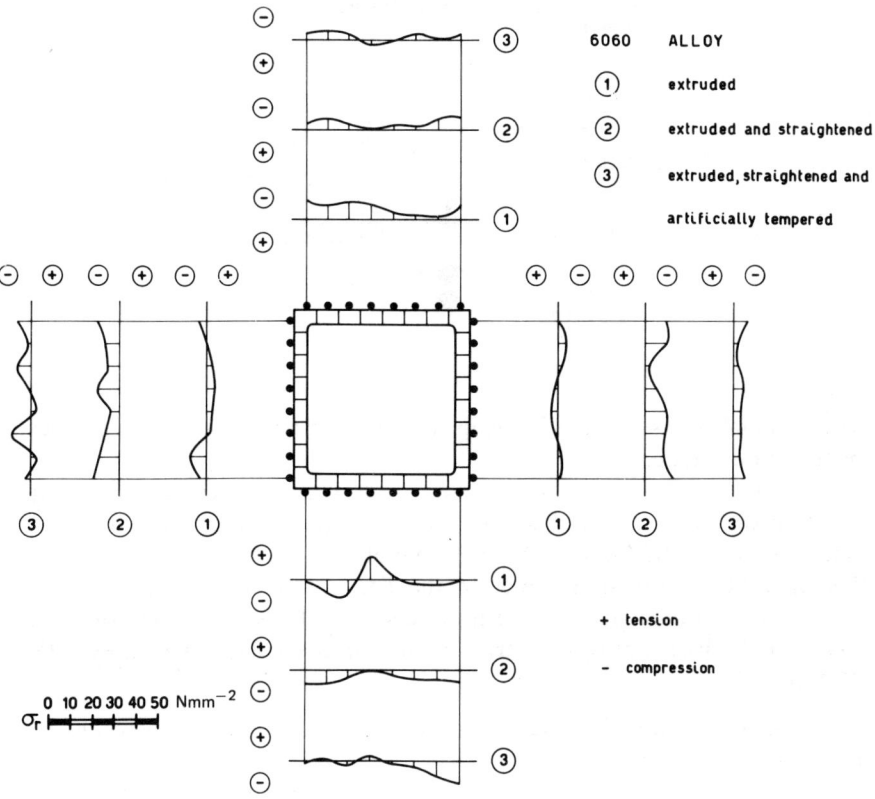

groove butt-welded joints. Two French alloys were examined (Fig. 2.35): non-heat-treated alloy A-G4MC H (corresponding to AlMg4.5Mn), and heat-treated alloy A-Z5G T6 (corresponding to AlZnMg1). The welding material was A-G5 (AlMg5).

Measurements taken using the sectioning method gave the results shown in Fig. 2.36, in which average residual stress distributions σ_r are drawn. The behavior of the two materials is very similar. Residual stress values are very close with peak values, corresponding to the weld, of 100 N mm^{-2} and with equilibrating compression values of 30–50 N mm^{-2} at the end of the members. For a range of structural welded shapes, the results of the international research carried out by Gatto, Mazzolani and Morri [16–18] for the ECCS committee are available.

In this research the profiles of Fig. 2.3 in AlSiMg alloy (6082 type) have been examined. The results of the sectioning tests (Figs 2.37–2.39)

Fig. 2.35

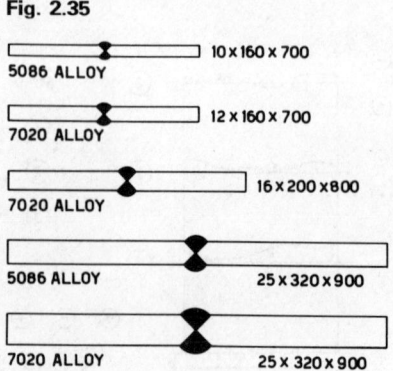

showed that:

The residual stress distribution is usually regular and is similar in profiles of the same shape.

The highest values of residual stresses are tension stresses corresponding to the position of fillet welds, and they are equal to about $140\,\text{N}\,\text{mm}^{-2}$, which is the elastic limit of the weld metal.

The highest compression values in the flanges are always smaller than $50\,\text{N}\,\text{mm}^{-2}$, whereas in the web they reach $120\,\text{N}\,\text{mm}^{-2}$ in the double T shape with fillet welds and are smaller in the other two cases (20–$40\,\text{N}\,\text{mm}^{-2}$).

These results confirm the theoretical estimates and the previous results on plate joints.

All these tests prove that residual stress distributions are characterized by tension regions close to the welds, where the highest values of tensile residual stresses are observed, and by equilibrating compression regions further from the welds. The highest values of compression residual stresses are always smaller than tension stresses.

The comparison with steel for a given structural profile (double T welded shape) in dimensionless form showed that in steel residual stresses in the weld can reach the yield limit of the parent metal. Even higher values occur because the weld metal usually has higher strength. In contrast to steel, the highest residual stresses in aluminum alloys, having the same strength as Fe360 (such as AlSiMg alloy), are less than 60 percent of the 0.2 percent elastic limit (Fig. 2.40).

This favorable difference is also valid for compression stresses at the ends of the flanges, which in aluminum profiles reach about 20 percent of the reference stress, compared with 70 percent for steel.

Hence it can be said that the effect of residual stresses in lowering the resistance of welded compression bars in aluminum alloys is smaller than

The structural material

Fig. 2.36

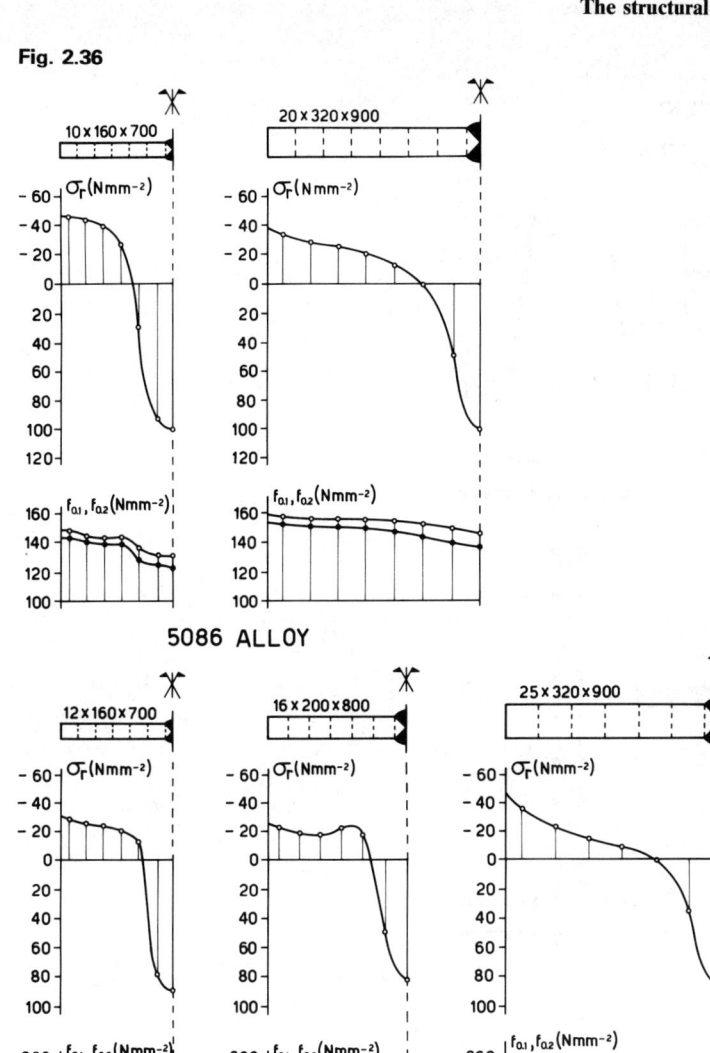

5086 ALLOY

7020 ALLOY

Aluminum alloy structures

Fig. 2.37

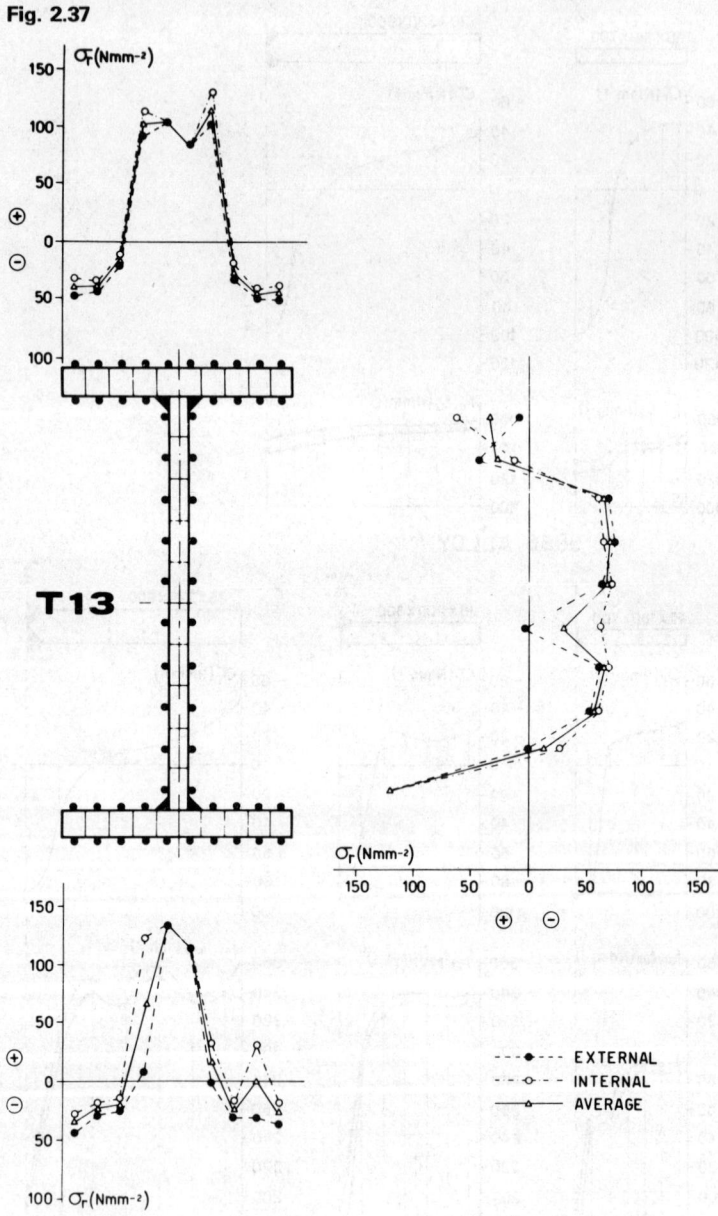

Fig. 2.37 (*contd*)

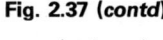

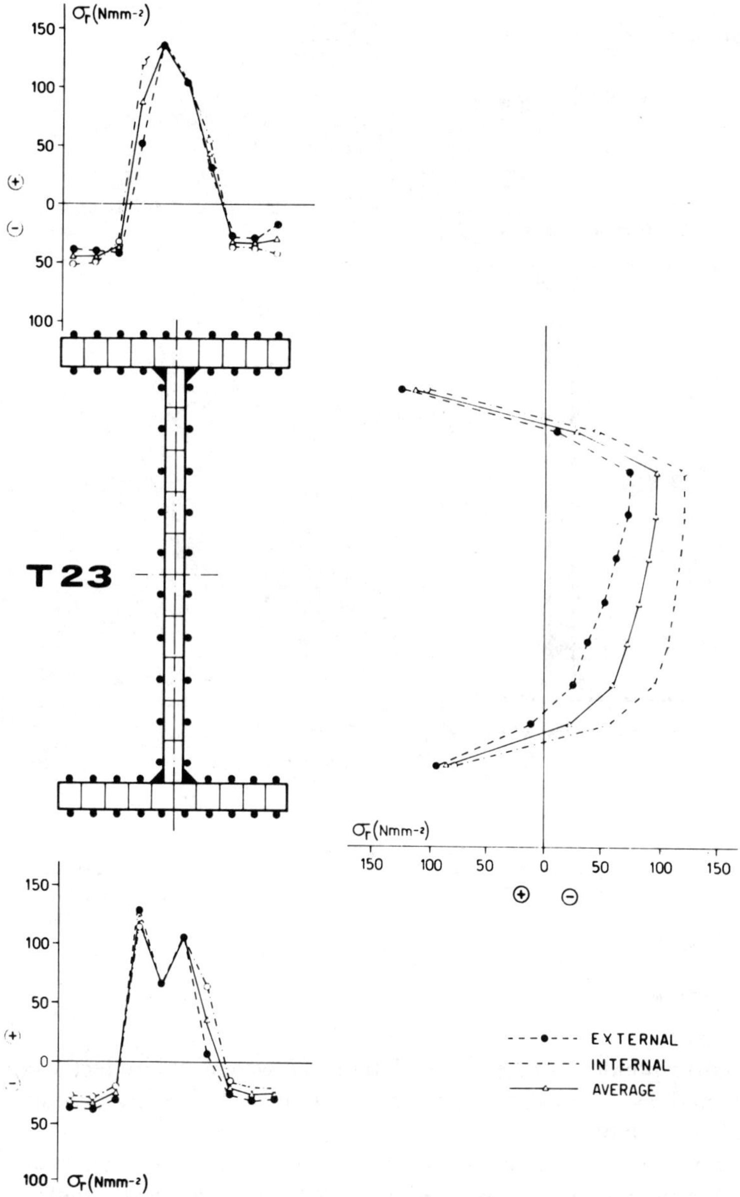

Fig. 2.38

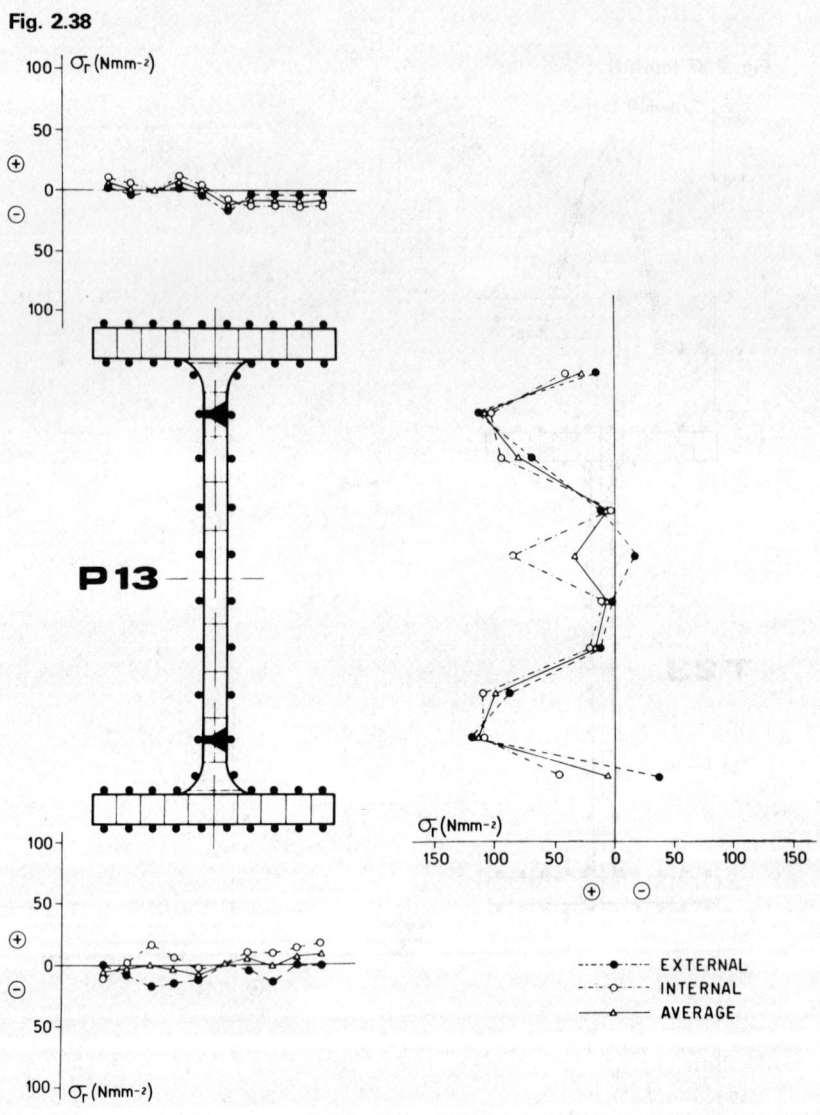

the corresponding effect in steel bars. However, this effect cannot be neglected as in extruded profiles, and must be taken account of in checking stability.

For this purpose it is necessary to convert experimental results into distribution models to be used as inputs to numerical simulations. It therefore seems reasonable to use the experimental ranges obtained for each profile in the most conservative way.

Fig. 2.38 (*contd*)

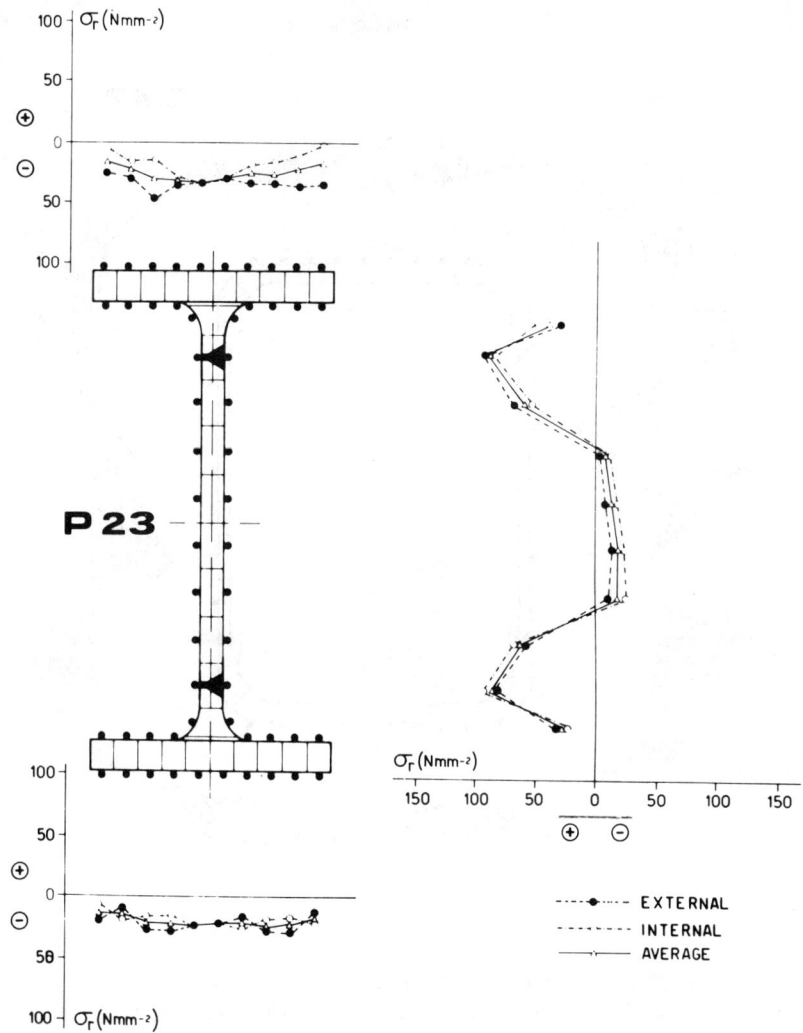

The distribution models defined in this way are given in Figs 2.41–2.43, where the symbols have the following meaning:

σ_f^+ highest tensile residual stress in the flanges
σ_f^- highest compressive residual stress in the flanges
σ_w^+ highest tensile residual stress in the web
σ_w^- highest compressive residual stress in the web

Aluminum alloy structures

Fig. 2.39

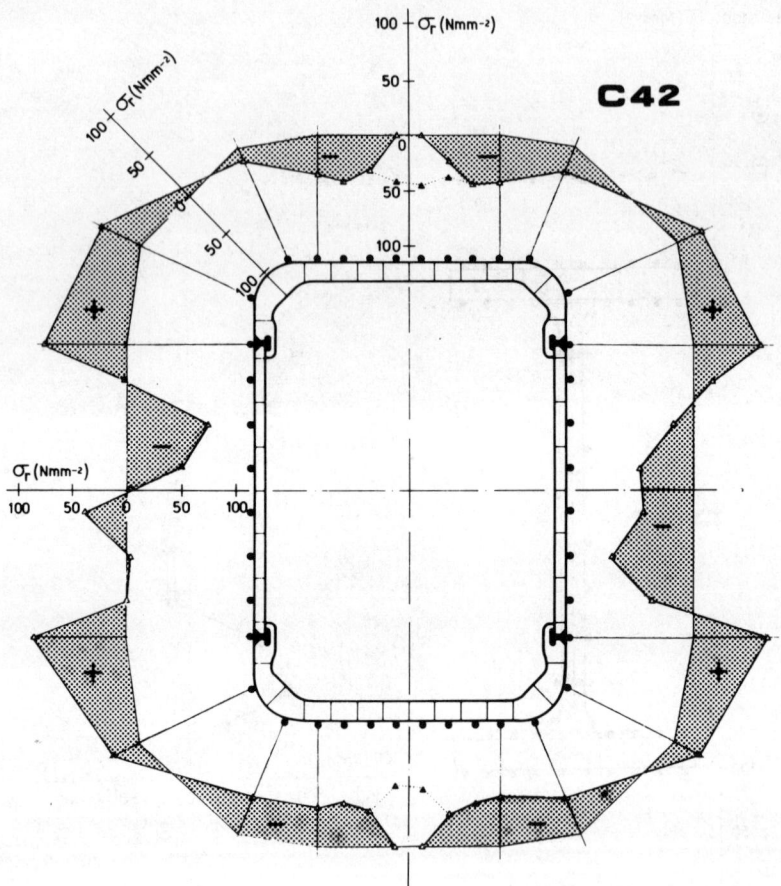

Double T profiles with fillet welds (Fig. 2.41) are characterized by a block distribution in the flanges (tension in the center and compression at the ends). In the web they are characterized by a distribution of constant compression in the center and linearly varying at the ends up to tension values corresponding to those in the fillet weld. This distribution can be quantified if the parameters χ and ξ, are known; these identify the point of inversion of stresses in the flanges and in the web, respectively. They can be obtained numerically from the equilibrium condition relative to the whole section:

$$2b(1-2\chi)t_f\sigma_f^+ + t_w\xi h\sigma_w^+ - 4t_f\chi b\sigma_f^- - t_w h\left(1-2\xi-\xi\frac{\sigma_w^-}{\sigma_w^+}\right)\sigma_w^- = 0 \quad (2.45)$$

Fig. 2.39 (*contd*)

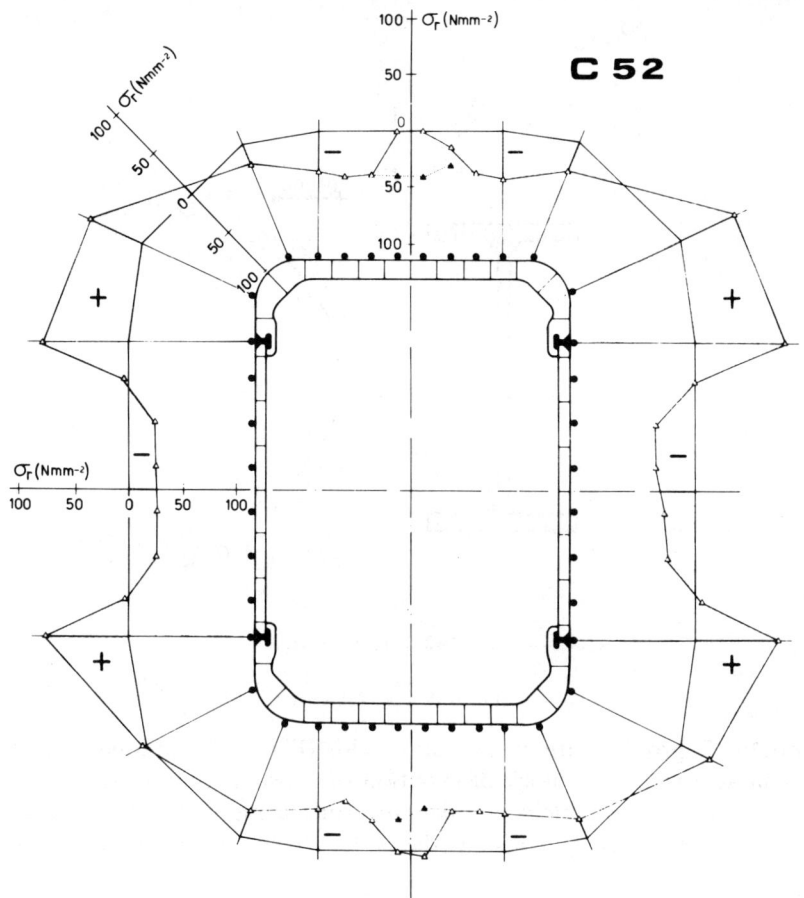

where b and h are the widths and t_f and t_w are the thicknesses of the flanges and of the web, respectively.

Assuming that the highest values which define the model coincide with the highest experimental values, we have:

$$\sigma_f^- = 50 \text{ N mm}^{-2}$$

$$\sigma_f^+ = \sigma_w^+ = 140 \text{ N mm}^{-2}$$

$$\sigma_w^- = 120 \text{ N mm}^{-2}$$

Equation 2.45 is satisfied by the values:

$$\chi = 0.30 \quad \xi = 0.157$$

Aluminum alloy structures

Fig. 2.40

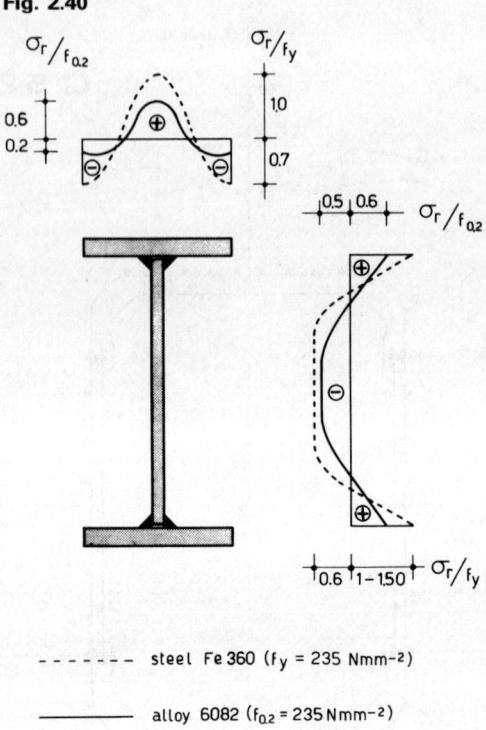

------- steel Fe 360 (f_y = 235 Nmm⁻²)

———— alloy 6082 ($f_{0.2}$ = 235 Nmm⁻²)

In double T profiles with groove butt welds (Fig. 2.42), residual stresses are characterized by a block distribution of constant compression in the flanges, and by three blocks in the web (one compression block at the center and two tension blocks at the welds) which are identified by the parameter ξ.

The equilibrium condition is expressed by

$$2\xi h t_w \sigma_w^+ - 2bt_f \sigma_f^- - (h - 2\xi h)t_w \sigma_w^- = 0 \qquad (2.46)$$

from which we get

$$\xi = \frac{2bt_f \sigma_f^- + h t_w \sigma_w^-}{2h t_w (\sigma_w^- + \sigma_w^+)} \qquad (2.47)$$

Assuming for the highest tensile stresses the values

$$\sigma_f^- = 30 \text{ N mm}^{-2}$$
$$\sigma_w^+ = 120 \text{ N mm}^{-2}$$
$$\sigma_w^- = 20 \text{ N mm}^{-2}$$

then from Eq. 2.47 we obtain

$$\xi = 0.254$$

The structural material

Fig. 2.41

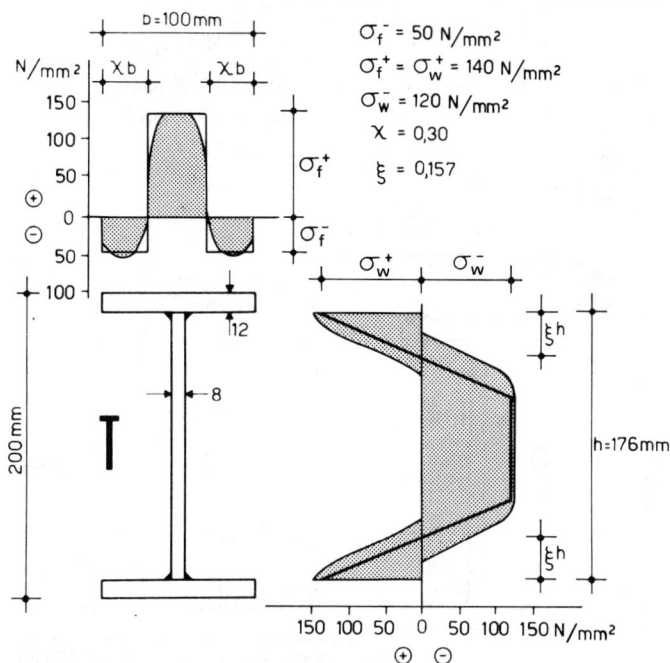

The model of distribution of residual stresses in the box section (Fig. 2.43), owing to the position of the weld, is qualitatively similar to the previous one and can be derived from the following equilibrium condition:

$$2\xi h \sigma_w^+(t_f+t_w) - 2bt_f\sigma_f^- - 2(h-2\xi h)t_w\sigma_w^- = 0 \quad (2.48)$$

which only partially takes into account the complex geometry of this profile. Assuming

$$\sigma_f^- = 40 \text{ N mm}^{-2}$$
$$\sigma_w^+ = 100 \text{ N mm}^{-2}$$
$$\sigma_w^- = 40 \text{ N mm}^{-2}$$

then Eq. 2.48 is satisfied by:

$$\xi = 0.25$$

The difference between the model and the experimental values observed in the web (see Fig. 2.43) is due to the fact that measurements were not taken on the interior surface, since it is a box section.

Aluminum alloy structures

Fig. 2.42

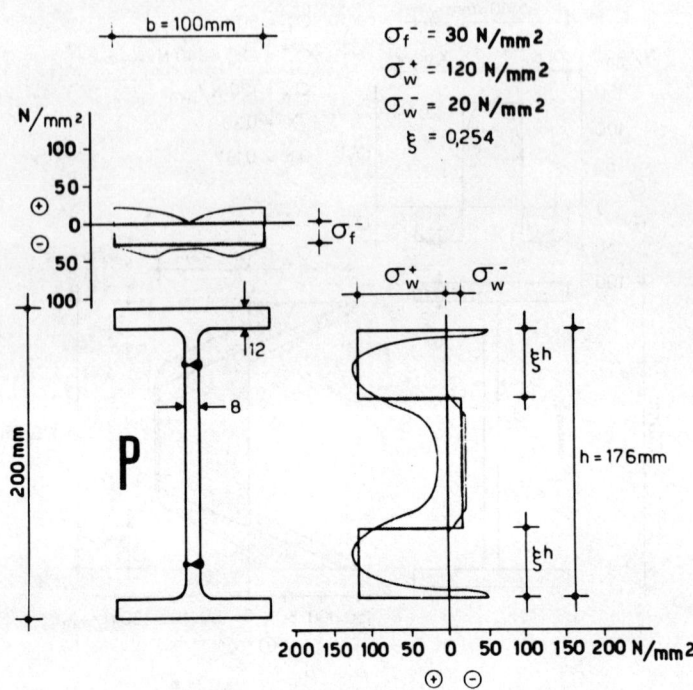

These models (Figs 2.41–2.43) have been used by the ECCS committee in a numerical simulation which led to a method for verifying the instability of welded columns; this has been introduced in the European recommendations.

2.5.2 Inhomogeneous Distribution of Mechanical Properties

2.5.2.1 Nominal parent metal values

The industrial fabrication processes (extrusion and welding) produce, albeit in different ways, an inhomogeneous distribution of mechanical properties. This is characterized by a scattered deviation between the measured values at different points of the cross-section of a profile and the nominal values.

For the structural behavior of members in particular it is important to know the distribution along the cross-section of the following properties:

E Young's modulus
$f_{0.2}$ elastic limit at 0.2 percent strain, conventionally assumed to be the yield limit

Fig. 2.43

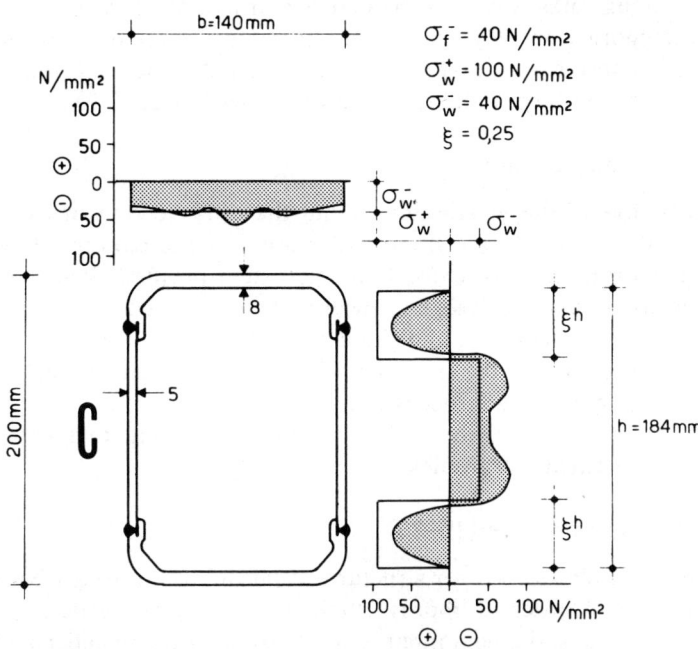

$f_{0.1}$ elastic limit at 0.1 per cent strain. Together with $f_{0.2}$ this is necessary to evaluate the exponent n of the Ramberg–Osgood law and thus to classify the different alloys from the mechanical point of view (see Section 2.2.5).

In the alloys commonly used in structural applications, Young's modulus varies between 68 000 and 75 000 N mm^{-2}. Most specifications usually assume the value $E = 70\,000$ N mm^{-2}. Nominal values of $f_{0.2}$ lie between 100 and 400 N mm^{-2}, depending on the chemical composition of the alloys.

From experiment, two classes of alloy commonly used in structural applications can be defined with respect to mechanical properties:

(a) *Non-heat-treated alloys* (e.g. AlMg, AlMgMn), in which $f_{0.2}$ varies between 100 and 200 N mm^{-2}
(b) *Heat-treated alloys* (e.g. AlSiMg, AlZnMg), in which $f_{0.2}$ varies between 200 and 400 N mm^{-2}

As pointed out by Steinhardt (see Section 2.2.5.2), owing to the direct relationship between $f_{0.2}$ (expressed in 10^{-1} N mm^{-2}) and the exponent n of

Aluminum alloy structures

the Ramberg–Osgood law (Eq. 2.33), the two classes show different strain-hardening rates; this is important for structural behavior.

These categories are even significant from the point of view of the mechanical imperfections examined here, especially when the imperfections are produced by a welding process (see Section 2.5.2.3).

2.5.2.2 Extruded profiles

The distribution of the elastic limit along the cross-section of extruded profiles in aluminum alloys is quite uniform and is not closely related to the manufacturing process as for hot-rolled steel profiles. Some experimental results obtained at Liège showed that the greatest differences are no more than a few percent, and these are not significant with regard to load-bearing capacity of the members. This is in contrast to steel structures, in which greater differences occur.

This is the reason why the ECCS committee decided to ignore this imperfection in extruded profiles.

2.5.2.3 Welded profiles [15–23]

Most aluminum alloys used for structural applications are heat treated or work hardened in order to improve their mechanical properties.

In welded profiles, the heat input removes some of the beneficial effects due to these treatments, and leads to a decrease in the elastic limit. The result is a distribution of strength varying along the cross-section of the profile, with a minimum at the welds. This imperfection is also related to a distribution of residual stress which, at the welds, is equal to the elastic limit of the annealed material (see Section 2.5.1.4).

The first experimental analyses carried out in the USA by Hill, Clark and Brungraber [19] were on plate joints with longitudinal welds at the center, using the 6000 series alloy. They led to the identification of a region close to the weld called the reduced-strength zone (Fig. 2.44), the

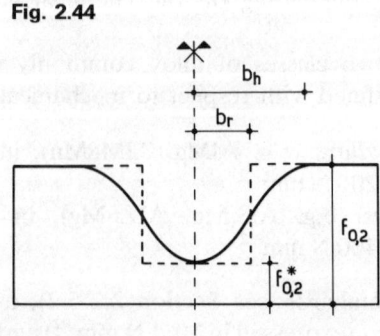

Fig. 2.44

Fig. 2.45

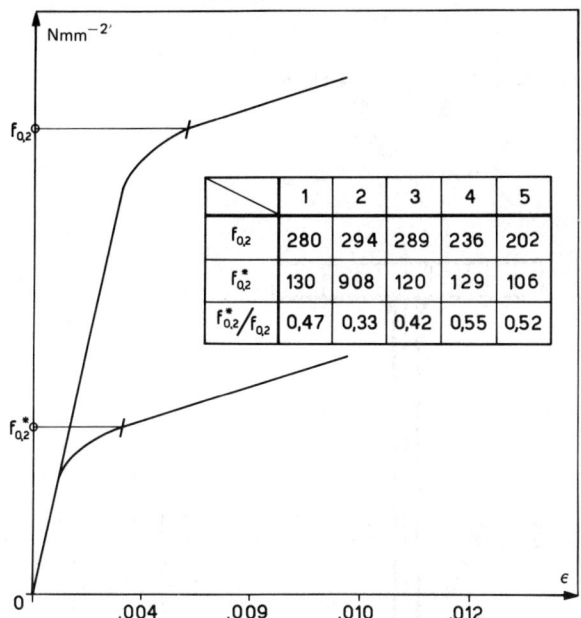

extent of which is equal to $2b_r$. This region corresponds to about one-half of the actual heat-affected zone, and a reduced, constant value, limit stress $f_{0.2}^*$ is attributed to this region.

As an upper bound to b_r, these researchers found a value of 0.74 in. On this basis, the French specifications (DTU régles Al, January 1971) give b_r values equal to or smaller than 25 mm.

The same researchers found the decreased values of the elastic limit $f_{0.2}^*$ given in Fig. 2.45. In comparison with the parent material, reductions of between 50 and 33 percent are observed.

This problem was later studied by Mazzolani from both theoretical [22] and experimental [15] points of view. Experiments carried out on the welded joints of Fig. 2.35 gave the average values given in Fig. 2.36, in which the variations of $f_{0.1}$ and $f_{0.2}$ are drawn along the different members.

As might be expected, the distributions depend greatly upon the heat treatment. Whereas for the non-heat-treated alloys (AlMg4.5Mn) the decrease of strength at the weld is about 10 percent, for the heat-treated alloys (AlZnMg1) the decrease reaches 40–50 percent. The extent of the heat-affected zones confirmed the American results.

Tests carried out at the Experimental Institute for Light Metals (ISML) in Novara (1977) [16, 17, 18] on welded profiles of 6082 alloy (Fig. 2.3)

Aluminum alloy structures

Fig. 2.46

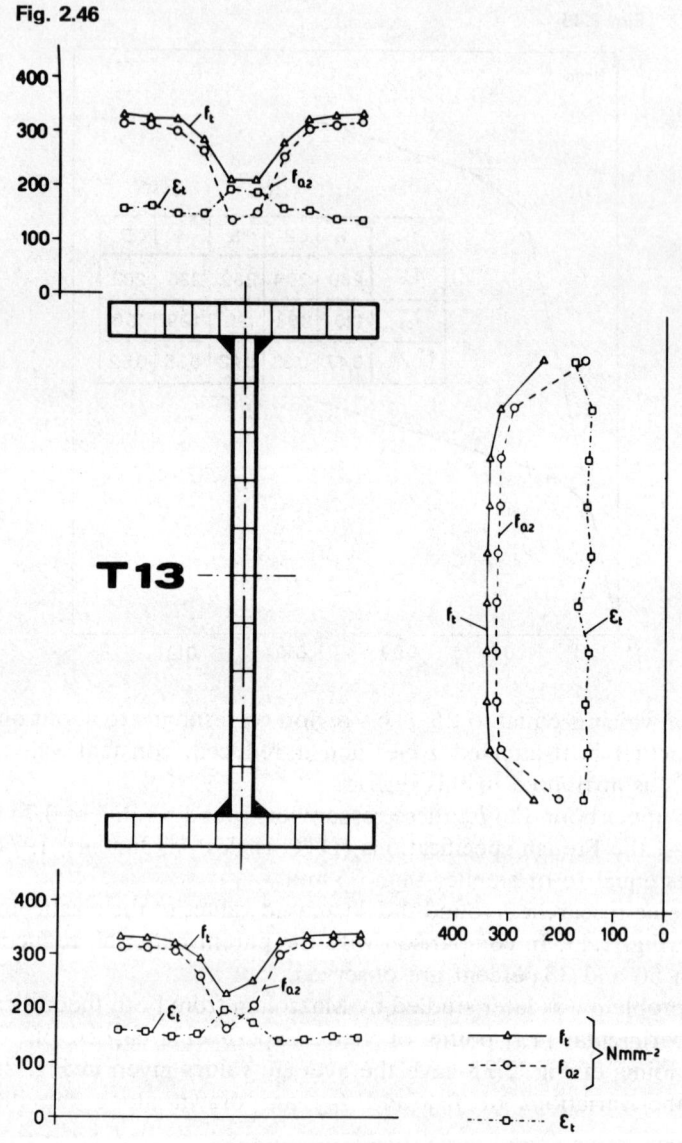

led to the results given in Figs 2.46–2.48 for the elastic limit distribution $f_{0.2}$, the strength at rupture f_t and the elongation at rupture ε_t, measured in tension. It can be observed that the distributions are usually homogeneous for each profile and have minimum values at the welds. These values correspond to the natural aging stage for the parent metal and to the annealed stage for the filler metal (AG5).

The structural material

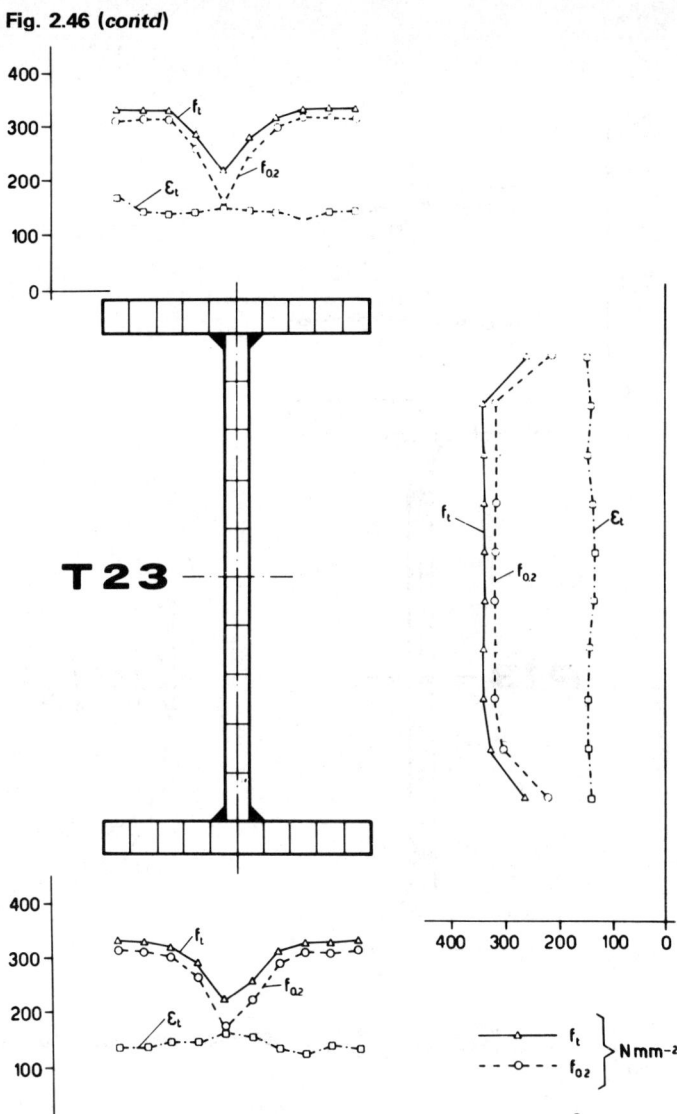

Fig. 2.46 (contd)

The extent of the decreased-strength region was always less than 20 mm on each side of the weld.

Three regions have been identified (Fig. 2.49):

A   Unaffected parent metal
B   Partially affected parent metal
C   Heat-affected zones around the weld metal

Aluminum alloy structures

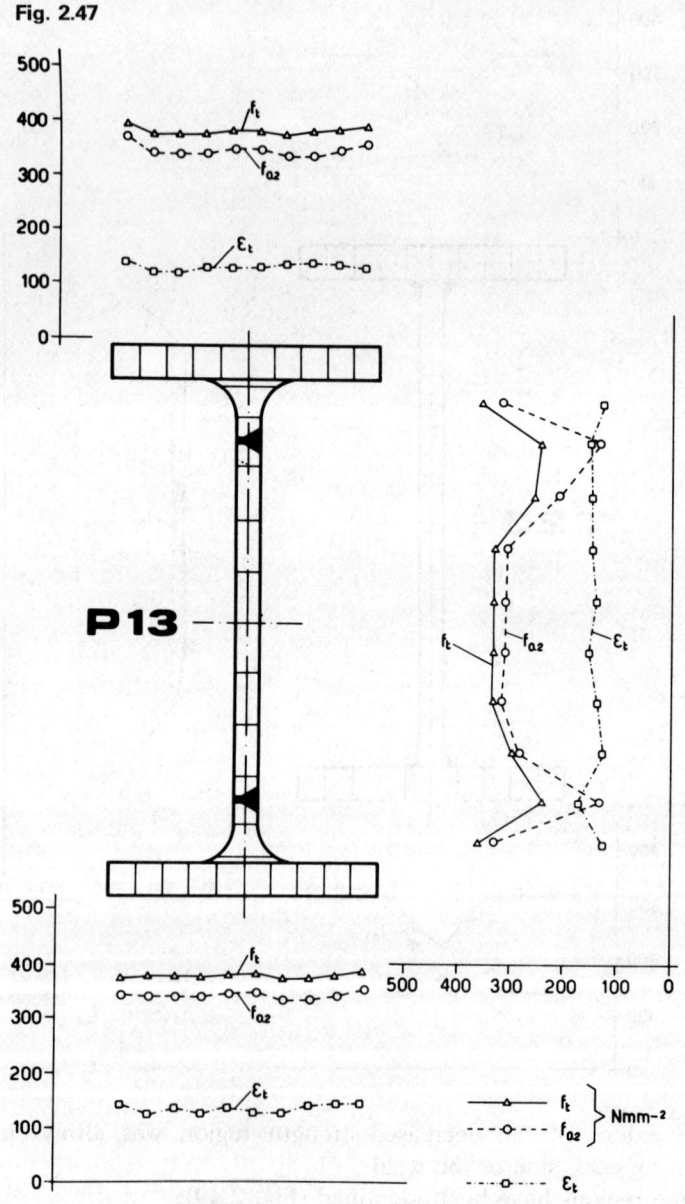

Fig. 2.47

Fig. 2.47 (*contd*)

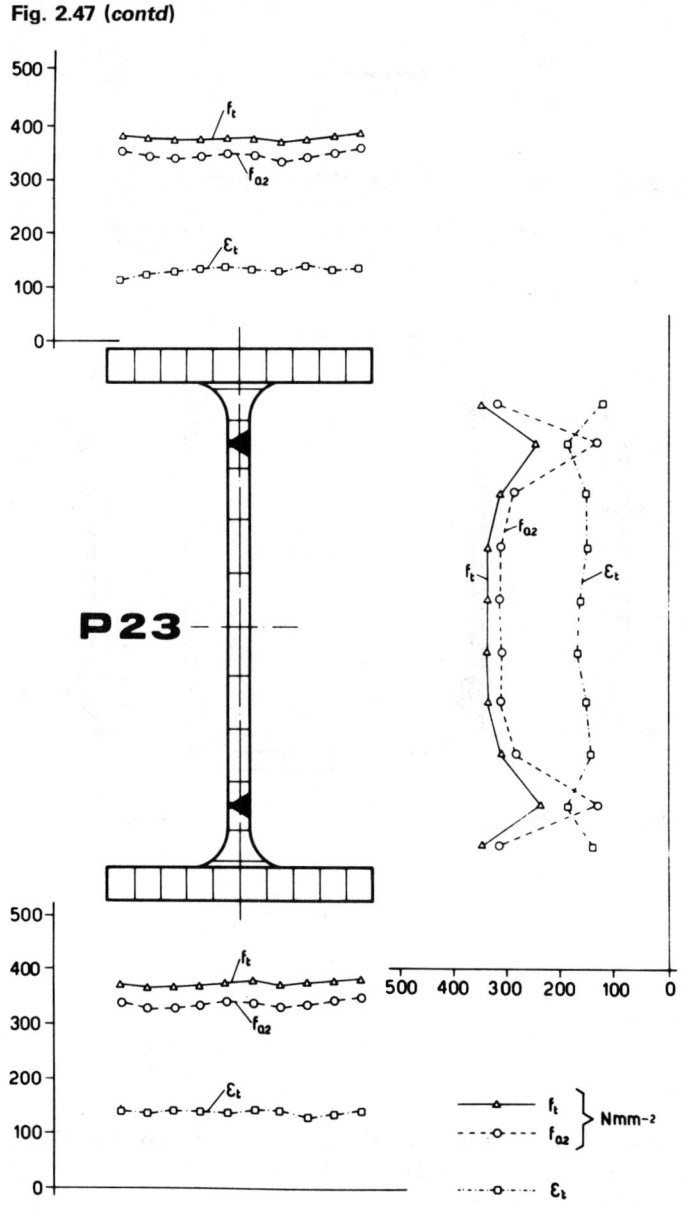

Aluminum alloy structures

Fig. 2.48

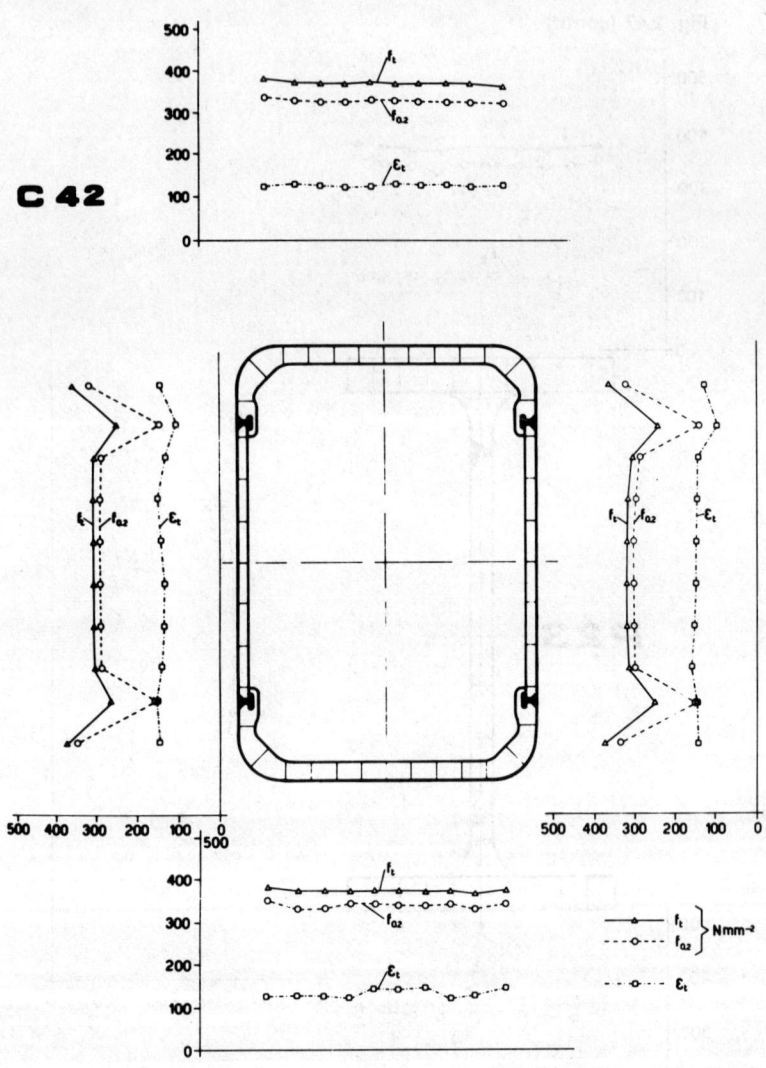

The average σ–ε diagrams for the three regions are compared in Figs 2.50–2.52 for the three profiles studied.

The comparison between the values of $f_{0.2}$ in compression and in tension shows that their elastic limits do not substantially differ, except for case B. However, the behavior of the curves in tension and in compression is completely different. In particular, after reaching the

Fig. 2.48 (*contd*)

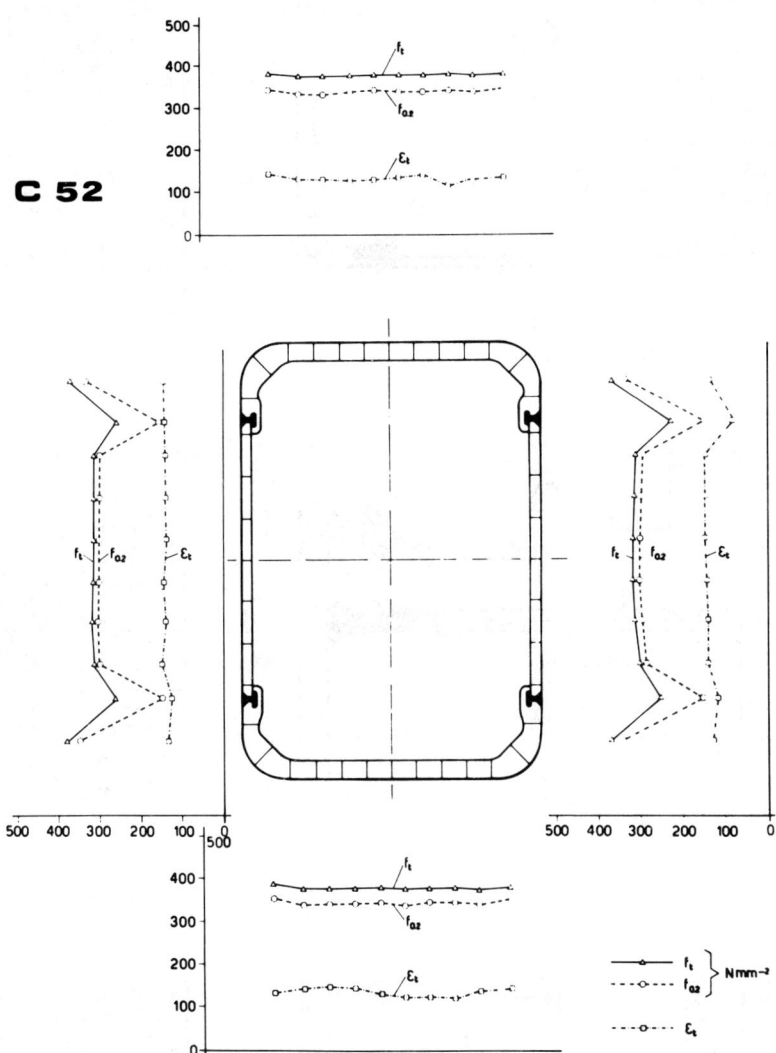

proportional limit the curves in compression join the knee more gradually than the curves in tension and they start at smaller values; in addition, the curves in tension are less gradual. Furthermore, it can be observed that in the range of small plastic deformations the compressive resistance is usually smaller than that in tension, whereas the contrary is true for larger deformations.

Aluminum alloy structures

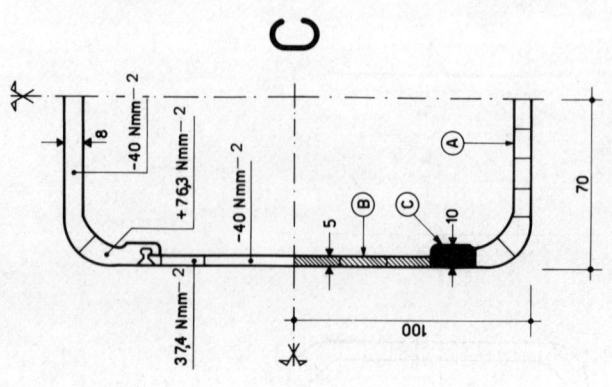

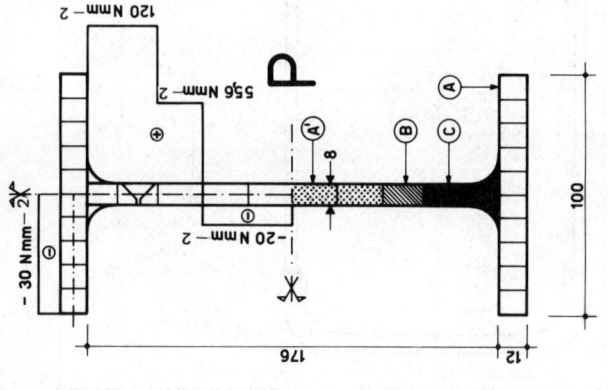

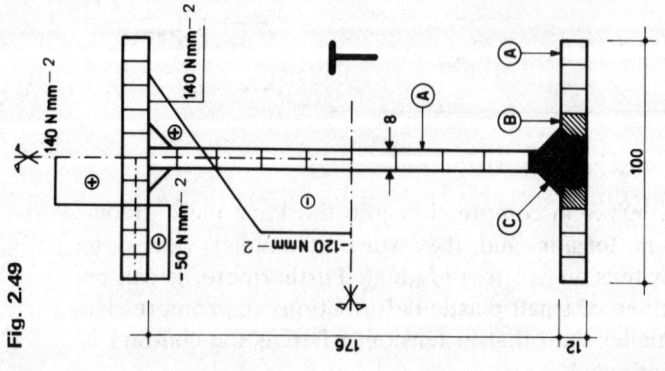

Fig. 2.49

Fig. 2.50

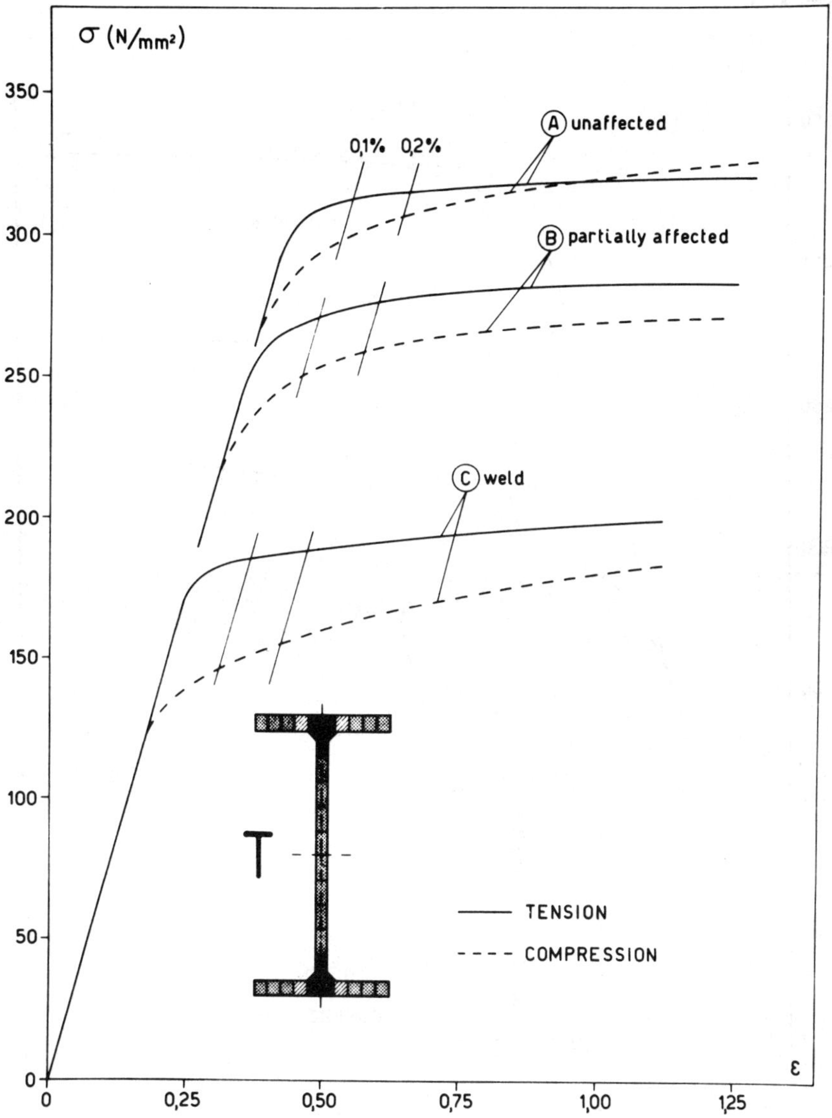

Fig. 2.51

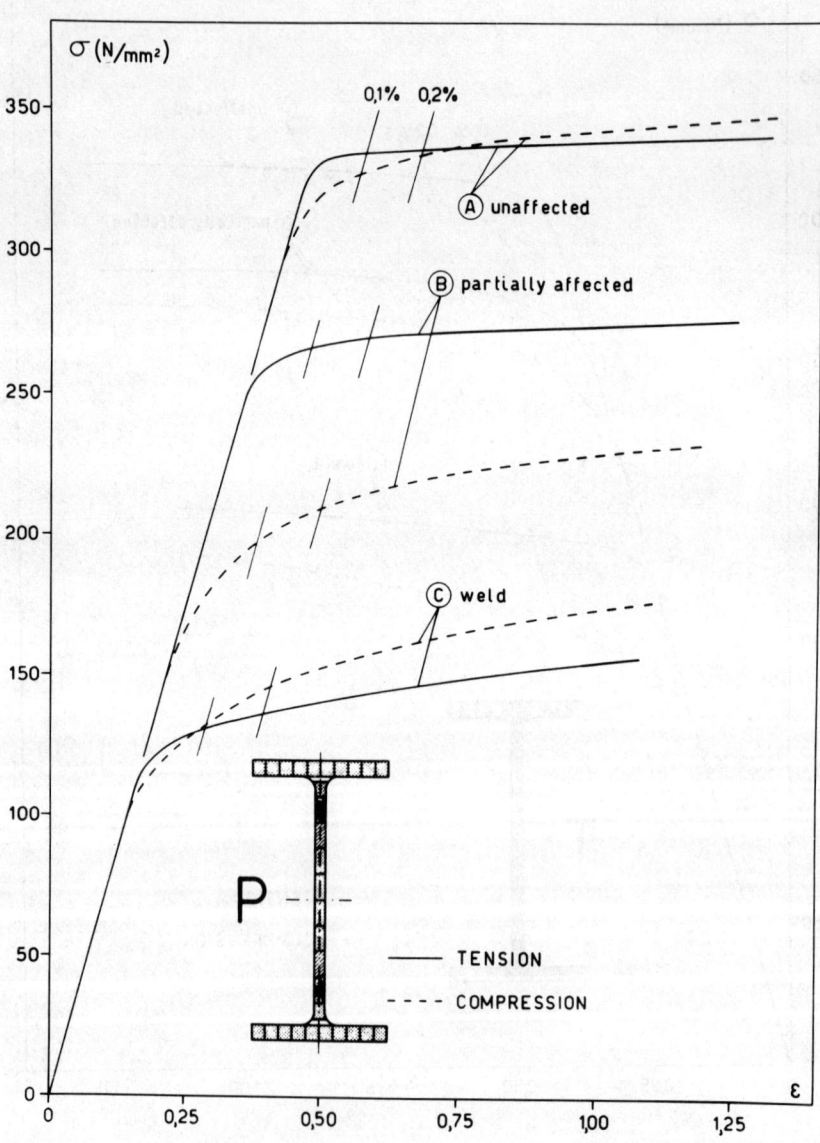

The structural material

Fig. 2.52

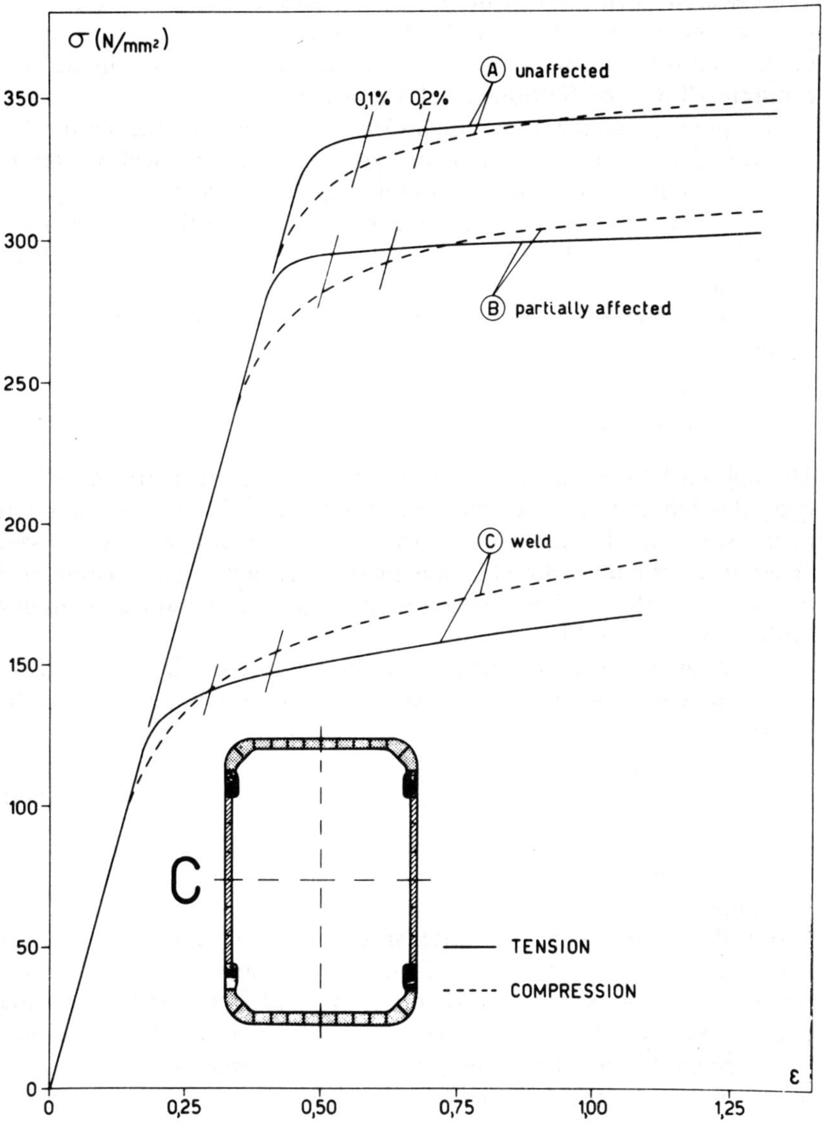

2.5.3 Models of Mechanical Imperfections

2.5.3.1 ECCS models [24]

The research carried out by the ECCS committees has been referred to in the previous sections. It led to the definition of the models of distribution of mechanical imperfections caused by welding processes in the most common alloys (see Sections 2.5.1.4 and 2.5.2.3).

The models related to residual stresses (a) and to the elastic limit variation $f_{0.2}$(b) are summarized in Figs 2.53 and 2.54, which refer respectively to butt-welded joints and built-up welded sections.

At the same time an alternative method for testing and analyzing mechanical imperfections has recently been developed at Cambridge University, UK (see Section 2.5.3.3). The comparison between the two different approaches is of great interest, and is therefore presented in Section 2.5.4.

2.5.3.2 Tendon force concept [25]

The approach adopted in Britain to determine residual stresses is based upon the definition of a conventional 'tendon force' which can be related to the shrinkage force F_s. This shrinkage force is a tension force which arises at welds and is caused by the greater resistance to elongation of the fibers close to the weld, which experience higher temperatures, than those further from the weld.

The shrinkage force F_s, which corresponds to the area AABB of Fig. 2.55, can be considered proportional to Q/v (heat input per unit length of weld):

$$F_s = k \frac{Q}{v} \tag{2.49}$$

k is a nondimensional constant which in practice is independent of the welding process.

In order to overcome the scatter in test data with respect to Eq. 2.49, the tendon force was introduced. This is insensitive to the width of the tension zone, plate dimensions, material yield stress and actual stress pattern on the cross-section. This force F_t is resisted by the whole cross-section of the plate and is defined by the area AACC (Fig. 2.55). When divided by the total plate area, it gives the compression stress on the whole plate.

This model is therefore equivalent to the two blocks of Fig. 2.55 which respectively represent the tendon force (tension block) and the compression stresses on the whole cross-section (area DDEE).

The structural material

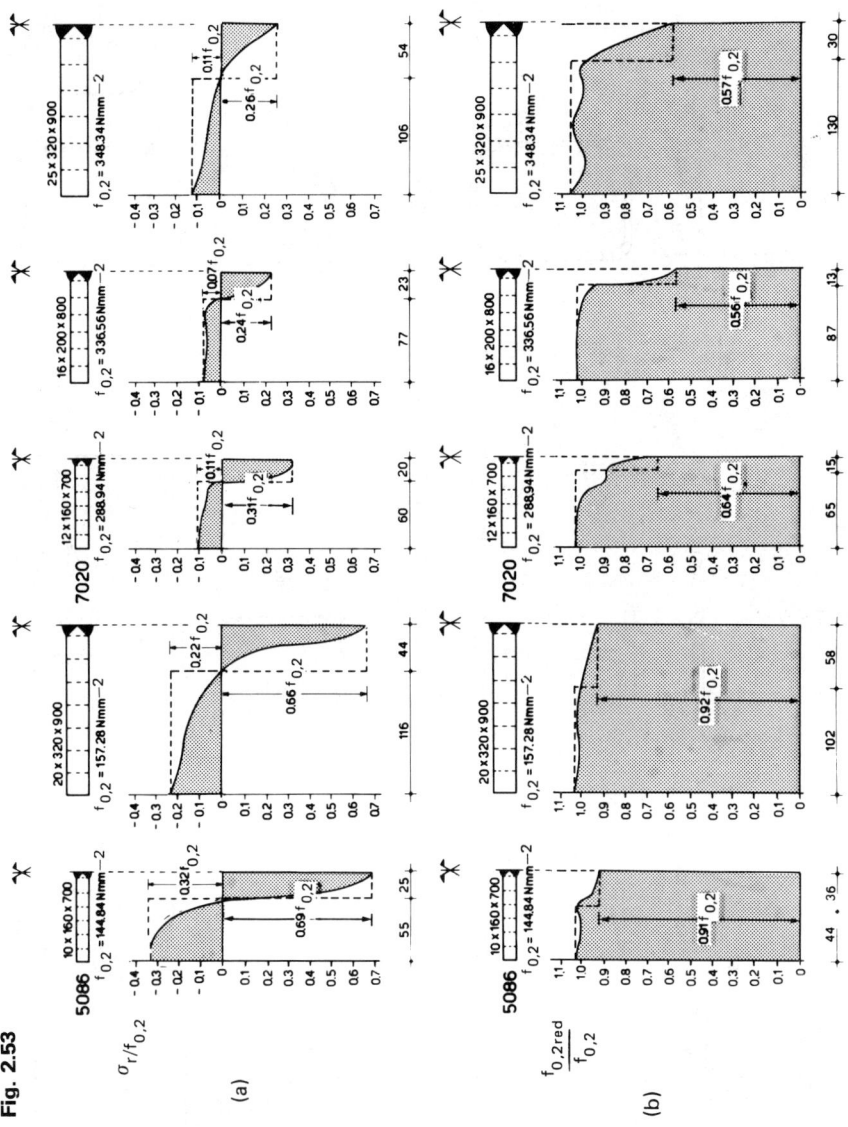

Fig. 2.53

Aluminum alloy structures

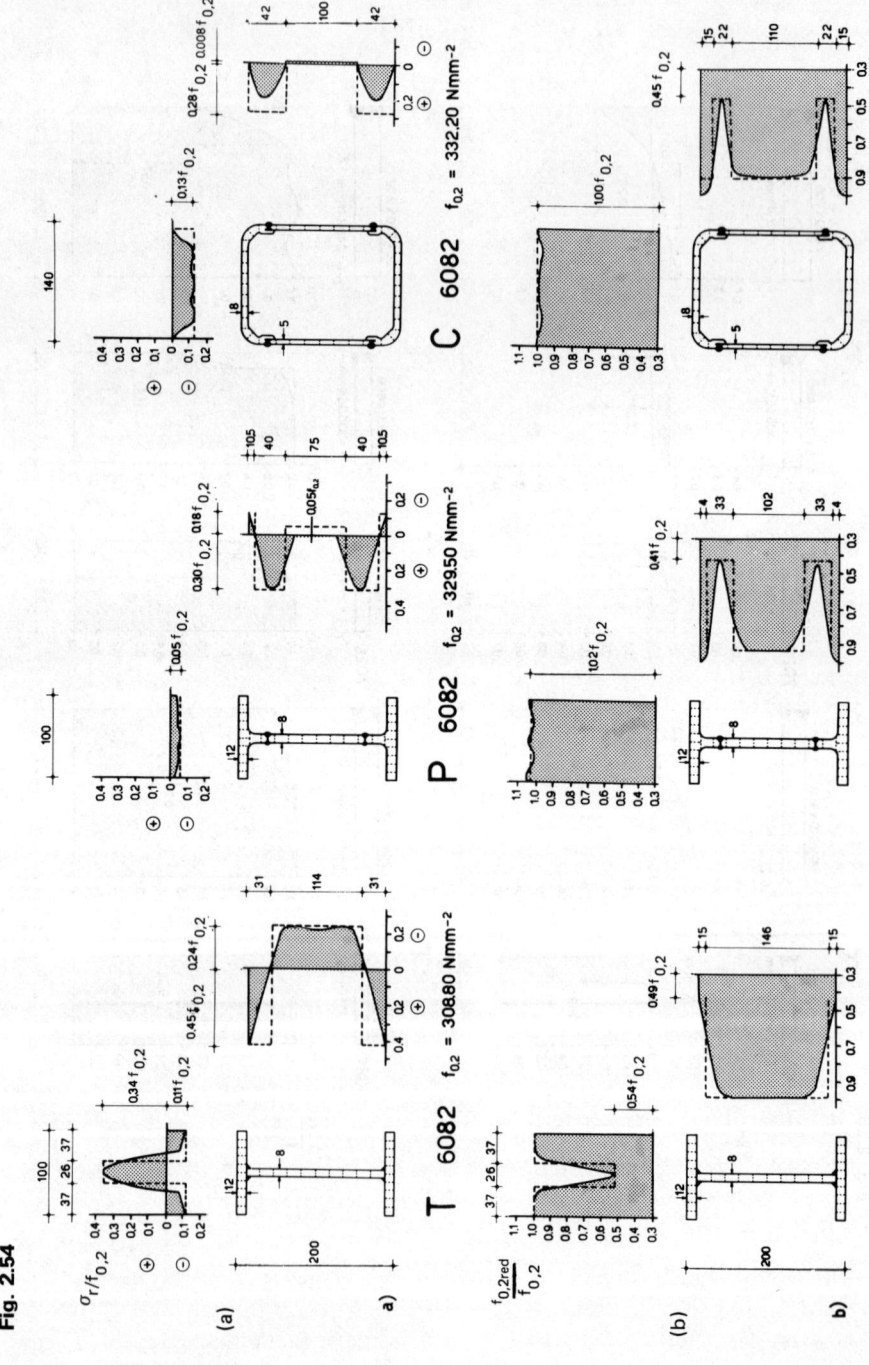

Fig. 2.54

The structural material

Fig. 2.55

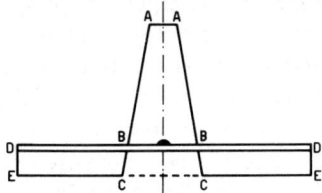

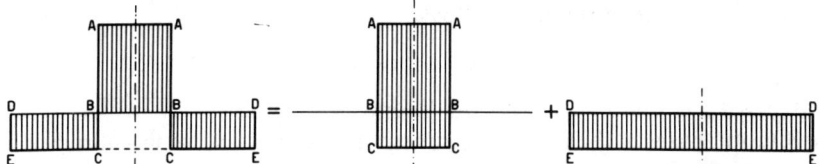

The following equations can be then obtained:

$$F_t = \sigma_{r,c} b t \qquad (2.50)$$

$$c = F_t / t(\sigma_{r,c} + \sigma_{r,t}) \qquad (2.51)$$

The tendon force can also be expressed, analogously to Eq. 2.49, by:

$$F_t = k_t \frac{Q}{v} \qquad (2.52)$$

or, alternatively, by

$$F_t = \psi A_w \qquad (2.53)$$

where

$$\psi = \chi k_t. \qquad (2.54)$$

A_w is the cross-sectional area of the weld deposit.

Equation 2.53 is simpler to use than Eq. 2.49. This advantage is countered by the wide range of variability in χ and ψ from process to process, in relation to the class of electrode used and its polarity.

The validity of Eq. 2.53 has been checked by comparing the results of its application with a finite difference method which simulates the thermoplastic phenomenon of residual stress formation.

2.5.3.3 Cambridge models [26–31]

The tendon force concept, originally developed for steel structures, has been extended to nonlinear materials such as aluminum alloys. It has been verified through experiment and numerical simulation.

Aluminum alloy structures

The draft of the British specifications for aluminum alloys (CP 118), now under revision, suggests an equation of type 2.53 to compute the tendon force. It provides two different values of the constant ψ for the cases of heat-treated and non-heat-treated alloys. In fact, the experimental results of tests carried out on specimens of 5000, 6000 and 7000 series alloys proved that the value of the tendon force in heat-treated alloys is almost one-half of the value for non-heat-treated alloys. The reason for this is connected with the greater degrading effect when welding the heat-treated alloys.

The corresponding equations are therefore:

for 5000 series alloys $\qquad F_t = 5 A_w \qquad$ (2.55)

for 6000 and 7000 series alloys $\quad F_t = 2.5 A_w \qquad$ (2.56)

where A_w is the area of the weld deposit.

On the basis of the same experimental research, an empirical relation is suggested:

$$A_h \cong 10 A_w \qquad (2.57)$$

which expresses the relationship between the weld deposit and the reduced-strength region A_h.

This equation is similar to the well-known 'one-inch' rule. In fact the experimental results available to date have shown that the semi-width of the reduced-strength region can be considered to be less than 25 mm, which corresponds to 1 in.

Equations 2.55 and 2.56 can then be rewritten:

for 5000 series alloys $\qquad F_t = A_h/2 \qquad$ (2.58)

for 6000 and 7000 series alloys $\quad F_t = A_h/4 \qquad$ (2.59)

In order to facilitate the use of these equations, A_h values are provided for different joints for various numbers of passes. The corresponding approximate formulas result:

Straight butt welds
Single pass $\qquad 2t^2$
Two pass (single V) $\quad 5t^2$
Two pass (double V) $\quad 2.5t^2$
n passes (single V) $\quad 7t^2/\sqrt{n}$
n passes (double V) $\quad 3.5t^2/\sqrt{n}$
T butt (one pass per side)
Consecutive passes $\quad 2.5t^2$
Simultaneous passes $\quad 5t^2$
Corner weld
(Depending on
 penetration) $\qquad 3t^2$ to $7t^2$

The structural material

Fillet welds	Automatic	Manual
Single fillet	$6t_1^2$	$10t_1^2$
T fillet (sequential)	$t_1(6t_1+t_2)$	$t_1(10t_1+t_2)$
T fillet (simultaneous)	$12t_1^2$	$20t_1^2$
Cruciform	$t_1(7t_1+t_2)$	$t_1(11t_1+t_2)$

For butt welds, when the thicknesses (t_1 and t_2) differ, replace t^2 by $t_1 t_2$.
For fillet welds, t_1 is the greater thickness and t_2 the lesser.

2.5.4 Comparison between ECCS and British Models [32]

2.5.4.1 Butt-welded joints

Residual stress distributions
The shrinkage force values have been computed (Fig. 2.56):

By integrating the experimental results (column 4)
By applying the block model of Section 2.5.3.1 (column 5)

The tendon force values have been computed by means of the block model (column 6) and by making use of the Cambridge method explained in Section 2.5.3.3 (column 7).

The comparison between the maximum experimental compressive stresses (column 3), which are equal to those of the block model (column 8), and the compressive stresses which are in equilibrium with the tendon force (column 9), show that the deviations are with one exception within 18 percent. However, in this exception the deviation reaches 48 percent. It should be noted, though, that the two models are practically equivalent since the values of compressive stresses are extremely small.

Elastic limit distribution
The area of the reduced-strength region, and hence its semi-width b_r can be computed by the approximate formulas given in Section 2.5.3.3.

The comparison between the values computed in this way and those of the ECCS (Section 2.5.3.1) is given in Fig. 2.57 (columns 6 and 7).

Fig. 2.56 Residual stress distribution values for butt-welded joints

No.	Max. residual stresses (N mm^{-2}) Tension	Compression	Test $\int \sigma \, dA$ (kN)	Model $F_c = F_t$ (kN)	Tendon force ECCS (kN)	Tendon force UK (kN)	$\sigma_{r,c}$ (N mm^{-2}) ECCS	UK	Scatter
1	101	45	38	49	72	88	45	55	18%
2	105	37	90	171	236	248	37	39	5%
3	91	30	33	44	58	63	30	33	9%
4	80	25	41	60	78	92	25	29	14%
5	89	46	101	240	364	193	46	24	48%

Aluminum alloy structures

Fig. 2.57 Elastic limit distribution values for butt-welded joints

No.	Specimen	Thickness (mm)	Passes	A_h (mm^{-2})	b_r (mm) ECCS	UK	Scatter	$f_{0.2}$ (N mm^{-2})	$f_{0.2}^*$ (N mm^{-2})	$f_{0.2}^*/f_{0.2}$
1	10×160×700 5086	10	4	175	36	9	75%	145	131	0.89
2	20×320×900 5086	20	8	495	58	13	78%	157	145	0.91
3	12×160×700 7020	12	4	252	15	11	30%	290	185	0.63
4	16×200×800 7020	16	6	365	13	12	12%	337	188	0.56
5	25×320×900 7020	25	8	773	30	16	48%	348	200	0.57

It should be noted that in the case of heat-treated alloys, except for case 5, there is good agreement between the ECCS model and the British one. In the case of non-heat-treated alloys (cases 1 and 2) large deviations result, and none of the models can be considered to closely interpret actual tests. The high values of the $f^*_{0.2}/f_{0.2}$ ratio (column 11) recorded in experiments suggest that this imperfection should be neglected in non-heat-treated alloys.

2.5.4.2 Built-up members

Residual stress distributions
In order to apply the tendon force method to the built-up members examined, three different models have been used to compute A_h and therefore F_t. These are:

In the case of T profiles, a T fillet-weld joint with consecutive passes
In the case of P profiles, a T butt-weld joint
In the case of C profiles, a straight butt-weld joint made from two plates, one with web and the other with flange thickness

The comparison, summarized in Fig. 2.58, can be considered satisfactory. It should be noted that the tendon force concept, when applied to T joints, yields equal compressive stresses in the flange and web:

$$\sigma_{r,c} = F_t/(b_f t_f + b_w t_w) \tag{2.60}$$

since it provides constant $\sigma_{r,c}$ on the whole joint.

This conservative hypothesis leads to a discrepancy between the Cambridge models and the experimental results of the T profiles, since the recorded compressive stresses were different in the flange and web.

Elastic limit distribution
The semi-width b_r of the reduced-strength regions can be obtained even in the case of built-up members through the approximate formulas for the

Fig. 2.58 Residual stress distribution values for built-up members

| Profile | Max. residual stresses (N mm^{-2}) | | | | $\sigma_{r,c}$ (N mm^{-2}) | | Scatter |
| | Compression | | Tension | | | | |
	Web	Flange	Web	Flange	ECCS	UK	
T	120	50	140	140	120	130	7%
P	30	20	120	30	20	20	0%
C	40	40	100	—	40	44	9%

Aluminum alloy structures

Fig. 2.59 Elastic limit distribution values for built-up members

Profile	A_h (mm^{-2})	b_r (mm) ECCS	UK	Scatter	$f_{0.2}$ (N mm^{-2})	$f_{0.2}^*$ (N mm^{-2})	$f_{0.2}^*/f_{0.2}$
T	480	13	15	20%	309	168	0.54
P	160	17	10	39%	330	135	0.41
C	200	11	16	31%	332	150	0.45

area A_h by making use of the previous models. In order to obtain b_r from the value of A_h, the following ideas have been used:

In the case of T profiles it has been assumed that the dimensions of the reduced-strength region are directly proportional to the web and flange widths.

In the case of P profiles the semi-width has been obtained by dividing the area A_h by the web thickness.

In the case of C profiles the semi-width can be obtained by dividing the area A_h by the average thickness of the web and the flange.

The comparison between the semi-widths of the reduced-strength regions computed by means of the block model (column 3) and the Cambridge model (column 4) is given in Fig. 2.59. The deviations are within 40 percent.

2.5.4.3 Concluding remarks

The comparison between the results of the ECCS research and the British research leads to the following conclusions:

(a) The magnitude of the residual stress values given by the two models is comparable even though the Cambridge model does not allow the actual distributions of the imperfections to be represented.
(b) The ECCS models are closer to the experimental results even though it is not possible to extrapolate their results to other cases.
(c) On the other hand, the British models have been developed in order to allow extrapolation to all types of joints even though rough approximations are necessary.
(d) If a model of imperfection has to be used in order to analyze the influence on the behavior of a structural element, it does not seem reasonable to adopt extremely simplified models even if on the conservative side.
(e) By the use of actual simulations in numerical programs, the limita-

tions existing in structural modeling have been eliminated. This allows all the imperfections of the industrial bar to be characterized.

2.5.5 Bauschinger Effect

2.5.5.1 Definitions

After being loaded in the inelastic range, strain-hardening metallic materials such as aluminum alloys exhibit a behavior different from that prior to inelastic loading. The most significant effect of this phenomenon, which is called the *Bauschinger effect*, can be observed in the uniaxial test. In fact if we carry out a compression test on a specimen which has already yielded in tension, we will normally find an elastic limit which is lower than that obtained in the tension test (which was not preceded by a test of opposite sign).

This difference of strength under compression and tension is another industrial imperfection which characterizes the industrial bar. There are different interpretations of this phenomenon from a crystalline point of view. However, it is important to have experimental results which allow quantification of the Bauschinger effect from the macroscopic viewpoint. This enables its effects on structural behavior to be considered.

It has not been possible to reach a single definition of the problem in the light of the experimental results obtained by Ferton (see Section 2.5.4.4) because there was considerable scatter in the results. They do not seem to follow any simple law, in contrast to the classical idealization of the Bauschinger effect into three regions: ideal, semi-ideal and with isotropic strain hardening. Furthermore, these results do not always agree with those obtained by Faella and Mazzolani (Section 2.5.4.4).

However, the classical idealizations of the Bauschinger effect have been used to explain the plastic behavior of the metals by means of the theory of 'slip' planes.

Under this theory the nonreversible plastic deformations are caused by the tangential stresses, and the plasticity condition between two planes is expressed by a τ–γ relationship of the rigid-hardening type with the Bauschinger effect. If the limiting tangential stress is expressed as a function of the history of the plastic slipping between two planes, this theory allows the σ–ε behavior of the material in the tension and compression tests to be interpreted.

The influence of the Bauschinger effect can be important for those aluminum alloy members in which the industrial manufacturing processes (extrusion, tractioning etc.) introduce an initial loading condition. The ultimate strength that they can carry under live loads is directly affected, very often in an adverse way, by the level of the initial stresses. This influence is even more complex in the case of alternating loads.

Aluminum alloy structures

2.5.5.2 Influence of tractioning [33, 34]

The fabrication of metallic profiles involves a technological phase to reduce the initial geometrical imperfections (longitudinal curvature) to within the tolerance limit required by erection procedures and specifications. These provisions usually agree in requiring the highest camber to be less than a limiting value of $L/1000$. This value of camber limits the load-bearing reduction the bar will undergo, especially when in compression.

In the case of aluminum alloy extruded profiles the straightening process is usually done by 'tractioning'. This consists of subjecting the bar to a high-tension force applied at its ends. The force is applied until the initial curvature of the bar is eliminated or reduced to the tolerance value. This force causes high plastic deformation ($\sim$1–3 percent). At the same time tractioning produces a strain hardening of the material which increases strength and relieves the residual stresses caused by heat treatment; these stresses can thus be neglected in extruded profiles (see Section 2.5.1.3). On the other hand, the Bauschinger effect reduces the load-bearing capacity under compression. Usually this effect counterbalances the beneficial effects of straightening the bar and relieving its residual stresses.

The final result on the resistance of the bar depends upon the interacting effect of these phenomena. Qualitatively the problem can be explained as follows.

Consider a bar of a material for which are known the σ–ε law in compression and in tension and the residual stress field due to heat treatment (Fig. 2.60). After extrusion the bar has a curvature higher than

Fig. 2.60

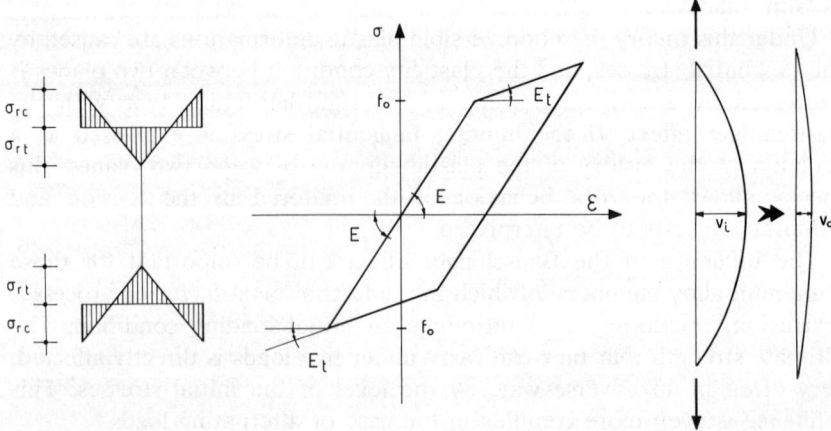

Fig. 2.61

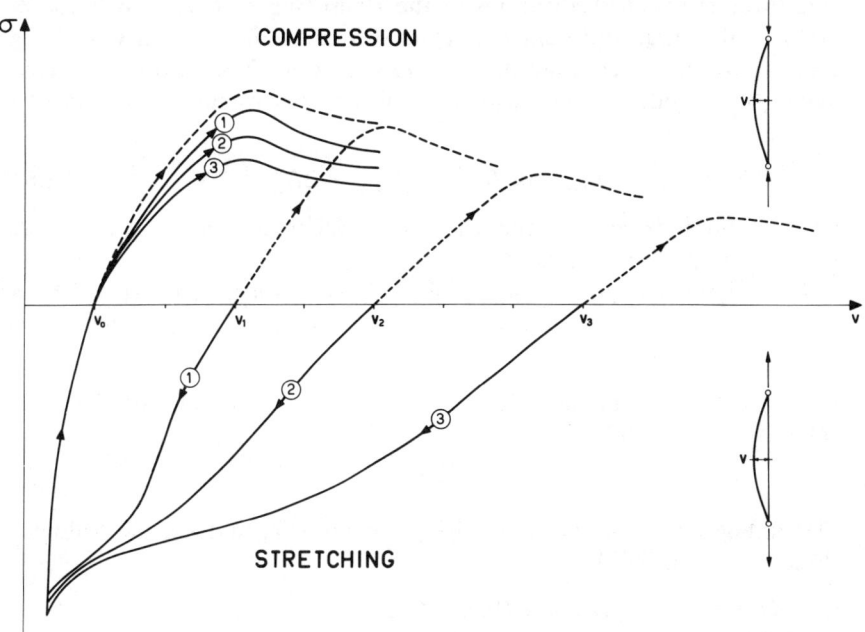

that allowed, the value of the camber being v_i. The magnitude of tractioning has to be such that this value is brought back within the tolerance limit v_0.

Now consider bars with different initial geometrical imperfections with cambers v_1, v_2 and v_3. Then apply different tractioning forces to produce the same camber v_0 in the different bars (Fig. 2.61). When these straightened bars, geometrically equal to each other, undergo compression, they will withstand a different maximum load dependent upon the magnitude of the imperfection before tractioning.

It is also interesting to compare the curves of the tractioned bars under compression and those of similar nontractioned bars. It is not always true that a reduction of the geometrical imperfections compensates for the introduction of structural imperfections due to the Bauschinger effect. Clearly the differences which are shown by this comparison are closely related not only to the amount of the geometrical imperfection, but also to the nature of the Bauschinger effect, to the intensity and distribution of initial residual stresses, and to the magnitude of the plastic deformation experienced.

2.5.5.3 Behavioral models [34, 35, 36]

The three classical idealizations of the Bauschinger effect, which correspond to the three different behaviors shown by the σ–ε curves of Fig. 2.62, apply to an elastic strain-hardening material. They are known as (a) ideal, (b) semi-ideal and (c) with isotropic strain hardening. Analytically it can be said that:

$$f_{\lim}^{-(+)} = k_1 f_{\lim}^{+(-)} + k_2 [f_{\lim}^{+(-)} - 2f_{0,\lim}] \tag{2.61}$$

$f_{0,\lim}$ limit strength of the virgin material (assumed equal in tension and in compression)

$f_{\lim}^{-(+)}$ limit strength in compression (tension) for a material which has been prestressed in tension (compression) to a $f_{\lim}^{+(-)}$ stress above the elastic limit

Owing to the assumed isotropy of the virgin material, the following relation is also valid:

$$k_1 + k_2 = 1 \tag{2.62}$$

Depending upon the values of the parameters k_1 and k_2, the following three cases can be obtained:

(a) *Ideal Bauschinger effect* (Fig. 2.62a)
    With the position

$$k_1 = 0 \qquad k_2 = 1 \tag{2.63}$$

Eq. 2.61 becomes

$$f_{\lim}^{-(+)} = f_{\lim}^{+(-)} - 2f_{0,\lim} \tag{2.64}$$

(b) *Semi-ideal Bauschinger effect* (Fig. 2.62b)
    With the position

$$k_1 = k_2 = 1/2 \tag{2.65}$$

Eq. 2.61 becomes

$$f_{\lim}^{-(+)} = -f_{0,\lim} \tag{2.66}$$

(c) *Bauschinger effect with isotropic strain-hardening* (Fig. 2.62c)
    With the position

$$k_1 = 1 \qquad k_2 = 0 \tag{2.67}$$

Eq. 2.61 becomes:

$$f_{\lim}^{-(+)} = -f_{\lim}^{+(-)} \tag{2.68}$$

The widely adopted hypothesis is of an elastic-hardening material behavior but this is not always close to the response of those materials

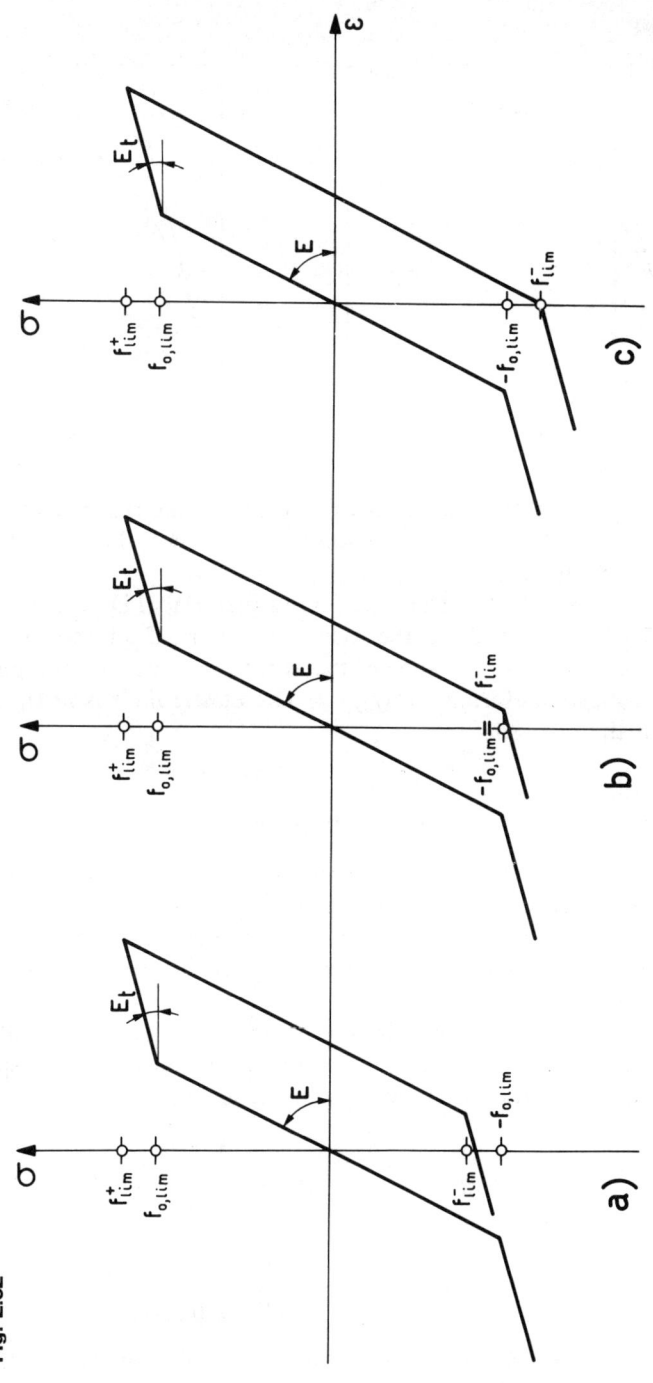

Fig. 2.62

Aluminum alloy structures

Fig. 2.63

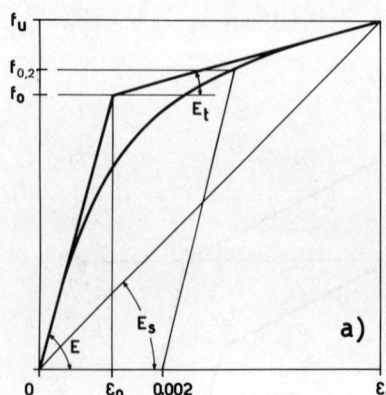
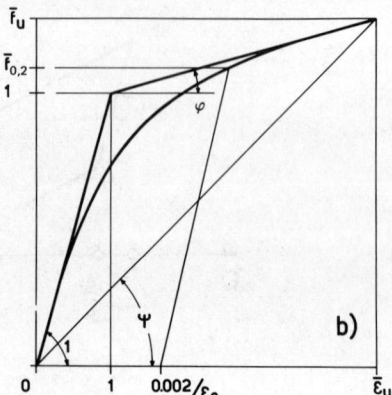

which have a continuous behavior. For these materials some criteria have to be defined to ensure correspondence between the discontinuous and the continuous model.

Consider an elastic strain-hardening material (Fig. 2.63a) with a $\sigma-\varepsilon$ law which is defined by the elastic modulus E at the origin, by the hardening modulus E_t, and by the secant modulus E_s computed at the limit of plastic deformation (f_u, ε_u). The elastic limit is at (f_0, ε_0).

With the position

$$\bar{\varepsilon} = \varepsilon/\varepsilon_0 \qquad \bar{\sigma} = \sigma/f_0$$
$$\varphi = E_t/E \qquad \psi = E_s/E \qquad (2.69)$$

we have the dimensionless form in the $\bar{\sigma}-\bar{\varepsilon}$ plane (Fig. 2.63b).

Correspondence with a continuous law, between the same bounds of stresses and strains is ensured by imposing the condition that the tangent modulus at the origin and at the point of maximum deformation are coincident.

For the continuous material the Ramberg–Osgood law in the form of Eq. 2.32 is assumed. It can be nondimensionalized with respect to the point (f_0, ε_0) of the discontinuous curve, giving:

$$\bar{\varepsilon} = \bar{\sigma} + \frac{1}{k}\left(\frac{\bar{\sigma}}{\bar{f}_{0.2}}\right)^n \qquad (2.70)$$

where

$$\bar{f}_{0.2} = f_{0.2}/f_0 \qquad k = \frac{f_0}{0.002E} \qquad (2.71)$$

The tangent condition is given by the same value of the modulus E. The tangent condition at the point $(\bar{f}_u, \bar{\varepsilon}_u)$ is given instead by the condition

$$\bar{\varepsilon}_u = \bar{f}_u + \frac{1}{k}\left(\frac{\bar{f}_u}{\bar{f}_{0.2}}\right)^n$$

$$\varphi = \frac{1}{1 + \frac{n}{k\bar{f}_u}\left(\frac{\bar{f}_u}{\bar{f}_{0.2}}\right)^n} \tag{2.72}$$

The parameters n and $\bar{f}_{0.2}$ can be obtained from Eq. 2.72. In this way the curve Eq. 2.70 is equivalent to the bilinear law following the proposed criterion.

From Eq. 2.72 we get:

$$\left(\frac{\bar{f}_u}{\bar{f}_{0.2}}\right)^n = \left(\frac{1}{\varphi} - 1\right)\frac{\bar{f}_u k}{n} \tag{2.73}$$

In addition,

$$n = \frac{1/\varphi - 1}{1/\psi - 1} \tag{2.74}$$

From Eq. 2.73 we get:

$$\bar{f}_{0.2} = \frac{\bar{f}_u}{\sqrt[n]{[k(\bar{\varepsilon}_u - \bar{f}_u)]}} \tag{2.75}$$

where

$$\bar{f}_u = 1 + \varphi(\bar{\varepsilon}_u - 1) \tag{2.76}$$

and

$$(\bar{\varepsilon}_u - \bar{f}_u) = (\bar{\varepsilon}_u - 1)(1 - \varphi) \tag{2.77}$$

Therefore

$$\bar{f}_{0.2} = \frac{1 + \varphi(\bar{\varepsilon}_u - 1)}{\sqrt[n]{[k(\bar{\varepsilon}_u - 1)(1 - \varphi)]}} \tag{2.78}$$

Eqs 2.74 and 2.78 characterize in dimensionless form the continuous curve corresponding to the bilinear law in the range defined by $\bar{\varepsilon}_u$.

The numerical correspondence between the parameters of the continuous curve $(n, \bar{f}_{0.2})$ and of the bilinear law (φ) is given in Fig. 2.64 for $k = 1$ and for the following range of variability: $\bar{\varepsilon}_u = 1.5, 2.0, 2.5, 3.0$ and $\varphi = 0.05, 0.10, 0.15$.

2.5.5.4 Experimental results [37–41]

The first interesting experimental results on this subject are those of Templin and Sturm [47] and Jackman [48] carried out on aluminum

Aluminum alloy structures

Fig. 2.64 Correspondence between parameters of continuous curve and bilinear law

$\bar{\varepsilon}_u$	φ 0.05	0.10	0.15	
1.5	0.683	0.700	0.717	ψ
	40.937	21.000	14.357	n
	1.044	1.091	1.141	$\bar{f}_{0.2}$
2.0	0.525	0.550	0.575	ψ
	21.000	11.000	7.667	n
	1.050	1.110	1.174	$\bar{f}_{0.2}$
2.5	0.430	0.460	0.490	ψ
	14.333	7.667	5.444	n
	1.049	1.105	1.172	$\bar{f}_{0.2}$
3.0	0.367	0.400	0.433	ψ
	11.016	6.000	4.327	n
	1.038	1.088	1.150	$\bar{f}_{0.2}$

alloys 2017 and 2024. These tests showed that the Bauschinger effect is more important for small values of the tractioning force (approximately 1–2 percent) (see Fig. 2.65) and that the effect is a function of the time interval between tempering and work hardening (e.g. 2017 alloy) and that it disappears after quenching (e.g. 2024 alloy). This research, however, was mainly devoted to aeronautical applications. Only in the 1970s was experimental research carried out on the alloys used in other structures.

Some relevant experimental research was carried out by the ECCS committee on extruded profiles (1972) [46], although study of the Bauschinger effect was not the major object. It did show that for the material of a given profile the values of $f_{0.2}$ from a tension test and a stub column test differ (see Fig. 2.66). Since residual stresses are negligible in extruded profiles, the difference between the two tests can be attributed to the Bauschinger effect. However, no quantitative explanation of the difference is possible because it was of opposite sign and the history of the manufacturing process was not known.

At a later date specific tests were carried out by Faella and Mazzolani [37, 38] in order to confirm the existence of a specific Bauschinger effect, its nature and the possibility of simulating its behavior. A non-heat-treated (AlMg4.5Mn, type 5086) and a heat-treated alloy (AlZnMg1, type 7020) were examined. The experimental procedure consisted of the following stages (Fig. 2.67):

(a) The specimens are prepared ($L = 200$ mm).

The structural material

Fig. 2.65

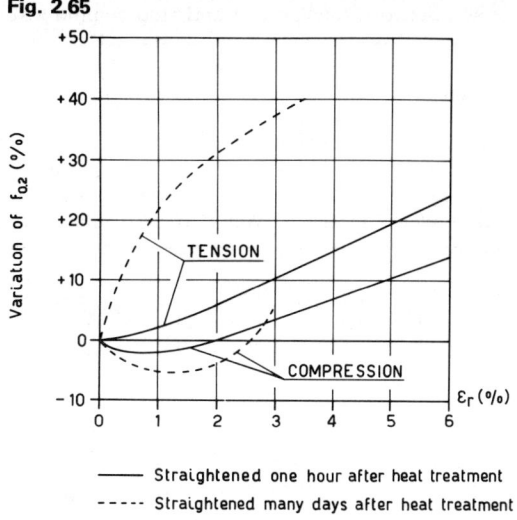

— Straightened one hour after heat treatment
---- Straightened many days after heat treatment

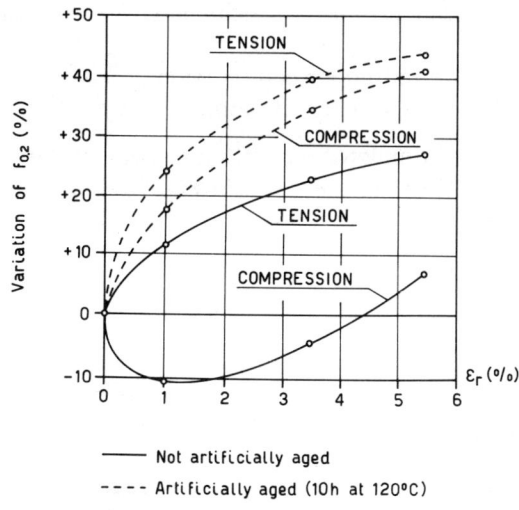

— Not artificially aged
---- Artificially aged (10h at 120°C)

(b) From each complete specimen is taken specimen A for the virgin compression test (with $L = 20$ mm) and specimen B for the virgin tension test (with $L = 170$ mm).
(c) Specimen A is tested under compression.
(d) Specimen B is tested under tension up to a given value of ε_r (~1–3 percent).

Aluminum alloy structures

Fig. 2.66 Comparison between tension test and stub column test

Shape	$f_{0.2}^+$ Tension (N mm^{-2})					$f_{0.2}^-$ Overall compression (N mm^{-2})	
	Standard specimens		Nonstandard specimens		Mean		$\|f_{0.2}^+\|-\|f_{0.2}^-\|$
Alloy	Flanges	Web	Flanges	Web	values	Mean values	(N mm^{-2})
Double T							
5083	182	170	166	157	170	158	+12
7020	320	305	300	290	305	335	−30
6060	290	270	260	260	272	282	−10
Tubes							
5052		190		167	180	163	+17
7020		305		305	305	340	−35
6051		290		270	280	298	−18
6061		236		244	240	245	−5

(e) Then from specimen B is obtained a specimen C for the compression test (with $L = 20$ mm) and a specimen D for the tension test (with $L = 140$).
(f) Specimen C is tested under compression.
(g) Specimen D is tested under tension.

The results of three complete tests with a tractioning force corresponding to $\varepsilon_r = 1.6, 2.0, 2.9$ percent are given in Fig. 2.68 and the corresponding curves are given in Figs 2.69–2.71.

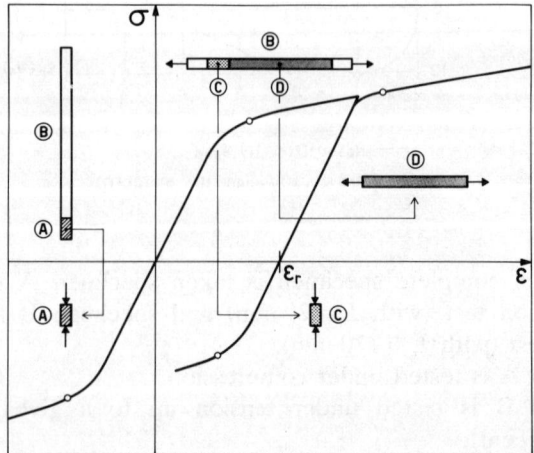

Fig. 2.67

The structural material

Fig. 2.68 Results of Faella and Mazzolani tests [37, 38]

Test	1 ε_r (%)	2 $f_{0.2}^+$ (N mm^{-2})	3 $f_{0.2}^-$ (N mm^{-2})	4 $f_{0.2}^+ - f_{0.2}^-$ (N mm^{-2})	5 Increment $\Delta f_{0.2}$ in tension (%)	6 Reduction $\Delta f_{0.2}$ in compression (%)	7 $\dfrac{f_{0.2}^+ - f_{0.2}^-}{(f_{0.2}^+ - f_{0.2}^-)_0}$
1	0 1.6	312 360	−370 −330	682 690	— +15	— −11	1 1.005
2	0 2.0	161 214	−214 −178	375 392	— +33	— −17	1 1.047
3	0 2.9	161 235	−252 −217	413 452	— +46	— −14	1 1.095

Of significance in Fig. 2.68 is the comparison between the conventional values of the elastic limit at 0.2 percent under tension (column 2) and under compression (column 3) for the virgin material ($\varepsilon_r = 0$) and after tractioning. The corresponding differences are given in Fig. 2.72, in which there is a constant increase in $f_{0.2}$ under tension due to tractioning. The increment of $f_{0.2}^+$ with respect to the initial value $f_{0.2}^0$ (column 5) varies between 15 and 46 percent depending upon the magnitude of the tractioning force and upon the slope of the strain-hardening portion of the σ–ε diagram. The reduction in $f_{0.2}$ (column 6) always lies between 11 and 17 percent.

Fig. 2.69

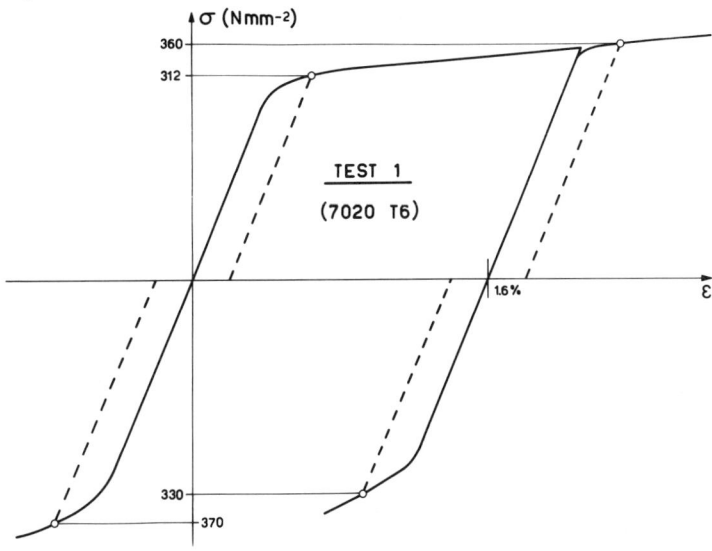

Aluminum alloy structures

Fig. 2.70

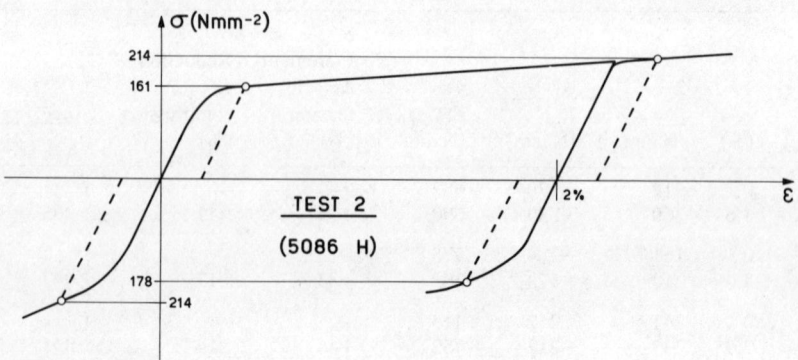

The significant difference between the elastic limits in tension and in compression of the virgin material are due to the testing procedure, which utilizes different specimens and different restraint conditions in tension and in compression.

The effect of tractioning on the extension of the conventional elastic portion (column 4) is given by the ratios of column 7. These values seem to support the hypothesis of an ideal behavior of the material with respect to the Bauschinger effect, especially for the specimen used in test 1 (Fig. 2.68) which is made of a material with a strain-hardening effect smaller than the other two.

This is confirmed by Eq. 2.69 for continuous diagrams, if the conventional elastic limit before and after tractioning ($f_{0,\text{lim}}$ and f_{lim}) are considered instead of the limit stresses $f_{0.2}$. If we nondimensionalize the equation (2.61)

$$\bar{f}^-_{\text{lim}} = -k_1 \bar{f}^+_{\text{lim}} + k_2(\bar{f}^+_{\text{lim}} - 2) \tag{2.79}$$

Fig. 2.71

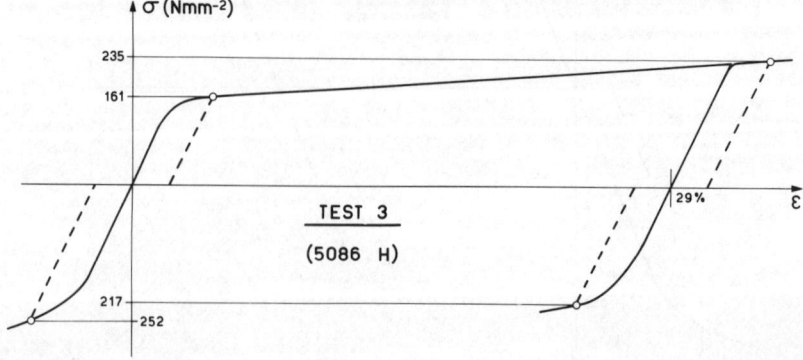

Fig. 2.72

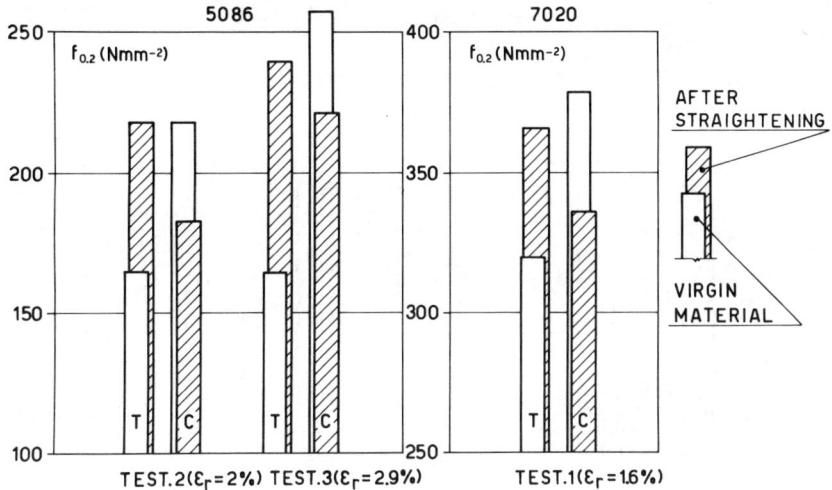

with the positions

$$\bar{f}_{\lim}^{+(-)} = \frac{f_{\lim}^{(+)-}}{f_{0,\lim}^{(+)-}} \qquad f_{0,\lim} = 1 \qquad (2.80)$$

Then, Eq. 2.79 together with Eq. 2.62 gives:

$$\begin{array}{lll} \text{Test 1} & k_1 = 0.112 & k_2 = 0.888 \\ \text{Test 2} & k_1 = 0.249 & k_2 = 0.751 \\ \text{Test 3} & k_1 = 0.349 & k_2 = 0.651 \end{array}$$

Because:

$$0 < k_1 < 0.5 \qquad 0.5 < k_2 < 1$$

the materials experience a Bauschinger effect bounded between the ideal case ($k_1 = 0$; $k_2 = 1$) and the semi-ideal one ($k_1 = k_2 = 1/2$).

Analogous research has been carried out independently at the same time in France by Ferton [40, 41]. The most important results are given in Figs 2.73–2.75 and can be summarized as follows. Strain hardening due to a tractioning force equivalent to a strain of a few percent for the annealed stage of the alloy A-G4MC (type 5086, $\varepsilon_r = 3.5$ percent) and immediately after tempering of the alloy A-SG (type 6081, $\varepsilon_r = 4.5$–5 percent) results in a variation of the elastic limits in tension and in compression which cannot be explained by the classical models of the Bauschinger effect.

Actually, this strain hardening causes an increase of the elastic limit of the alloys A-G4MC and A-SG, tempered and naturally aged, in three

Aluminum alloy structures

Fig. 2.73

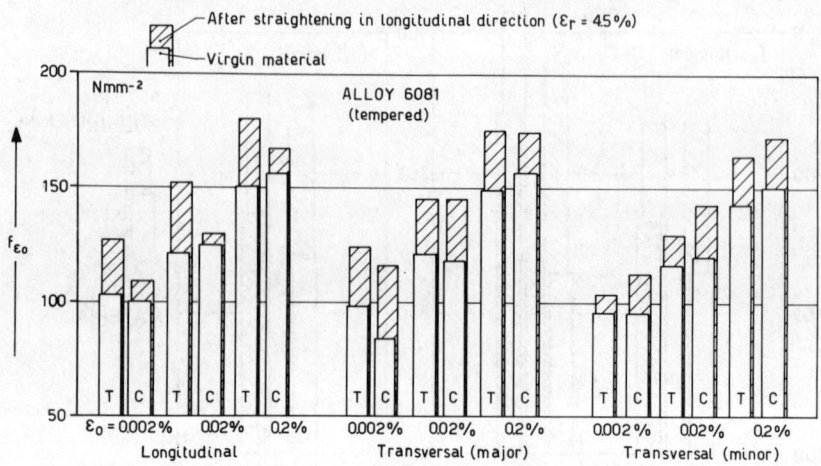

directions (the longitudinal and two transverse directions). This increment is present in both tension and compression. In the tempered A-SG and artificially aged alloy the effect of work hardening is different in the three directions of drawing. In the longitudinal direction, which is coincident with the tractioning direction, the elastic limit is increased in tension and decreased in compression. In the two transverse directions the elastic limit is decreased in both tension and compression.

Fig. 2.74

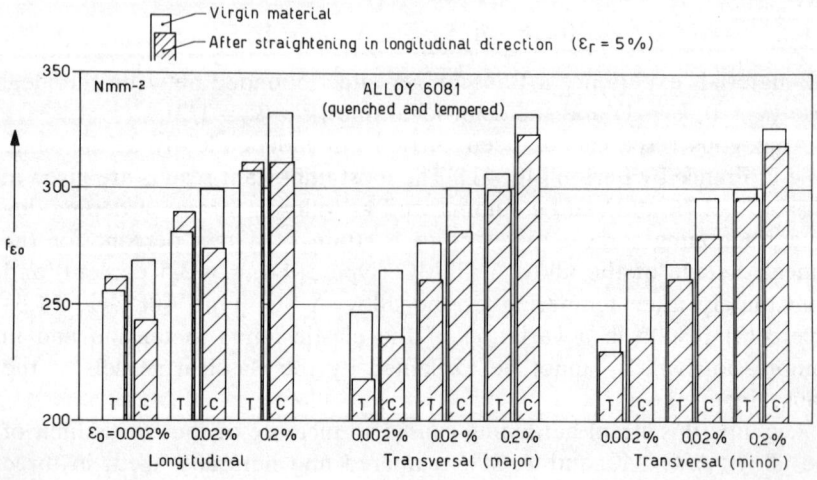

Fig. 2.75

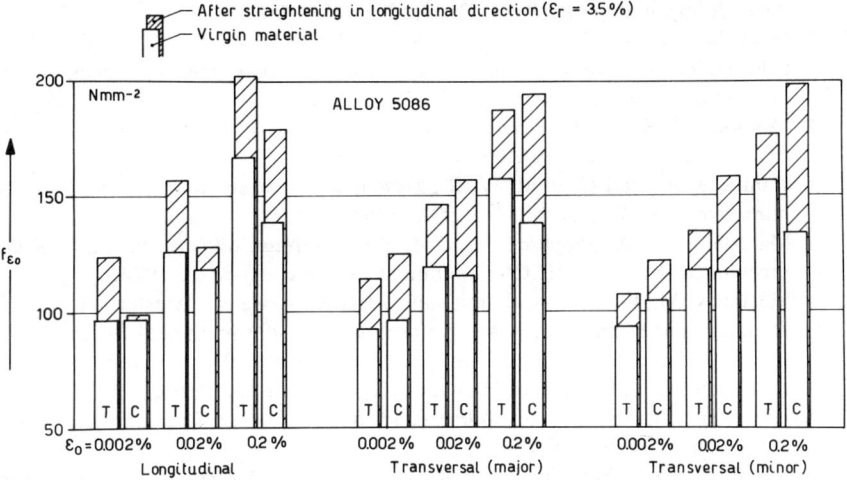

2.5.5.5 Practical consequences

The apparent contrast between the restricted number of known experimental results does not allow extrapolation to a general law. For this reason the ECCS committee does not take into account the Bauschinger effect in its recommendations. This decision seems to be a compromise between two opposite effects due to tractioning: the positive effect of relieving residual stresses, and the negative one of the Bauschinger effect on the load-bearing capacity under compression.

However this is the most common treatment in different specifications. Only the Aluminum Association (USA) specifications consider different elastic limits in tension and compression. The values allowed in compression are always smaller for non-heat-treated alloys and are related to the magnitude of the treatment (alloys of the 3000 and 5000 series), whereas for heat-treated alloys these values are almost equal to those in tension for the 2000 series or exactly equal for the 6000 series.

The variations are always within 20 percent of the minimum values guaranteed for the elastic limit.

References

1. Sutter, K., Die theoretischen Knickdiagramma bei Aluminiumlegierungen (Theoretical stress–strain diagram for aluminum alloys), *Techniske Rundschau, Bern*, 1959, Nos. 20–24.
2. Baehre, R., *Trycktastravorav elastoplastikt material-nagrafragestallningar*

(*Comparison between structural behavior of elastoplastic materials*), Tekn. Dr Arne Johnson Ingenjorsbyra, Report No. 16, 1966.
3. Mazzolani, F. M., La caratterizzazione della legge $\sigma-\varepsilon$ e l'instabilità delle colonne di alluminio (Characterization of the $\sigma-\varepsilon$ law and buckling of aluminum columns), *Costr. Metall.*, 1972, No. 3.
4. Osgood, W. R. and Holt, M., *The column strength of two extruded aluminum alloy H-sections*, NACA Report 656, 1939.
5. Ramberg, W. and Osgood, W. R., *Description of stress–strain curves by three parameters*, NACA Techn. No. 902, 1943.
6. Mazzolani, F. M., *Proposal to classify the aluminum alloy on the basis of the mechanical behavior*, ECCS Committee 16, Doc. 16-74-2, 1974.
7. Marincek, M., Ductility and limit states, *IABSE Congress*, Amsterdam, 1972.
8. Daddi, I. and Mazzolani, F. M., Determinazione sperimentale delle imperfezioni strutturali nei profilati di acciaio (Experimental determination of structural imperfections in steel H-shapes), *Costruzioni Metalliche*, 1972, No. 5.
9. ECCS-IABSE-SSCR-CRC, *Introductory Report*, Second Int. Itinerant Colloquium on Stability, 1976.
10. SSRC, *Guide to Stability Design Criteria for Metal Structures*, B. J. Johnston (Ed.) Wiley Interscience, New York, 1976.
11. Mazzolani, F. M., Analisi Sperimentale delle Tensioni Residue nei Profilati Metallici (Experimental Analysis of Residual Stresses in Metal Profiles), 1st *National Congress AIAS*, Palermo, 1972.
12. Mazzolani, F. M., *Residual stress tests on French profiles* $63 \times 63 \times 4$ *in aluminum alloys*, ECCS Committee 16, Doc. 16-74-3, 1974.
13. Mazzolani, F. M., *Residual stress tests alu-alloy Austrian profiles*, ECCS Committee 16, Doc. 16-75-1, 1975.
14. Hill, H. N., Residual welding stresses in aluminum alloys, *Metal Progress*. ALCOA, 1961, August.
15. Mazzolani, F. M., Les imperfections structurales dans les assemblages soudés en aluminium (Structural imperfections in aluminum welded joints), *Revue de l'Aluminium*, 1974, No. 431.
16. Gatto, F., Mazzolani, F. M. and Morri, D., Analisi sperimentale delle tensioni residue e delle caratteristische meccaniche nei profili saldati di lega Al–Si–Mg (tipo 6082), *Ingegnaria Meccanica*, 1979, Nos. 6, 7, 8.
17. Gatto, F., Mazzolani, F. M. and Morri, D., Etude expérimentale des contraintes résiduelles et des caractéristiques mécaniques dans les profilés soudés en alliage Al–Si–Mg (type 6082), *Soudage et Technique Connexes*, 1979, July–August.
18. Gatto, F., Mazzolani, F. M. and Morri, D., *Experimental analysis of residual stresses and of mechanical characteristics in welded profiles of Al–Si–Mg (type 6082)*, ECCS Committee 16, Doc. 16-77-5, 1977; *Italian Machinery and Equipment*, **11,** 1979, No. 50, March.
19. Hill, H. N., Clark, J. W. and Brungraber, R. J., Design of welded aluminum structures, *Trans. ASCE*, **127**, 1962, Part II.
20. Task Committee on Light Weight Alloys, Suggested specifications for structures of aluminum alloys 6061-T6 and 6062-T6, *J. Struct. Div. Proc. ASCE*, 1962, December, pp. 3341–2.

21. Brungraber, R. J. and Clark, J. W., Strength of welded aluminum columns, *Trans. ASCE*, **127**, 1962, Part II.
22. Mazzolani, F. M., Il comportamento inelastico dei profili in aluminio saldati (Inelastic behavior of welded aluminum profiles), *Costruzioni Metalliche*, 1971, No. 5.
23. Steinhardt, O., *Der Einfluss der Fugetechnik auf Gestaltung und Normung von Aluminium Konstruktionen des Bauweser (The Influence of Connection Techniques on the Design of Aluminum Constructions)*, Deutscher Verlag fur Schweisstechnik (DVS) GmbH, Dusseldorf, 1974.
24. Mazzolani, F. M., The influence of mechanical imperfection on the structural behavior of welded aluminum alloy members, *2nd International Conference on Aluminum Weldments*, Munich, May 1982.
25. Gibson, G. S., An approximate method for calculating the distortion of welded members, *Welding J.*, **17**, No. 7, July 1938.
26. White, J. D., *Longitudinal shrinkage of a single pass weld*, Rep. CUED/C. Struct/TR 57, 1977.
27. White, J. D., Leggatt, R. H. and Dwight, J. B., Weld shrinkage stresses in plated structures, *Int. Conf. on Stability of Steel Structures*, Liège, April 1977.
28. White, J. D., *Longitudinal stresses in welded T-sections*, Rep. CUED/D-Struct./TR 60, Cambridge, 1977.
29. Wong, M. P., *Shrinkage effects in bead-on-plate welds in 5083 aluminum*, Rep. CUED/D-Struct./TR 88, Cambridge, 1981.
30. Wong, M. P. and Dwight, J. B., Longitudinal weld shrinkage in materials having a rounded stress–strain curve, *Tecseide Conference*, April 1981.
31. Dwight, J. B. and Mofflin, D. S., *Local buckling of aluminum*, Preliminary Proposals, Cambridge, March 1982.
32. De Luca, A. and Mazzolani, F. M., Modelli di distribuzione delle imperfezioni meccaniche nelle sezioni saldate in lega d'alluminio (Models of distribution of mechanical imperfections in welded aluminum alloy section), *10th AIAS Congress*, Vibo Valentia, September 1982.
33. Faella, C. and Mazzolani, F. M., L'effetto-trazionamento sulla capacità portante delle colonne metalliche (Stretching effect on load bearing capacity of metal columns), *Giornate Italiane della Costruzione in Acciaio CTA*, Firenze, October 1975.
34. Faella, C. and Mazzolani, F. M., Analysis of high strength steel bars under repeated axial loading, *Colloquium on Stability of Steel Structures*, Liège, 1977.
35. Como, M. and D'Agostino, S., Strain-hardening plasticity with Bauschinger effect, *Meccanica*, **IV**, No. 2, 1969.
36. Como, M., D'Agostino, S. and Grimaldi, A., Influence of the offset on the experiments yield surfaces of metals: a theoretical evaluation, *Archives of Mechanics*, Warsaw, **25**, No. 4, 1973, pp. 685–93.
37. Faella, C. and Mazzolani, F. M., L'influenza dell'effetto Bauschinger sul comportamento di aste metalliche sotto cicli di elongazioni, *Colloquium RILEM-CISM*, Udine, October 1974; *La Ricerca*, 1974, May–August.
38. Faella, C. and Mazzolani, F. M., *The influence of Bauschinger effect on the*

behaviour of metal bars under elongation cycles, ECCS Committee 16, Doc. 16-74-4, 1974.
39. Laurent, P. and Ferry, M., *Fatigue et Effet Bauschinger*, Société Française de Métallurgie, October 1946.
40. Ferton, D., *Influence d'un ecrovissage par traction sur les characteristiques mechaniques en traction et en compressiondes alliages A-SG et A-GAMC. Effect Bauschinger*, Aluminum Pechiney, Report. No. 879, Voreppe, February 1974.
41. Ferton, D., *The influence of stretching on the mechanical properties of alloys A-SG and A-GAMC under tension and compression. Bauschinger effect*, ECCS Committee 16, Doc. 16-74-5, 1974.
42. ECCS, *European Recommendations for Aluminum Alloy Structures*, First edn, 1978.
43. Steinhardt, O., Aluminum in Konstruktiven Ingenierbau (Aluminium constructions in civil engineering), *Aluminium*, **47,** 1971, pp. 131–9; 254–61.
44. DIN 4113 Part 1, *Aluminiumkonstruktionen unter vorwiegend ruhender Belastung: Berechnung und bauliche Durchbildung*, 1975.
45. DIN 4113 Part 2, *Aluminiumkonstruktionen unter vorwiegend ruhender Belastung: Geschwaisste Konstruktionen; Berechnung und bauliche Durchbildung*, 1975.
46. Bernard, A., Frey, F., Janss, J. and Massonnet, Ch., *Recherches sur le comportement au flambement de barres en aluminium (Research on the buckling behaviour of aluminium columns)*, Report CIDA, Liège–Paris, May 1971; IABSE Mem; Vol. 33-I, Zurich, 1973.
47. Templin, R. L. and Sturm, R. G., Some stress–strain studies of metals, *J. Aer. Sci.*, **7**(5), March 1940.
48. Jackman, K. R., Super aluminium alloys for aircraft structures, *Aviation*, August–October 1943.

3 Safety principles

3.1 Trends

3.1.1 Philosophy of Specifications

The principal scope of specifications is to provide general principles and computational methods in order to verify safety of structures. The 'safety factor', which according to modern trends is homogeneous and independent of the nature and combination of the materials used, can usually be defined as the ratio between the conditions which yield failure of the construction and the worst predictable working conditions. This ratio is also proportional to the inverse of the probability (risk) of failure of the structure.

Failure has to be considered not only as overall collapse of the structure but also as unserviceability or, according to a more precise, common definition, as the reaching of a 'limit state' which causes the construction not to accomplish the task it was designed for. There are two categories of limit state:

(a) *Ultimate limit state*, which corresponds to the highest value of the load-bearing capacity. Examples include local buckling or global instability of the structure; failure of some sections and subsequent transformation of the structure into a mechanism; failure by fatigue; elastic or plastic deformation or creep that cause a substantial change of the geometry of the structure; and sensitivity of the structure to alternating loads, to fire and to explosions.
(b) *Service limit states*, which are functions of the use and durability of the structure. Examples include excessive deformations and displacements without instability; early or excessive cracks; large vibrations; and corrosion.

Computational methods used to verify structures with respect to the different safety conditions can be separated into:

(a) *Deterministic methods*, in which the main parameters are considered as nonrandom parameters.

Aluminum alloy structures

(b) *Probabilistic methods*, in which the main parameters are considered as random parameters.

Alternatively, with respect to the different use of factors of safety, computational methods can be separated into:

(a) *Allowable stress method*, in which the stresses computed under maximum loads are compared with the strength of the material reduced by given safety factors.
(b) *Limit states method*, in which the structure may be proportioned on the basis of its maximum strength. This strength, as determined by rational analysis, shall not be less than that required to support a factored load equal to the sum of the factored live load and dead load (ultimate state).

The stresses corresponding to working (service) conditions with unfactored live and dead loads are compared with prescribed values (service limit state). From the four possible combinations of the first two and second two methods, we can obtain some useful computational methods. Generally, two combinations prevail:

Deterministic methods, which make use of allowable stresses
Probabilistic methods, which make use of limit states

In particular, with respect to the computation of aluminum alloy structures, the first (classical) method is still adopted in the standard specifications of several countries (Austria, West Germany, UK, Italy, Sweden and the USA). The second (modern) method is adopted by more advanced specifications (France, Canada and the ECCS).

The main advantage of probabilistic approaches is that, at least in theory, it is possible to scientifically take into account all random factors of safety, which are then combined to define the safety factor. Probabilistic approaches depend upon:

Random distribution of strength of materials with respect to the conditions of fabrication and erection (scatter of the values of mechanical properties throughout the structure)
Uncertainty of the geometry of the cross-sections and of the structure (faults and imperfections due to fabrication and erection of the structure)
Uncertainty of the predicted live loads and dead loads acting on the structure
Uncertainty related to the approximation of the computational method used (deviation of the actual stresses from computed stresses)

Furthermore, probabilistic theories mean that the allowable risk can be

Safety principles

based on several factors, such as:

Importance of the construction and gravity of the damage by its failure
Number of human lives which can be threatened by this failure
Possibility and/or likelihood of repairing the structure
Predicted life of the structure

All these factors are related to economic and social considerations such as:

Initial cost of the construction
Amortization funds for the duration of the construction
Cost of physical and material damage due to the failure of the construction
Adverse impact on society
Moral and psychological views

The definition of all these parameters, for a given safety factor, allows construction at the optimum cost. However, the difficulty of carrying out a complete probabilistic analysis has to be taken into account. For such an analysis the laws of the distribution of the live load and its induced stresses, of the scatter of mechanical properties of materials, and of the geometry of the cross-sections and the structure have to be known. Furthermore, it is difficult to interpret the interaction between the law of distribution of strength and that of stresses because both depend upon the nature of the material, on the cross-sections and upon the load acting on the structure. These practical difficulties can be overcome in two ways. The first is to apply different safety factors to the material and to the loads as well as to the risk, without necessarily adopting the probabilistic criterion. The second is an approximate probabilistic method which introduces some simplifying assumptions (semi-probabilistic methods).

Torroja's [36] proposal can be considered in the first group. He indicated the different values of the safety factors with respect to the amount of damage (serious and less serious), to the nature of loads (live and dead load), to the structural component (floor system and beams), and to the type of construction (buildings and bridges). A more detailed evaluation [36] was proposed in the UK by the Institution of Structural Engineers, which assumed the safety factor $\nu = \nu_1 \nu_2$. The coefficient ν_1 is closely related to the three safety factors connected to the procedure of erection, evaluation of loads and approximation of computational methods; each of these is divided into four classes of different quality (excellent, sufficient, good, bad). The coefficient ν_2 takes account of injuries to people or things, the effects being judged as light, severe or very severe.

A method which subdivides the level of safety even further is that

Fig. 3.1

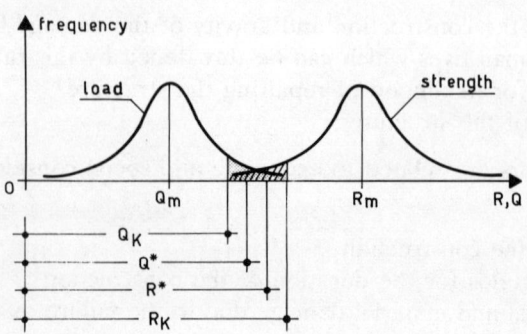

implied in USSR standard specifications [36]. In this case in order to take into account uncertainty and irregularity factors, three corrective coefficients are provided for the corresponding nominal values (standard values). The first is related to the uncertainty of predicted loads, the second depends upon the approximations of the computational method, the concentrations of stresses and the erection procedures, and the third takes into account the scatter of the material properties.

The same philosophy is followed by the instructions CNR-UNI 10012-67 on loading hypotheses [37]. In these instructions a safety factor ν_1 is required for the main loading condition (principal actions added in the most unfavorable way) and a coefficient ν_2 for the secondary loading condition (principal and secondary actions added in the most unfavorable way). In particular, CNR-UNI 10012 for steel construction uses $\nu_1 = 1.5$ and $\nu_2 = \nu_1/1.125$.

Among semi-probabilistic methods, the most common, adopted by CEB-FIP, ECCS and ISO, is based upon the definition of 'characteristic values' of the main quantities. These are resistance R and subsequent stress S (CEB-FIP), or resistance R and load Q (ECCS, ISO), assuming a normal (Gaussian) distribution of frequency (Fig. 3.1). The greater realism is offset, even in the case of simple semi-probabilistic theories, by a more laborious computational procedure; however, this is now made easier by the increased use of computers of different sizes. The alternative of using the various safety factors (the course adopted by most standard specifications) can be considered as a compromise between the deterministic and the probabilistic approach. It is, indeed, justified by the practical simplifications it provides.

3.1.2 Materials Considered in Different Specifications

All types of commercial alloys, and their physical, mechanical and chemical properties, are defined in specific codes provided in each country.

Safety principles

Examples are given as follows:

NBN [1–4]   (Belgium)
DIN [5, 6]   (Germany)
BS [7]   (UK)
UNI-UNIMET [8–12] (Italy)
NEN [13–23]   (Netherlands)
SIS-SVR [24, 26]   (Sweden)
NF-AFNORM [27, 28] (France)
VSM   (Switzerland)

Specifications concerning structural computations do not consider all commercial materials. They consider, instead, the most significant alloys from the structural point of view [29–33]. However, it is possible to extrapolate to other alloys, using the same principles, provided that the same safety factor is assured.

European recommendations [34, 35] follow the same principle. They consider six alloys: heat-treated alloys (7020, 6082, 6061, 6060) and non-heat-treated (5083, 5454), with the properties given in Fig. 3.2.

The choice of these alloys, even though they were used in different European countries with slightly different nominal values of mechanical properties, represented an attempt to unify production (see Fig. 3.3).

French specifications [33] make a substantial distinction between the behavior of non-heat-treated alloys and that of heat-treated alloys. They provide mechanical properties of several alloys of the 2000, 5000, 6000 and 7000 series, which are also referred to in some dimensional tables concerning the ultimate load of columns.

Fig. 3.2

Alloy	Product Sheet and plate	Tube	Extruded	Condition: Aluminum Association designation	$f_{0.2}$ (N mm^{-2})	f_t (N mm^{-2})
7020	×	×	×	T6 or T5	280	340
6082	×	×	×	T6	260	310
6061	×	×	×	T6	240	260
6060		×	×	T5	150	200
5083	×	×	not recommended	0	120	270
5454	×	×	×	H24	200	270

Aluminum alloy structures

Fig. 3.3

Numerical designation	Switzerland	France	Germany	Italy	Netherlands	Spain	Sweden	UK
7020 (T5 or T6)	×	×	×	×		×	×	×
6082 (T6)	×	×	×	×	×		×	×
6061 (T6)	×	×	×	×	×			×
6060 (T5)	×	×	×	×		×	×	×
5083 (F or 0 or H111)	×	×	×	×	×	×	×	×
5454 (H24)		×		×				

Fig. 3.4

		Plates and sheets		Tubes		Profiles	
	Product: alloy	f_t	$f_{0.2}$	f_t	$f_{0.2}$	f_t	$f_{0.2}$
1	AlZnMg1F36	360	280	360	280	360	280
2	AlMgSi1F32	320	260	320	260	320	260
3	AlMgSi1F28	280	200	280	200	280	200
4	AlMgSi0.5F22	—	—	220	160	220	160
5	AlMg4.5MnF30	300	210	—	—	—	—
6	AlMg4.5Mnw/F28	280	125	280	160	280	160
7	AlMgMnF23	230	140	230	140	—	—
8	AlMgMnF20	—	—	200	100	200	100
9	AlMgMn w/F18 AlMg3 w/F18	180	80	180	80	180	80

British specifications [7] refer mainly to the three alloys 5083, 6060 and 6082, which are also covered in the ECCS European recommendations [34, 35].

Aluminum alloys considered in German specifications [5, 6] are in accordance with those given in the ECCS European recommendations (see Fig. 3.4, in which minimum guaranteed values of $f_{0.2}$ and f_t, in N mm^{-2}, are also provided for tubes and profiles). The materials are divided into different categories with respect to the manufacturing process.

In Italy the XIII UNIMET committee [12] has recently identified the alloys given in Fig. 3.5 as structurally interesting. This choice allows the

Fig. 3.5

Alloy	National designation	Numerical designation	Code UNI	Product: Plate	Product: Extruded	Condition	Mechanical properties: $f_{0.2}$(N mm^{-2})	f_t(N mm^{-2})	ε_t(%)
	P-AlMg2.5	5052	3574	$t \le 6$ mm	—	H20	175	215	8
	P-AlMg2.7Mn	5454	7789	$t \le 6$ mm	—	H20	175	245	8
Al–Mg	P-AlMg4.4	5086	5452	$t = 8$ to 75 mm	—	HL	110	255	12
	P-AlMg4.4	5086	5452	$t \le 6$ mm	—	H15	215	285	12
	P-AlMg4.5	5083	7790	—	×	Hp	110	265	12
	P-AlMgSi	6060	3569	—	$t \le 12$ mm	TaA16	145	195	11
	P-AlSi1MgMn	6082	3571	$t \le 4$ mm	—	TA16	245	295	11
Al–Si	P-AlSi1MgMn	6082	3571	—	×	TA16	265	315	10
	P-AlMg1SiCu	6061	6170	$t \le 6$ mm	—	TA16	245	295	10
	P-AlMg1SiCu	6061	6170	—	×	TA16	235	265	9
	P-AlZn4.5Mg	7020	7791	$t \le 12$ mm	×	TaN	215	315	10
Al–Zn	P-AlZn4.5Mg	7020	7791	$t \le 12$ mm	×	TaA	275	355	8
	P-AlZn5.8MgCu	7075	3735	$t \le 13$ mm	—	TA	450	520	8
	P-AlZn5.8MgCu	7075	3735	—	$t \le 38$ mm	TA	480	540	7
	P-AlCu4.5SiMnMg	2014	3581	—	$t = 9$ to	TA	370	410	7
Al–Cu	P-AlCu4.5SiMnMg	2014	3581	—	20 mm forged and pressed	TA	345	445	9
	P-AlCu4.5MgMn	2024	3583	$t \le 20$ mm	—	THN	275	425	8

Aluminum alloy structures

use of other alloys provided that the same safety factor is assured. Physical and mechanical properties of the structural alloys used in Italy are given in [8].

3.2 Loading conditions

3.2.1 Definitions

Loads consist of:

Concentrated and distributed forces (direct actions)
Imposed deformations (indirect actions)

A load is assumed as a single load if it is not related to any other load or imposed deformation acting on the structure. In practice more than one single load acts on the structure, although it is convenient to consider each load separately.

Loads are random processes; more precisely, they are stochastic processes. However, in order to match the requirements of the methods of calculation actually used in most structural specifications (allowable stresses and semi-probabilistic methods), each load is also characterized by the parameters representative of the different computational methods.

The loads can be classified with respect to their effect on the structure (static or dynamic) or with respect to their variation of intensity. Loads can also be classified with respect to some particular aspect, such as limited or not limited, having long or short duration, dependent or not on human activities etc.

3.2.2 Classifications

3.2.2.1 Classification of loads with respect to the structural response

A distinction is made between two types of load according to the response of the structure:

(a) *Static loads*, which are applied to the structure without causing significant accelerations of the structure or of structural elements
(b) *Dynamic loads*, which cause significant accelerations of the structure

The same load might be static or dynamic depending upon the structure to which it is applied. Generally loads can be considered static loads, provided that the dynamic effects are taken into account by an increment of the intensity of the loads. In other cases a dynamic analysis is necessary.

3.2.2.2 Classification of the loads with respect to the variation in time of their intensity

If we indicate by Δt_s the life of the structure, which can be assumed equal to 50 years (unless there are different provisions), the loads are divided in the following way:

(a) *Dead loads* act on the structure for the whole of its life with negligible variations of intensity. They include:
The weight of the structure
The weight of each superstructure
Forces caused by the pressure of the ground (except from the effects of moving loads applied on the ground)
Deformations imposed by the fabrication and erection processes
Actions caused by shrinkage of concrete and the distortions due to welding
Loads due to the pressure of water when it is constant in time
Loads due to the displacements of supports
The actions due to prestressing

(b) *Live loads* act on the structure with instantaneous values which can be noticeably different from each other. These actions can be further divided into loads of long and short duration:
(i) *Variable long-duration loads* act on the structure, and their duration is of the same order as the life of the structure Δt_s. Their intensity is significant. Examples of loads which fall into this category are:
The weight of the non-structural elements which form the construction (floor, plaster, roof covering etc.)
The weight of furniture, stocked merchandise, parked cars etc.
Loads related to the fabrication or erection or with some of their phases
(ii) *Variable short-duration loads* act on the structure for a period that is short compared with the life of the structure Δt_s. Typical examples are:
Moving loads due to people, moving cars etc.
The action of the wind
Ice formation
Earthquakes in seismic areas
(iii) Some actions can be considered of short or long duration depending upon the particular case. Among these are:
Snow loads
The effects due to variation of water level in a tank
The effects due to variation of temperature

(c) *Exceptional loads* are those loads which are very unlikely to act on the structure, such as those due to:

Collisions
Explosions
Fires
Displacement of the ground
Earthquakes in nonseismic areas

3.2.2.3 Classification of loads with respect to their variation of intensity along the structure

We distinguish throughout the structure between:

(a) *Fixed loads* with a known spatial distribution, such that the magnitude and position of the load is known on all of the structure when it is known at a given point
(b) *Nonfixed loads*, which can have (within certain bounds) any distribution throughout the structure.

Many loads can be considered to consist of a fixed and a nonfixed portion. It is often necessary to separate a load into a fixed portion and other portions acting in a random way on the structure. The analysis of nonfixed loads implies consideration of different loading conditions, each of them being defined when the magnitude and position of all the nonfixed loads are given.

3.3 Allowable stress method

The allowable stress method verifies the structure by checking that in the most highly stressed point the stress field, expressed by an appropriate stress σ_{id} (see Section 3.5.1), does not exceed the value which has been defined as allowable. Hence it has to be checked that

$$\sigma_{id} \leq \sigma_{adm} \tag{3.1}$$

The allowable (or admissible) stress is given by the ratio

$$\sigma_{adm} = \frac{f_{lim}}{\nu} \tag{3.2}$$

between a stress defined as limit stress because it corresponds to a fixed dangerous situation (f_{lim}) and an appropriate safety factor $\nu > 1$. ν should take into account all uncertainties related to the unavoidable imperfections which come into play when analyzing actual structures in a particular way. The most important of these uncertainties are:

Evaluation of loads
Determination of strength
Fabrication tolerances

Safety principles

Specifications for aluminum alloy structures usually agree in assuming, for the limit stress, the conventional elastic limit $f_{0.2}$ which corresponds to the yield limit in steel.

However, it has to be taken into account that, as has already been observed (Section 1.7.1), the ratio between the ultimate strength f_u and the elastic limit $f_{0.2}$ is smaller than that of steel and is different in each alloy. In mild steels a safety factor is required against yielding, and the further margin with respect to the ultimate strength is safely ignored. However, in aluminum alloys the actual value of f_u has to be considered either by introducing it in the definition of σ_{adm}, as it is done in British specifications, or by defining two safety factors:

$\nu_{0.2}$ with respect to reaching the elastic limit
ν_u with respect to reaching rupture

as it is done in American specifications [30]. The values assumed for the safety factors are different in each case.

British specifications (BS CP 118) [7] define the allowable stresses in the following way. For axial loads without instability problems, they are:

$$\sigma_{adm} = 0.44 f_{0.2} + 0.09 f_u \tag{3.3}$$

For flexural loads, without instability problems, they are:

$$\sigma_{adm} = 0.44 f_{0.2} + 0.14 f_u \tag{3.4}$$

where $f_{0.2}$ and f_u are the minimum required values for the elastic limit and the ultimate strength respectively.

Equations 3.3 and 3.4 are valid both for compression and for tension stresses. Both take into account the difference between the elastic limit $f_{0.2}$ and the ultimate strength f_u and therefore the different slope of the hardening portion of the σ–ε diagram. The application of these equations to the most common aluminum alloys leads to conventional values of the safety factor $\nu_{0.2}$, referred to the elastic limit $f_{0.2}$, varying between 1.5 and 1.9.

German specifications (DIN 4113) [5, 6] establish the different safety factors with respect to the loading conditions, the values being equal to:

$\nu_{0.2} = 1.70$–1.80 for the principal loading condition (H)
$\nu_{0.2} = 1.50$–1.60 for the secondary loading condition (HZ)

These values correspond to safety factors with respect to failure which are equal to:

Aluminum alloy structures

$$\nu_u = 2.20\text{--}4.00 \text{ for the condition (H)}$$
$$\nu_u = 1.90\text{--}3.60 \text{ for the condition (HZ)}$$

USA specifications (Aluminum Association) [30, 31] also adopt different safety factors with respect to the type of structure (building or bridge) and the type of loading, and they also define a different margin with respect to the elastic limit or to rupture.

In particular, for axial and flexural loading in the absence of instability problems, the following values are required:

Buildings

$$\nu_{0.2} = 1.65 \qquad \nu_u = 1.95$$

Bridges

$$\nu_{0.2} = 1.85 \qquad \nu_u = 2.20$$

These specifications also take into account the different values of $f_{0.2}$ and f_u in compression and in tension, as already observed in Section 2.5.5.5.

The value of allowable (or admissible) stress is assumed as:

$$\sigma_{adm} = \text{lesser of} \begin{cases} f_{0.2}/\nu_{0.2} \\ f_u/\nu_u \end{cases}$$

This formulation can be considered a general one among the allowable stress methods applied to aluminum alloy structures [12].

Analogous criteria are used, for example, in the definition of allowable stress for high-strength steel structures, in which the σ–ε law is continuous and is qualitatively close to that of aluminum structures.

3.4 Semi-probabilistic method

3.4.1 Characteristic Values

If we consider as random both the loads acting on the structure and its strength, the interpretation of these values from a probabilistic point of view leads to a definition of characteristic values [38].

The *characteristic value of strength* is defined by

$$f_k = f_m(1 - k\delta) \tag{3.5}$$

and represents the value which has the probability p_k (for example, 5 percent) of not being achieved or has the probability $1 - p_k$ (therefore 95 percent) of being exceeded, where:

f_m arithmetic average of the strengths obtained from different experimental results
δ coefficient of variation, which depends upon the shape of the probability function $p(f)$; it is given by

$$\delta = s/f_m$$

where s is the standard deviation
k a coefficient dependent upon the value p_k of the assumed probability

The *characteristic value of the load* is given by

$$F_k = F_m(1 + k\delta) \tag{3.6}$$

and represents the value which has an assumed probability p_k of not being exceeded in an unfavorable way during a reference period determined on the basis of the life of the structure, where:

F_m mean value
δ variation coefficient
k coefficient dependent upon the value p_k of the assumed probability

The check against a given limit state has to show that the stress S in a given section, determined by the loads F_k (Eq. 3.6) acting on the structure, is less than the resistance R of the same section of a material with a characteristic strength f_k (Eq. 3.5):

$$S(F_k) \le R(f_k) \tag{3.7}$$

3.4.2 Design Values

Actually, in order to take into account those remaining causes of uncertainty that are not interpreted by the definition of characteristic values of strength and loads, the semi-probabilistic method transforms the characteristic values into so-called computational or design values by using appropriate coefficients.

The *design strength* results:

$$f_d = \frac{f_k}{\gamma_m} \tag{3.8}$$

where

$$\gamma_m = \gamma(\gamma_1, \gamma_2)$$

γ_1 a coefficient which takes into account the possible reduction of strength of the material in the structure with respect to the values measured on the specimens

Aluminum alloy structures

γ_2 a coefficient which takes into account further local reductions of strength, including the dimensional tolerances

The *design load* is defined by the value $\gamma_F F_k$, where:

$$\gamma_F = \gamma(\gamma_1, \gamma_2, \gamma_3)$$

γ_1 a coefficient which takes into account the possibility that the load will reach values more unfavorable than the characteristic ones, owing to abnormal or unpredictable situations

γ_2 a coefficient which takes into account the reduced probability that the different loads acting on the structure all have their characteristic value at the same time

γ_3 a coefficient which takes into account the possibility of unfavorable modifications of the stresses, owing to computational assumptions being different from reality (simplified supports; effects neglected because difficult to evaluate; erection imperfections, such as out-of-straightness and eccentricities in geometrical shape)

Since in reality there are always a number of load types acting on the structure at the same time, it is not reasonable to consider a design load. It is better to refer to a *design combination* given by the summation of the design values $\gamma_F F_k$ of the single loads. For example,

$$F_d = \sum \gamma_{F_i} F_i \qquad (3.9)$$

The following symbols are defined:

G_k characteristic value of dead loads
q_{k_i} characteristic value of the single live load
γ_G coefficient of dead loads
γ_Q coefficient of the dominant live load
$\psi_i \gamma_Q$ coefficient of the other live loads (with $\psi_i \leq 1$)

The design combination of loads F_d is usually expressed in the form

$$F_d = \gamma_G G_k + \gamma_Q \left(Q_{k_1} + \sum_{i=2}^{n} \psi_i Q_{k_i} \right) \qquad (3.10)$$

Q_{k_1} is the base load. Q_{k_i} represents the remaining $n-1$ live loads, known as accompanying loads; the probability of these acting at the same time depends upon the coefficients ψ_i.

Equation 3.7 thus becomes:

$$S(F_d) \leq R(f_d) \qquad (3.11)$$

If there are n live loads, there are n possible design combinations like Eq. 3.10, obtainable by varying the loads and assuming each of them to be the principal. In reality, most of these combinations can be neglected since they are on the safe side.

It is necessary to know the values of the coefficients (γ_m related to strength and γ_G, γ_Q, ψ_i related to loads) in order to define Eqs 3.8 and 3.9 and therefore to apply Eq. 3.11 in safety controls. These coefficients mainly depend upon the shape of the distribution of loads and of strength, upon the probable contemporaneity of loads, upon the assigned level of probability and upon the structural typology.

The corresponding numerical values are given in different specifications which are based upon the semi-probabilistic method (such as the French specifications and the European recommendations).

3.4.3 Limit States

3.4.3.1 Definitions

The load-bearing capacity of a structure as a whole, or of part of it, has to be guaranteed by several safety controls against specific situations, called *limit states*, beyond which the structure cannot satisfy its design requirements.

A general definition is given in the ISO 2394 report [38], valid for all construction materials, followed also by the ISO/TC 167 Commission for Metallic Materials 'Steel and Aluminium Structures' (1979), which divides limit states into two:

(a) Ultimate limit states, which correspond to the maximum load-bearing capacity or to a permanent deformation considered as critical
(b) Service limit states, which correspond to the criteria governing normal use of a construction

For these states some simplifications are provided.

Ultimate limit states can correspond to:

Overturning of the structure or of a part of it considered as a rigid body
Failure of a critical section of the structure because the material strength is exceeded (in some cases reduced by repeated loading or deformation)
Transformation of the structure into a mechanism
Loss of stability (buckling)
Elastic, plastic or creep deformations leading to a change in the geometry which necessitates replacing the structure

Service limit states can correspond to:

Deformations which prevent the correct use of the structure or which affect the appearance of the structure or of noncarrying members
Vibrations producing discomfort, especially when they are in resonance

Aluminum alloy structures

with the noncarrying members or with equipment
Local damage which reduces the durability of the structure or affects the efficiency and appearance of structural members as well as of noncarrying members

For our purpose, we consider that more valid definitions are those closely related to the behavior of metallic materials and to corresponding typologies. These definitions are as follows.

3.4.3.2 Service limit states

Service limit states correspond to the loss of functionality of the structure, and in most cases they can be identified as follows:

(a) *Deformation limit states* correspond to excessive residual deformations when the loads which caused the limit state are removed from the structure.
(b) *Displacement limit states* correspond to excessive displacements of the structure with respect to functionality, durability and compatibility conditions.
(c) *Vibration limit states* correspond to dynamic effects caused by natural events (wind) or by machines. They also can cause loss of functionality. They can be approximately verified by limiting displacements of the structure.
(d) *Slipping limit states of joints* can cause considerable displacements of the structure due to movement of joints. This applies particularly to bolted joints due to bolt-hole clearance.
(e) *Other service limit states* can be defined for structures designed to accomplish special functions.

3.4.3.3 Ultimate limit states

Ultimate limit states correspond to failure of the structure and, in most cases, can be identified as one of the following limit states:

(a) *Overturning limit states* relate to loss of equilibrium of a part or of the whole structure considered as a rigid body.
(b) *Ultimate states of a connection* correspond to failure of a connection with the subsequent collapse of the connected component due to loss of support.
(c) *Ultimate limit states of a structural element* correspond to failure of a section of a rectangular element when there is insufficient ductility to redistribute loads and stresses on the structure.
(d) *Instability limit states* correspond to overall instability of the structure or to local and/or global instability of a single member.

(e) *Plastic collapse limit states* correspond, in the case of ductile structures, to the transformation of the structure into a mechanism.
(f) *Fatigue limit states* can even cause collapse of the structural element due to alternating loads.

To date there are few specifications that provide ultimate computational methods for a prediction of the limit states of the cross-section and of the structure.

Most specifications, and in particular all those concerning aluminum structures, check against ultimate limit states in a conventional way by applying elastic computational methods.

In these cases, the ultimate limit state is considered as corresponding to the conventional elastic limit state of the cross-section, which requires the elastic limit of the material in the most highly stressed fiber to be reached. In this case $\gamma_m = 1$ in Eq. 3.8 is allowed because the analysis of the stress field is carried out in a very conservative way.

More details on the ultimate limit states and on the post-elastic behavior of aluminum alloy structures are given in Chapter 6.

3.4.4 Design Combination of Loads

The design combination of loads (Eq. 3.10) is quantified when the combination coefficient values γ_G, γ_Q, ψ_i are defined. If it is not proposed to compute more exact values by probabilistic computations, the values first proposed by CEB (Comité Européen du Béton), and later used by ECCS in its recommendations on steel structures (1975) and on aluminum structures [34], have to be used.

Some values of ψ_i proposed by CEB are given here as an example:

Actions (working loads)	ψ_0	ψ_1	ψ_2
Domestic and office buildings, stores	0.5	0.8	0.4
Parking	0.6	0.7	0.6
Highway bridges	0.5	0.3	0.0
Wind and snow	0.5	0.2	0.0

The general philosophy given here is followed by all semi-probabilistic specifications.

The following symbols are defined:

G_1 weight of the structure
G_2 dead load (except the weight of the structure)
Q_i independent live loads

In the case in which the effects of dead loads G_1 and G_2 are to be added to those of live loads Q, the combination of loads can be written as:

Aluminum alloy structures

$$\gamma_{G_1}G_{1,\max} + \gamma_{G_2}G_{2,\max} + \gamma_Q\left(Q_1 + \sum_{i=2}^{n}\psi_i Q_i\right) \quad (3.12)$$

In the opposite case:

$$\gamma_{G_1}G_{1,\min} + \gamma_{G_2}G_{2,\min} + \gamma_Q\left(Q_1 + \sum_{i=2}^{n}\psi_i Q_i\right) \quad (3.13)$$

When *checking service limit states*, it can be assumed that:

$$\gamma_{G_1} = \gamma_{G_2} = \gamma_Q = 1 \quad \psi_i = 0.8 \quad (3.14)$$

It has to be checked which live load Q_i causes the most unfavorable situation with respect to the limit state that is being considered. If this load is Q_1, the combination is equal to:

$$G_1 + G_2 + Q_1 + 0.8\sum_{i=2}^{n} Q_i \quad (3.15)$$

In this equation are not considered those live loads Q that give a positive contribution to reaching the limit state being analyzed.

When *checking ultimate limit states*, the determination of the stress field, in some cases, is computed by nonlinear equations. In these cases it is necessary to make several attempts in order to identify the most unfavorable combination of loads. If the following values are assumed:

$$\gamma_{G_1} = 1.30 \quad \gamma_{G_2} = \gamma_Q = 1.50 \quad \psi_i = 0.6 \quad (3.16)$$

then the most severe combinations from the following has to be found:

$$\begin{aligned}
1.3G_1 + 1.5G_2 + 1.5Q_1 + 0.6(Q_2 + Q_3 + \cdots + Q_n) \\
1.3G_1 + 1.5G_2 + 1.5Q_2 + 0.6(Q_1 + Q_3 + \cdots + Q_n) \\
\cdots \quad \cdots \quad \cdots \quad \cdots \\
1.3G_1 + 1.5G_2 + 1.5Q_n + 0.6(Q_1 + Q_2 + \cdots + Q_{n-1})
\end{aligned} \quad (3.17)$$

Those loads Q_i are neglected when they make a positive contribution to reaching the limit state under analysis. Furthermore, if the contribution of dead loads G_1 and G_2 is positive, their minimum values $G_{1,\min}$ and $G_{2,\min}$ are used and multiplied by the coefficients

$$\gamma_{G_1} = \gamma_{G_2} = 1$$

The values given for γ_{G_1} and γ_Q are the same as those given in [34]. However, in these recommendations the ψ_i values are not given.

The ECCS recommendations consider another combination of loads (known as accidental) which are due to exceptional loading conditions Q_e

Safety principles

(see Section 3.2.2.2). This can be written in the form:

$$\gamma_{G_1} G_1 + \gamma_{G_2} G_2 + \gamma_e Q_e + \gamma_Q \psi_1 Q_1 + \sum_{i=2}^{n} \psi_i Q_i \qquad (3.18)$$

In the case in which the effect of dead loads is to be added to that due to other loads,

$$G_1 = G_{1,\max} \qquad G_2 = G_{2,\max}$$
$$\gamma_{G_1} = \gamma_{G_2} = 1.1 \qquad (3.19)$$
$$\gamma_e = \gamma_Q = 1$$

In the case in which the effect of dead loads reduces the effect due to other loads,

$$G_1 = G_{1,\min} \qquad G_2 = G_{2,\min}$$
$$\gamma_{G_1} = \gamma_{G_2} = 0.9 \qquad (3.20)$$
$$\gamma_e = \gamma_Q = 1$$

The use of combination coefficients (known as enhancement coefficients) is also used in French specifications [33]. In these specifications, in contrast to the general semi-probabilistic method, enhancement coefficients are directly applied to stresses because in most cases stresses are proportional to loads (linear behavior).

Under the hypothesis of elastic linear behavior of structures, French specifications require the computation of stresses for each loading condition. These stresses are defined as:

σ_g dead load stresses
σ_{gt} (σ'_{gt})   temperature stresses to be added to (subtracted from) σ_g
σ_q (σ'_q)   live load stresses to be added to (subtracted from) σ_g
σ_{qs} (σ'_{qs})   snow stresses to be added to (subtracted from) σ_g
σ_{qw} (σ'_{qw})   wind stresses to be added to (subtracted from) σ_g
σ_{qsr} (σ'_{qsr})   snow stresses to be added to (subtracted from) σ_g, reduced if concomitant with wind
σ_{qse} (σ'_{qse})   exceptional snow stresses to be added to (subtracted from) σ_g
σ_{qsre} (σ'_{qsre})   exceptional snow stresses to be added to (subtracted from) σ_g, reduced if concomitant with wind
σ_{qwe} (σ'_{qwe})   exceptional wind effect to be added to (subtracted from) σ_g

Specifications therefore define the following enhanced stress states. The worst conditions have to be chosen from these stresses:

Aluminum alloy structures

(a) $1.7\sigma_q + 1.5(\sigma_{gt}+\sigma_g)$
$1.7\sigma'_q + 1.5\sigma'_{gt} - \sigma_g$
$1.7\sigma_{qs} + 1.5(\sigma_{gt}+\sigma_g)$
$1.7\sigma'_{qs} + 1.5\sigma'_{gt} - \sigma_g$
$1.7\sigma_{qw} + 1.5(\sigma_{gt}+\sigma_g)$
$1.7\sigma'_{qw} + 1.5\sigma'_{gt} - \sigma_g$

(b) $1.6(\sigma_{qsr}+\sigma_{qw}) + 1.5(\sigma_{gt}+\sigma_g)$
$1.6(\sigma'_{qsr}+\sigma'_{qw}) + 1.5\sigma'_{gt} - \sigma_g$
$1.6(\sigma_{qw}+\sigma_q) + 1.5(\sigma_{gt}+\sigma_g)$
$1.6(\sigma'_{qw}+\sigma'_q) + 1.5\sigma'_{gt} - \sigma_g$
$1.6(\sigma_q + \sigma_{qs}) + 1.5(\sigma_{gt}+\sigma_g)$
$1.6(\sigma'_q + \sigma'_{qs}) + 1.5\sigma'_{gt} - \sigma_g$

(c) $1.5(\sigma_q + \sigma_{qsr} + \sigma_{qw} + \sigma_{gt} + \sigma_g)$
$1.5(\sigma'_q + \sigma'_{qsr} + \sigma'_{qw} + \sigma'_{gt}) - \sigma_g$

(d) $1.1(\sigma_q + \sigma_{qsre} + \sigma_{qwe} + \sigma_{gt} + \sigma_g)$
$1.1(\sigma_q + \sigma_{qse} + \sigma_{gt} + \sigma_g)$
$1.1(\sigma'_q + \sigma'_{qsre} + \sigma'_{qwe} + \sigma'_{gt} - \sigma_g)$
$1.1(\sigma'_q + \sigma'_{qse} + \sigma'_{gt} - \sigma_g)$

The most unfavorable stress state has to be checked against the design strength, assumed equal to the elastic limit $f_{0.2}$ (if statistical values are not available, the minimum required from specifications is used).

Specifications for metallic structures commonly use $\gamma_m = 1$ in Eq. 3.8. When computational methods based upon elasticity theory are used the cross-section is checked against the conventional elastic limit. This criterion is also followed by the ECCS European recommendations.

The order of magnitude of enhancement coefficients is different in French and European specifications. Figure 3.6 gives a concise, if incomplete, comparison. It is immediately seen that French specifications are more conservative, providing a safety factor about 15 percent higher than the European ones.

There are specific reasons for these differences. Combination coefficients of loads, in fact, govern the statistical distribution of loads and are therefore independent of the material. This is the reason why ECCS adopted the same values for steel and aluminum of the coefficients γ_F. This philosophy has also been followed by the ISO/TC 167 Committee.

On the other hand the different σ–ε behavior of the two materials has to be taken into account. It has already been observed that the inelastic range is usually higher in steel than in aluminum alloys. Since strength is usually checked against the conventional elastic limit, it is reasonable to assume a larger safety factor in aluminum alloys than in steel structures. This is the reason why most specifications based upon allowable stresses (see Section 3.3) use safety factors $\nu_{0.2}$, with respect to the elastic limit,

Fig. 3.6

Actions	ECCS	DTU
Permanent loads		
normal conditions	1.3 (1.0)	1.5 (1.0)
exceptional conditions	1.1 (0.9)	1.1
Basic variable loads		
service	1.5	1.7
snow and wind	1.5	1.7

equal to 1.5–2.0 in aluminum alloys, whereas the corresponding range in steel structures is 1.3–1.5. This philosophy has probably been followed by French specifications in using enhancement coefficients for aluminum alloys different from those used in steel structures.

3.5 Strength of the base metal

3.5.1 Failure Criteria

Among different failure criteria, the one which is closest to the behavior of isotropic and homogeneous materials is the Huber–Hencky–Von Mises criterion. This criterion is universally used in steel structures as well as in aluminum alloy structures.

The corresponding reference stress, called the ideal stress, is given by the general expression:

$$\sigma_{id} = \sqrt{(\sigma_1^2 + \sigma_2^2 + \sigma_3^2 - \sigma_1\sigma_2 - \sigma_2\sigma_3 - \sigma_1\sigma_3)} \tag{3.21}$$

where σ_j ($j = 1, 2, 3$) are the principal stresses.

Stress fields more common in metal constructions are usually one or two dimensional. In the two-dimensional case the ideal stress is:

$$\sigma_{id} = \sqrt{(\sigma_1^2 + \sigma_2^2 - \sigma_1\sigma_2)} \tag{3.22}$$

or, if we consider the special components of stress:

$$\sigma_{id} = \sqrt{(\sigma_x^2 + \sigma_y^2 - \sigma_x\sigma_y + 3\tau_{xy}^2)} \tag{3.23}$$

The most common case is that in which we have bending and shear. In this case, since $\sigma_y = 0$, we have:

$$\sigma_{id} = \sqrt{(\sigma_x^2 + 3\tau_{xy}^2)} \tag{3.24}$$

In the case of shear only, since $\sigma_x = \sigma_y = 0$, we have:

Aluminum alloy structures

$$\sigma_{id} = \tau\sqrt{3} \tag{3.25}$$

The ideal stress is used in verifying the material for either the allowable stress method or the limit state method.

3.5.2 Practical Application of Calculation Methods

In conclusion, a summary is given of the application of the two main methods – the allowable stress method (Section 3.3) and the semi-probabilistic method (Section 3.4).

3.5.2.1 Allowable stress method

The specifications for computations of aluminum alloy structures are based upon the allowable stress method in the following countries:

Austria [29]
West Germany – DIN 4113 [5, 6]
UK – BS CP 118 [7]
Sweden – SVR [26]
USA – Aluminum Association [30]

The allowable stress procedure has the following steps:

(a) The loading condition is defined.
(b) The internal actions due to loading conditions (a) are computed.
(c) The stresses at critical points are computed.
(d) The stresses are combined in order to get the reference ideal stress σ_{id} (Section 3.5.1).
(e) The ideal stress is checked against the allowable stress

$$\sigma_{id} \leq \sigma_{adm}$$

where σ_{adm} is defined by Eq. 3.2.

3.5.2.2 Semi-probabilistic method

The semi-probabilistic method was first used in France for aluminum alloy structures in the DTU régles Al [33]. Later, Canadian specifications followed French specifications in the preparation of the new Canadian code [32], in which two existing specifications merged: CSA S157 (1969), 'The structural use of aluminum in buildings' and CSA S190 (1968), 'Computation of small dimension aluminum components'.

The ECCS committee on aluminum alloy structures decided to adopt the semi-probabilistic approach, already used in steel, from the beginning of its work in 1970. The European recommendations for aluminum alloy structures [34] make use of the semi-probabilistic approach in the first chapter 'Bases for design'.

The ISO/TC 167 committee also follows this approach (1980), because this committee decided to use most of the results of the ECCS T2 committee.

The steps usually followed when using a semi-probabilistic method are:

(a) The design combination of loads F_d is defined (Eq. 3.10).
(b) The internal actions $S(F_d)$ due to the loading condition as defined in (a) are computed.
(c) The stresses in the critical sections of the structure are computed.
(d) The reference stress σ_{id} (see Section 3.5.1) corresponding to the stresses defined in (c) is computed.
(e) Check

$$\sigma_{id} \leq f_d$$

where f_d is the design strength (Eq. 3.8), to be taken equal to the characteristic value at 0.2 percent with the assumption $\gamma_m = 1$.

The value f_k has to be experimentally justified on the basis of a statistical evaluation which requires at least twelve tests for each type of profile. In this case it is assumed that:

$$f_k = f_m - 2s \tag{3.26}$$

where f_m and s are the mean value and standard deviation, respectively.

If statistical values are not provided, it is commonly (and conservatively) assumed that f_k has the minimum guaranteed value given in the specifications for materials.

References

1. NBN 480, *Couvertures et parois en feuilles métalliques ondulées*, 1958
2. NBN 437, *Aluminiums et alliages d'aluminium d'usage courant pour produits corroyés*, 1958.
3. NBN 1-01, *Conditions générales applicables aux charpentes métalliques*, 1968.
4. NBN P21-101, *Aluminiums et alliages d'aluminium moulés*, 1972.
5. DIN 4113 Part 1, *Aluminiumkonstruktionen unter vorwiegend ruhender Belastung: Berechnung und bauliche Durchbildung*, 1975.
6. DIN 4113 Part 2, *Aluminiumkonstruktionen unter vorwiegend ruhender Belastung: Geschwaisste Konstruktionen; Berechnung und bauliche Durchbildung*, 1975.
7. BS CP 118, *The structural use of aluminum*, 1969.
8. UNI 7876, *Strutture in alluminio: Selezione dell'alluminio e delle leghe di alluminio per impieghi strutturali*, 1978.

Aluminum alloy structures

9. UNI 7427, *Alluminio e leghe di alluminio primaria da lavorazione plastica: prospetto delle qualità normalizzate*, 1978.
10. UNI 8278, *Alluminio e leghe di alluminio da lavorazione plastica*, 1981.
11. UNI 8209, *Giunti saldati di allumino e leghe di alluminio sollecitati staticamente: Istruzioni per il calcolo*, 1978.
12. UNI 8634, *Strutture in leghe d'alluminio: Istruzioni per il calcolo e l'esecuzione*, 1984.
13. NEN 3854, *Regulations for the calculation of building structures. Design of aluminum structures*, 1982.
14. NEN 6029, *Aluminum in wrought products. Classification, chemical composition and forms*, 1967.
15. NEN 6026, *Aluminum alloy castings. Classification and mechanical properties*, 1967.
16. NEN 6022, *Aluminum alloy ingots. Classification and testing*, 1965.
17. NEN 6021, *Result ingots and pigs of unalloyed aluminum. Classification and testing*, 1964.
18. NEN 5255, *Testing audic oxidation coatings on aluminum or aluminum alloys*, 1966.
19. NEN 6032, *Aluminum sheet and strip. Tolerances*, 1971.
20. NEN 6031, *Aluminum plate, sheet and strip. Mechanical properties*, 1969.
21. NEN 6035, *Aluminum tube. Mechanical properties*, 1971.
22. NEN 6039, *Aluminum rod. Mechanical properties*, 1973.
23. NEN 6040, *Aluminum rod. Tolerances*, 1973.
24. SVR, *Aluminium Konstruktioner*, 1966.
25. SVR, *Aluminium Konstruktioner: stabilites problem*, 1970.
26. SVR, *Svetsade aluminiumkonstruktioner*, 1971.
27. AFNORM, *Recueil de normes françaises sur les métaux légers. Volume I: Produits corroyés en aluminium et alliages d'aluminium—caractéristiques et dimension*, 1973.
28. AFNORM, *Recueil de normes françaises sur les métaux légers. Volume II: Produits en aluminium et alliages d'aluminium—méthodes d'essais et soudage*, 1973.
29. Osterreichischer Stahlbauverband, *Richtlinien für die Verwendung von Aluminiumlegierungen für tragende Konstructionen im Ingenieurbau*, 1977.
30. Aluminum Association, *Specification for aluminum structures*, 1976.
31. Aluminum Association, *Engineering data for aluminum structures*, 1975.
32. CSA S157-M, *Strength design in aluminum*, Fourth Draft, 1980.
33. DTU, *Règles de conception et de calcul des charpentes en alliages d'aluminium*, First edn., 1971; Second edn., 1976.
34. ECCS, *European recommendations for aluminum alloy structures*, First edn, 1978.
35. ECCS, *European recommendations for the stressed skin design of aluminum structures*, 1977.
36. Donato, L. F. and Sanpaolesi, L., *Gliacciai e la Sicurezza delle Costruzioni*, Vol. 1, Collana Italsider, 1970.
37. CNR-UNI, *Azioni sulle Costruzioni*, 1967.
38. ISO 2394, *General principles for verifying safety of structures*, 1968.

4 Welded connections

4.1 Technology of welded connections

4.1.1 Welding Procedures

Welding is a joining procedure widely used in aluminum alloy structures. Different methods are used to weld aluminum alloys, such as fusion welding, pressure welding and other special procedures.

Usually, in the welding of structures, two fusion welding procedures in protective atmospheres are used:

TIG – tungsten inert gas (Fig. 4.1a)
MIG – metal inert gas (Fig. 4.1b)

In both cases fusion is due to the high temperature in the electric arc between the electrode and the metal to be welded. The fusion is protected by an inert gas (argon or helium).

It should be noted that, even though aluminum has a melting point lower than steel (650 °C against 1500 °C), the heat necessary for a weld of a given dimension is almost the same. This is due to the greater heat dispersion in aluminum because of its higher conductivity (see Fig. 1.15).

The TIG procedure is a manual semi-automatic procedure; sometimes it is automatic with alternating current. It makes use of a permanent tungsten electrode (Fig. 4.1a), and the voltage does not pass through the weld metal.

The MIG procedure is automatic or semi-automatic with continuous current. The electrode consists of the weld metal which is continuously supplied (Fig. 4.1b).

In the case of a continuous weld, the higher the welding velocity the smaller the extent of the heat-affected zone. This is a beneficial effect since the reduced-strength regions are limited in this way (see Section 2.5.2.3).

Some examples of aluminum welded joints obtained by MIG or TIG procedures are given in Fig. 4.2. They are full penetration joints with or without preparation of the edges and fillet welds.

Aluminum alloy structures

Fig. 4.1

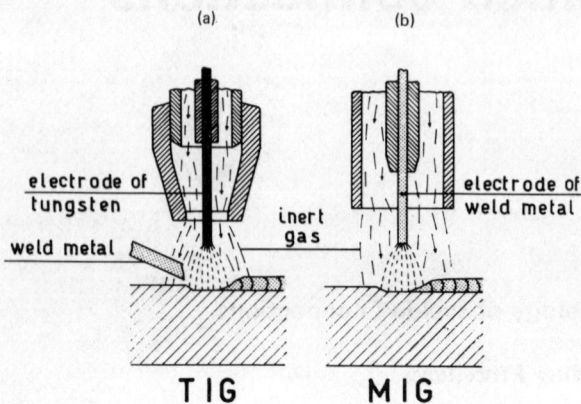

Tack welds can be made using TIG or MIG procedures as well as by using electric arc welding, which is used in plates 0.2 to 5 mm thick.

Some examples of typical electric arc tack welds are given in Fig. 4.3, in which a classification of quality of joints is also given.

4.1.2 Weldable Alloys

The weldability of aluminum alloys has been studied over several decades. Experimental research carried out by Brenner [11] led to the acceptance as weldable of some alloys of the 5000, 6000 and 7000 series, particularly AlMg5, AlMgSi1, AlZnMg1 (as well as of pure aluminum). In these alloys the effect of welding on mechanical properties ($f_{0.2}, f_t, \varepsilon_t$) was evaluated by checking the values given in Fig. 4.4.

Fig. 4.2

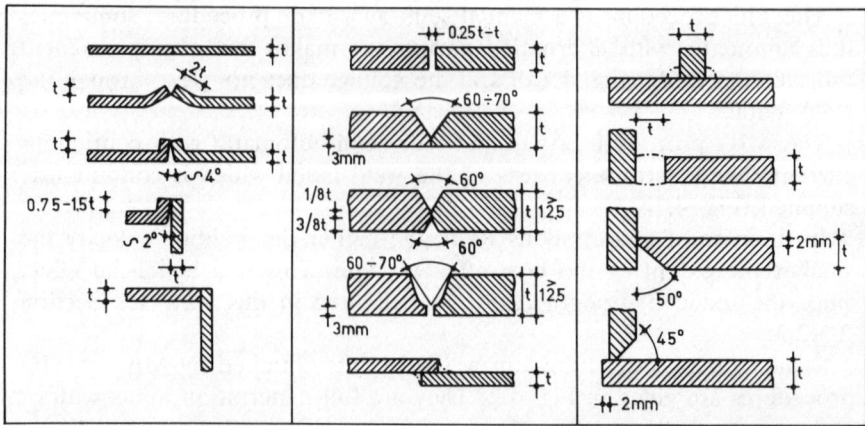

Fig. 4.3

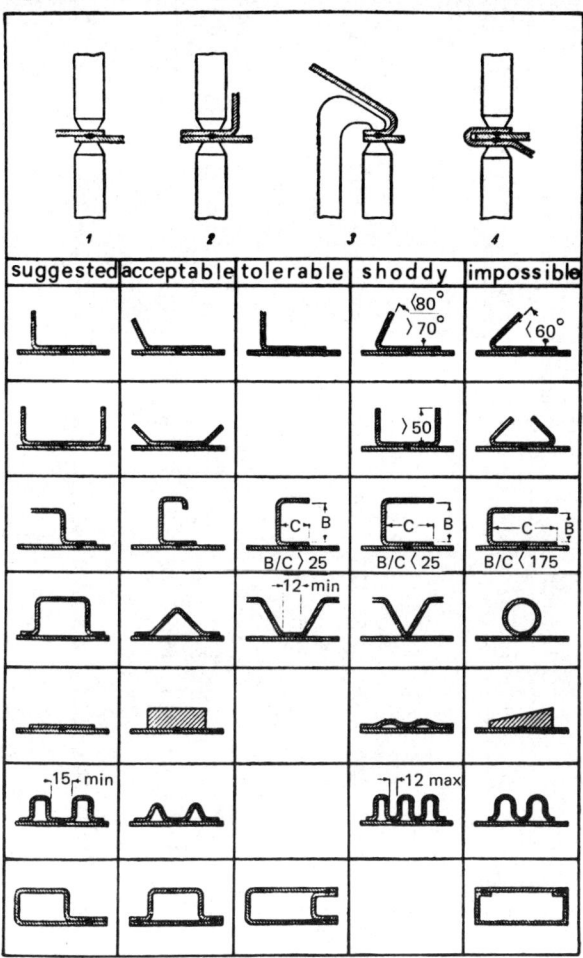

AlMg5 does not significantly change its mechanical properties after welding, but the same is not true of the other alloys.

In the heat-affected zone the following properties are drastically reduced, by the amounts given:

Elastic limit (50 percent in AlZnMg1 and 55 percent in AlMgSi1)
Ultimate strength (30 percent in AlZnMg1 and 45 percent in AlMgSi1).
Ultimate elongation (25 percent in AlZnMg1 and 60 percent in AlMgSi1).

Brenner's experiments also showed that the alloys recover some of their original strength through an aging process (see Fig. 4.5). After six months

Fig. 4.4 Effect of welding on mechanical properties (values of $f_{0.2}$ and f_t in kgf mm^{-2})

Material	Heat treatment	Not welded			Welded or aged														
					0 hours			1 month			3 months			6 months			2 days + 24 hours at 20°C / 2 days + 8 hours at 160°C		
		$f_{0.2}$	f_t	ε_t	$f_{0.2}$	f_t	ε_t	$f_{0.2}$	f_t	ε_t	$f_{0.2}$	f_t	ε_t	$f_{0.2}$	f_t	ε_t	$f_{0.2}$	f_t	ε
AlZnMg1 (7020)	Annealed with solution treatment; tempered; 2 days at room temperature; 48 hours at 160°C	33.3	39.7	10.8	16.2	26.5	8.3	22.5	31.7	4.4	25.6	34.1	5.6	26.2	34.6	5.6	26.1	33.4	2.8
AlMgSi1 (6082)	Annealed with solution treatment; tempered; 8 hours at 160°C	30.5	33.5	9.9	13.7	19.0	4.1	16.0	19.2	3.1	16.6	19.7	3.2	17.5	21.0	2.4	20.8	22.8	1.8
AlMg5 (5050)	Softened and slowly cooled between 400 and 150°C	11.4	28.1	24.3	12.1	28.1	14.1	12.0	27.8	11.5	11.9	25.2	8.8	—	—	—			
AlMg5 (5056)	Softened and slowly cooled between 400 and 150°C; cold worked	17.3	30.0	20.5	16.9	30.4	17.7	16.8	27.5	18.0	15.9	29.1	13.0	17.2	30.1	17.6			

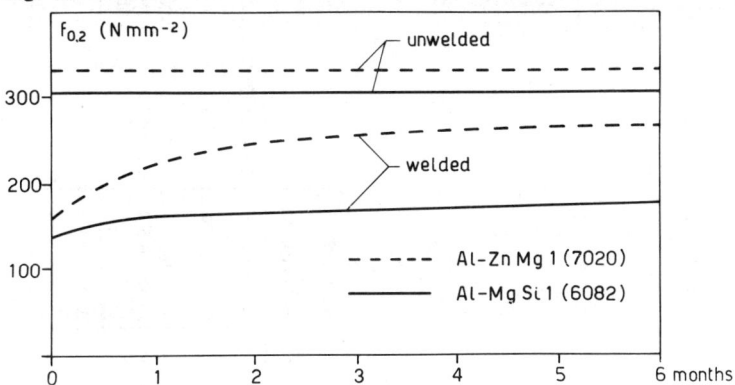

Fig. 4.5

of natural aging the elastic limit reached 80 percent of the original value in AlZnMg1, but in AlMgSi1 only 10 percent of the elastic limit is recovered.

Even better results can be obtained if artificial aging is carried out in the manner outlined in Fig. 4.4.

Ultimate elongation values remain very low, and can even decrease in the case of artificial aging. Such small ε_t values, even though they are limited to the region close to the weld, have to be considered as very significant because they drastically decrease the ductility of the material.

Nowadays, the wrought alloys most used in welded structures are those given in Fig. 4.6.

Fig. 4.6 Wrought alloys for welded structures

Series	Chemical composition	International designation
1000	Al99.0	1200
	Al99.5	1050A
5000	AlMg2.5	5052
	AlMg2.7Mn	5054
	AlMg3.5	—
	AlMg4.4	5086
	AlMg4.5	5083
6000	AlMgSi	6060
	AlSi1MgMn	6082
	AlMg1SiCu	6061
7000	AlZn4.5Mg	7020

Fig. 4.7 Possible combinations between base metals and weld metals

Base metal											
Conventional designation	Numerical	1080 A / 1070 A	1050 A / 1200	3004 / 3003	5052	5054	—	5083	5086	6060 / 6082 / 6061	7020
P-Al99.8 (UNI 4509) / P-Al99.7 (UNI 4508)	1080 A / 1070 A	S-Al99.8									
P-Al99.5 (UNI 4507) / P-Al99.0 (UNI 3567)	1050 A / 1200	S-Al99.5 / S-Al99.5Ti	S-Al99.5 / S-Al99.5Ti								
P-AlMg1.2Mg (UNI 6361) / P-AlMg1.2Co (UNI 7788)	3004 / 3003	S-Al99.5Ti	S-Al99.5Ti	S-Al99.5Ti (AlMn)							
P-AlMg2.5 (UNI 3574)	5052	(S-AlMg5)	(S-AlMg5)	(S-Al99.5Ti) / (S-AlMg5)	S-AlMg3Mg / S-AlMg3.5 / S-AlMg5(1)						
P-AlMg2.7Mn (UNI 7789)	5054	(S-AlMg5)	(S-AlMg5)	(S-AlMg6)	(S-AlMg5)	S-AlMg3.5 / S-AlMg5(1)					
P-AlMg3.5 (UNI 3375)	—	(S-AlMg5)	(S-AlMg5)	(S-AlMg5)	(S-AlMg5)	(S-AlMg5)	S-AlMg3.5 / S-AlMg5				
P-AlMg4.5 (UNI 7790)	5083	(S-AlMg5)	(S-AlMg5)	(S-AlMg5)	S-AlMg5	S-AlMg5	S-AlMg5	S-AlMg5 / S-AlMg4.5Mn			
P-AlMg4.4 (UNI 5452)	5086	(S-AlMg5)	(S-AlMg5)	(S-AlMg5)	S-AlMg5	S-AlMg5	S-AlMg5	S-AlMg5	S-AlMg5 / S-AlMg4.5Mn		
P-AlMgSi (UNI 3569) / P-AlSi1MgMn (UNI 3571) / P-AlMg1SiCu (UNI 6170)	6060 / 6082 / 6061	(S-AlSi5)	(S-AlSi5)	(S-AlSi5)	S-AlMg5	S-AlMg5	S-AlMg5	S-AlMg5	S-AlMg5	S-AlMg5 (2) / S-AlSi5 (3)	
P-AlZn4.5Mg (UNI 7791)	7020	—	(S-Al99.5Ti)	(S-Al99.5Ti)	(S-AlMg5)	(S-AlMg5)	S-AlMg5 / S-AlMg4.5Mg	S-AlMg5 / S-AlMg4.5Mg	S-AlMg5 / S-AlMg4.5Mg	S-AlMg5 / S-AlSi5 (3)	S-AlMg5 / S-AlMg 4.5 Mn
Base material numerical or conventional designation		1080 A / 1070 A	1050 A / 1200	3004 / 3003	5052	5054	P-AlMg3.5 (UNI 3575)	5083	5086	6060 / 6082 / 6061	7020

(1) Combination advisable for metallurgical reasons. (2) Combination recommended for mechanical reasons when no heat treatments are provided after welding. (3) Combination recommended for mechanical reasons when an appropriate heat treatment is provided after welding.

4.1.3 Choice of Base Metal and Weld Metal

A summary of possible combinations between base metals and weld metals is given in Fig. 4.7. The designation of alloys to be used as weld metal is preceded by the letter S. The combinations not commonly used in practice are given in parentheses. When using TIG and MIG procedures, some possible combinations are given in Fig. 4.8.

ECCS European recommendations consider the combinations given in Fig. 4.9, with the limitation that 7020 alloy is not used together with 4043 in the case of butt welds.

When joining different base metals, the combinations given in Fig. 4.10 are recommended.

4.1.4 Approval of Welding Procedures

For a welding procedure to be approved, preliminary test welds have to be completed. These tests are to determine, taking into account the aging time:

That the procedure is suited to the principal type of joints of the structure, either from an esthetic viewpoint or from consideration of internal defects. The latter are ascertained by radiographic or ultrasonic examination or by destructive testing.

Fig. 4.8 Metal combinations for TIG and MIG welding

Parent metal	Filler metal
5052	5554
	5654
5454	5356*
5086	5356
5083	
6060	5356†
6082	
6061	4043‡
7020	5356

* Suitable for metallurgical reasons.
† Recommended for strength reasons, when no thermal treatment after welding.
‡ Recommended for strength reasons, when an appropriate thermal treatment follows welding.

Aluminum alloy structures

Fig. 4.9 ECCS recommendations for welding metal combinations

Parent metal	7020	6082 6061	6060	5083	5454
7020	5356 4043 5183				
6082 6061	5356 4043 5183	5356 4043 5183			
6060	5356 4043	5356 4043	5356 4043 5754		
5083	5356 5183	5356 5183	5356	5356 5183	
5454	5356 5754 5183	5356 5183	5356 5754	5356 5754 5183	5356 5754 5183

That the strength of butt welds, measured on specimens transverse to the joint, is not smaller than that provided by specifications.

That the deformation capacity of butt joints, measured on specimens transverse to the joint by means of standard tests, is satisfactory.

Welding, both in the shop and in the field, has to be undertaken by specialized operatives who have passed examinations indicated by the appropriate specifications, with respect to the class of procedure (TIG or

Fig. 4.10 Joining different base metals

First parent metal	Second parent metal	Weld metal
5083 5086	6061 6082	5356† 4043
5083 5086	7020	5356
6082 6061	7020	5356† 4043‡

† Recommended for strength reasons, when no thermal treatment after welding.

‡ Recommended for strength reasons, when an appropriate thermal treatment follows welding.

MIG), to the product (plates or tubes) and to the thicknesses and operating conditions.

4.2 Strength of welded joints [1–10]

4.2.1 Reduced-Strength Zones

There are two main categories of welded joint (Fig. 4.11) – butt welds and fillet welds. In both cases the effect of heat of welding has to be considered in computations of the reduction in strength of the base metal. (see Section 2.5.2.3).

Mechanical properties $(f_{0.2}, f_t)$ gradually decrease near to the weld and reach a minimum at the center of the weld (see Figs. 2.46–2.48). This effect is taken into account through the definition of a reduced-strength zone. This zone is characterized by a width b_r which allows for the variation of strength between the center of the weld, where the strength corresponds to that of the affected metal, and the unaffected zones of parent metal.

Fig. 4.11

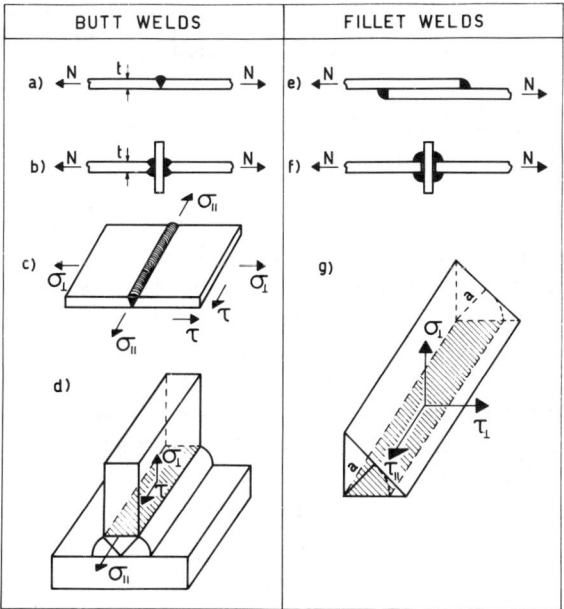

Aluminum alloy structures

Fig. 4.12

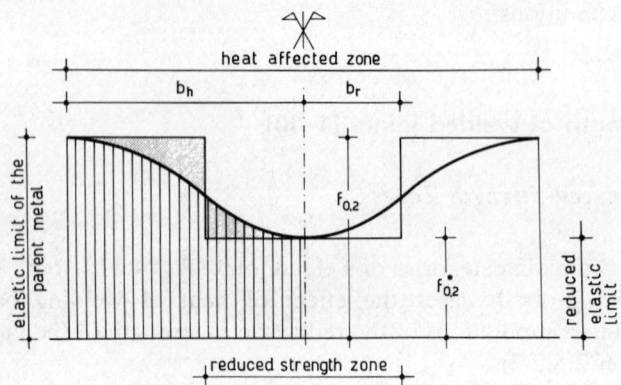

With respect to Fig. 4.12 we have:

$$f^*_{0.2} b_r = \int_0^{b_h} f(x)\,dx$$

$f_{0.2}$ elastic limit of the unaffected base metal
$f^*_{0.2}$ elastic limit of the material in the welded region
b_h semi-width of the heat-affected zone, from which the semi-width b_r of the reduced-strength zone is derived:

$$b_r = \left[\int_0^{b_h} f(x)\,dx\right] \Big/ f^*_{0.2}$$

The width of the heat-affected zone, and therefore b_r, depends upon the welding procedure (velocity, voltage, number of passes, thickness of the joint).

Amongst common welding procedures, experiments carried out to determine b_r led to quite uniform results. These tests allow the use of a constant value of 25 mm from each side of the axis of the weld (Fig. 4.13). This assumption can be considered valid for butt welds and fillet welds made with MIG or TIG procedures. If annealed alloys are used, b_r equals zero.

When calculating the strength of welded built-up members, the reduced-section modulus of the overall cross-section A_{red} has to be defined by considering the presence of the reduced-strength zones. The reduced cross-section is therefore given by

$$A_{red} = A - (1-\beta)\sum_i b_{r,i} t_i \qquad (4.1)$$

Fig. 4.13

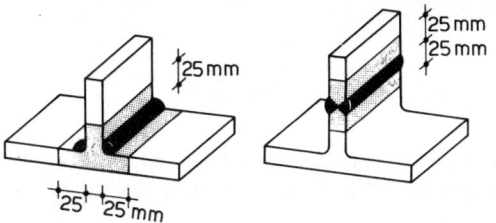

▨ reduced strength zone

A   geometric area of the cross-section (base metal and weld metal)
$b_{r,i}$ represents half of the generic reduced-strength zone
t_i average thickness of the base metal in that region
β a metallurgical efficiency factor

β can be expressed by the ratio

$$\beta = \frac{f_{d,red}}{f_d}$$

between the design strength $f_{d,red}$ of the base metal adjacent to the joint in the reduced-strength zone and the design strength of the unaffected base metal f_d.

$f_{d,red}$ is conventionally assumed equal to the elastic limit $f_{0.2}$ referred to the following stages:

Annealed (for 1000 and 5000 series)
Tempered in water and naturally aged (for 6000 series)
Tempered in the atmosphere and naturally aged (for 7000 series)

By analogy with Eq. 4.1, the following can be evaluated:

The reduced static moment

$$S_{red} = S - (1-\beta) \sum_i b_{r,i} t_i y_i \qquad (4.2)$$

The reduced moment of inertia

$$I_{red} = I - (1-\beta) \sum_i b_{r,i} t_i y_i^2 \qquad (4.3)$$

S, I nominal values
y_i distance of the center of gravity of the ith area with reduced strength and the center of gravity of the cross-section of the member

Aluminum alloy structures

As a consequence, the reduced-section modulus of the cross-section is given by:

$$W_{red} = \frac{I_{red}}{y_{max}} \quad (4.4)$$

If the extent of the reduced-strength zones is smaller than 10 percent of the entire cross-section, then by convention it is approximately assumed that:

$$A_{red} = A$$
$$S_{red} = S \quad (4.5)$$
$$I_{red} = I$$

4.2.2 Full Penetration Butt or T joints

4.2.2.1 Design strength

A butt joint in which the stresses are perpendicular to the axis of the joint has an effective section with a length L equal to the weld length and a thickness t equal to:

For butt welds, the smaller of the thicknesses of the connected plates which are at a distance less than or equal to one-half of the width b_r of the reduced-strength zone measured from the axis of the joint (Fig. 4.14a).

For T welds, the smallest thickness of the web which is at a distance less than or equal to one-half of the width of the reduced-strength zone b_r, measured from the surface of the flange (Fig. 4.14b).

When computing the thickness t of the effective section, the indications given in Fig. 4.15 can be useful.

The design strength $f_{d,w}$ of this effective section can be taken as equal

Fig. 4.14

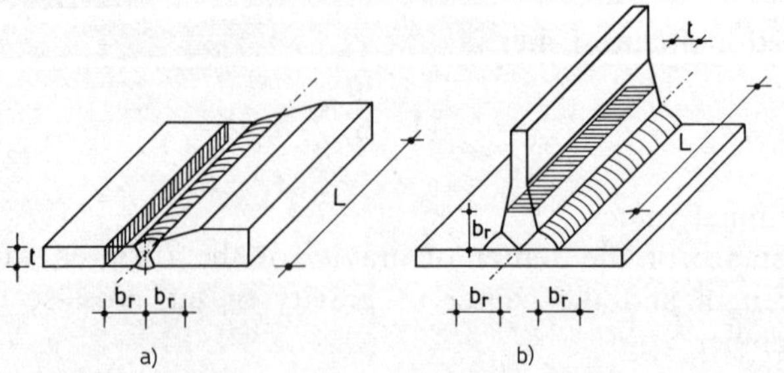

Welded connections

to:

$$f_{d,w} = \eta_w \times \text{lesser of} \begin{cases} f_{d,red} \\ f_{d,0} \end{cases} \quad (4.6)$$

$f_{d,red}$ strength of the material in the reduced-strength zone as defined in Section 4.2.1
$f_{d,0}$ design strength of the weld metal
η_w efficiency coefficient of the joint

This coefficient can be related to quality classes, which are defined by national standards. As an example, we can assume for the three hypothetical weld classes:

$\eta_w = 1$ in the case of first-class welds
$\eta_w = 0.85$ in the case of second-class welds
$\eta_w = 0.70$ in the case of third-class welds

Fig. 4.15

Type	Diagram	Dimensions
Butt welds		
T butt weld		$t = t_1$ $(t_1 \leq t_2)$
V butt weld with back recovering	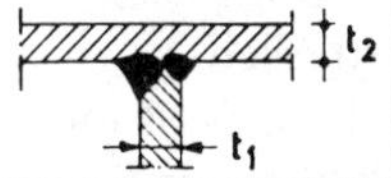	
V full penetration weld	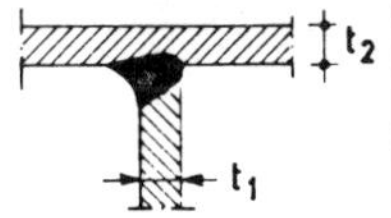	
K partial penetration weld	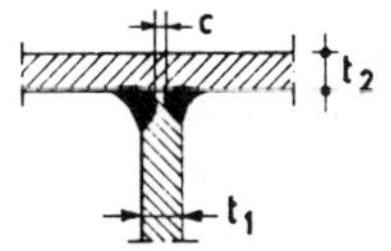	$t = t_1$ $c \begin{cases} \leq 1/5\, t_1 \\ \leq 3\,mm \end{cases}$
V partial penetration weld with back recovering	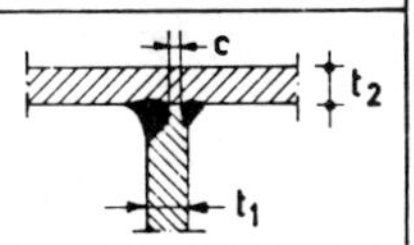	
V partial penetration weld	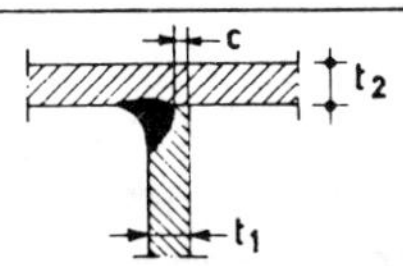	$t = t_1 - c$ $c \begin{cases} \leq 1/5\, t_1 \\ \leq 3\,mm \end{cases}$

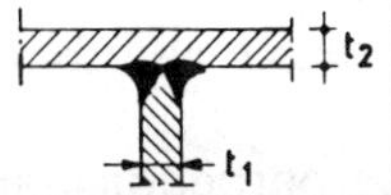

Fig. 4.16

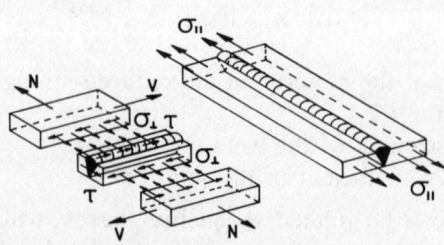

The following stresses can be identified in the effective section (see Fig. 4.16):

$\sigma_\perp$ normal stress (tensile or compressive) which acts in a direction perpendicular to the axis of the weld

τ shear stress which acts in a direction parallel to the axis of the weld

$\sigma_\parallel$ normal stress (tensile or compressive) which acts in a direction parallel to the axis of the weld

The following simple and combined stress states have to be checked.

4.2.2.2 Stresses perpendicular to the axis of the weld

If the joint is stressed by a simple tensile or compressive force N perpendicular to the axis of the weld (Fig. 4.17), where tL is the effective section, the nominal stress is equal to:

$$\sigma_\perp = \frac{N}{tL} \qquad (4.7)$$

Fig. 4.17

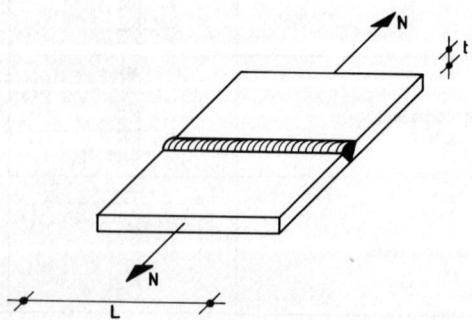

Fig. 4.18 $f_{d,red}$ and $f_{d,0}$ for common metal combinations

Parent metal	$f_{d,red}$ (N mm^{-2})	Weld metal	$f_{d,0}$ (N mm^{-2})	$\sigma_{adm,w}$ (N mm^{-2})		
				First class	Second class	Third class
AlMg2.5 or AlMg2.7Mn	80	AlMg3Mn or AlMg5	90 120	53 53	45 45	37 37
AlMg4.5 or AlMg4.4	120	AlMg5	120	80	68	56
AlSi1MgMn or AlMg1SiCu	110	AlMg5 AlSi5	120 70	73 46	62 39	51 33
AlMgSi	70	AlMg5 AlSi5	120 70	46 46	39 39	33 33
AlZn4.5	200	AlMg5	120	80	68	56

It has to be checked that:

$$\sigma_\perp \leq f_{d,w} \text{ (limit state method)}$$
$$\sigma_\perp \leq \sigma_{adm,w} \text{ (allowable stresses method)}$$

where $f_{d,w}$ is the design strength of the weld defined by Eq. 4.6. $\sigma_{adm,w} = f_{d,w}/\nu$ is the allowable stress in the weld, where ν is the safety factor.

Figure 4.18 gives some values of $f_{d,red}$ and $f_{d,0}$ for some common combinations of base metals and weld metals. From these the smallest (on the basis of Eq. 4.6) is assumed as the design strength $f_{d,w}$ for joints of the first class (since $\eta_w = 1$). The values of $\sigma_{adm,w}$ given in Fig. 4.18 are computed for $\nu = 1.50$, and take into account different values of η_w for the different quality classes of the joint (Section 4.2.2.1).

If a shear force V acts on the joint (Fig. 4.19) a shear stress arises in the effective section tL equal to

$$\tau = \frac{V}{tL} \qquad (4.8)$$

Fig. 4.19

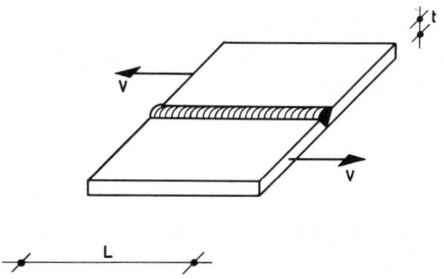

Aluminum alloy structures

In this case it has to be checked that

$\tau \leq \chi f_{d,w}$ (limit state method)

$\tau \leq \tau_{adm,w}$ (allowable stress method)

where

$$\tau_{adm,w} = \frac{\chi f_{d,w}}{\nu} \tag{4.9}$$

and χ is a coefficient which depends upon the failure criterion assumed. If we consider the von Mises criterion, we have:

$$\chi = 1/\sqrt{3} = 0.577$$

Some experimental results have given:

$$\chi = 1/1.64 = 0.61$$

which has been adopted by the European recommendations.

4.2.2.3 Stresses parallel to the axis of the weld

The joint is loaded by a tensile or compressive force N parallel to the axis of the weld. Since A_{red} is the effective section defined by Eq. 4.1, the nominal stress is equal to:

$$\sigma_{\|} = \frac{N}{A_{red}} \tag{4.10}$$

When the joint is loaded by a bending moment M, the corresponding stresses are given by

$$\sigma_{\|} = \frac{M}{I_{red}} y \tag{4.11}$$

The maximum values are equal to

$$\sigma_{\|,max} = \frac{M}{W_{red}} \tag{4.12}$$

where I_{red} and W_{red} are defined by Eqs 4.3 and 4.4, respectively.

It has to be checked that

$\sigma_{\|} \leq f_d$ (limit state method)

$\sigma \leq \sigma_{adm}$ (allowable stress method)

where f_d is the design strength of the base metal and $\sigma_{adm} = f_d/\nu$ is the allowable stress of the base metal.

4.2.2.4 Combined stresses

In butt and full penetration T joints, without internal defects, the stress field can be considered equal to that occurring in continuous metal, and therefore it is commonly accepted that the Von Mises criterion is used to calculate these types of weld. Under biaxial loading this criterion gives:

$$\sigma_{id} = \sqrt{(\sigma_\perp^2 + \sigma_\parallel^2 - \sigma_\perp \sigma_\parallel + 3\tau^2)} \qquad (4.13)$$

It has to be checked that:

$$\sigma_{id} \leq f_{d,w} \text{ (limit state method)}$$

$$\sigma_{id} \leq \sigma_{adm,w} \text{ (allowable stress method)}$$

where $f_{d,w}$ is the design strength of the welds and $\sigma_{adm,w} = f_{d,w}/\nu$ is the allowable strength of the welds.

The ECCS European recommendations use Eq. 4.13 assuming $\sigma_\parallel = 0$. French specifications follow the same philosophy (DTU régles Al).

Actually, the problem of taking into account $\sigma_\parallel$ in verifying welds has been widely discussed in international organizations (IIW, ISO, ECCS). Those who are in agreement with the inclusion of $\sigma_\parallel$ in calculating welds under combined stresses say that this method is more homogeneous with the criterion used for base metal.

Some experimental tests on welded steel joints showed that even high values of $\sigma_\parallel$ do not decrease the load-bearing capacity of joints under static loads in the case of ductile materials. Although the trend to neglect $\sigma_\parallel$ is almost universally followed in steel structures, it may be prudent to take into account $\sigma_\parallel$ stresses in aluminum alloy joints. Experimental results are awaited to confirm what has already been recorded for steel structures. This philosophy has been followed by German (DIN 4113) and Italian (UNI 8209) specifications.

4.2.3 Fillet-welded Joints

4.2.3.1 Design strength

In contrast to butt welds, in which the stress field is very close to that of a continuous plate, the stress distribution in fillet welds is very complex. The stress field, in fact, changes from point to point; the stresses deviate greatly in joints between different plates. Because of this there is a high stress concentration, especially at the end of the fillet.

However, as the loads increase, if there is sufficient ductility to achieve complete plasticity, the concentration of stresses tends to reduce and their distribution becomes closer to a uniform distribution.

These considerations justify the assumption of uniform stresses in the

Fig. 4.20

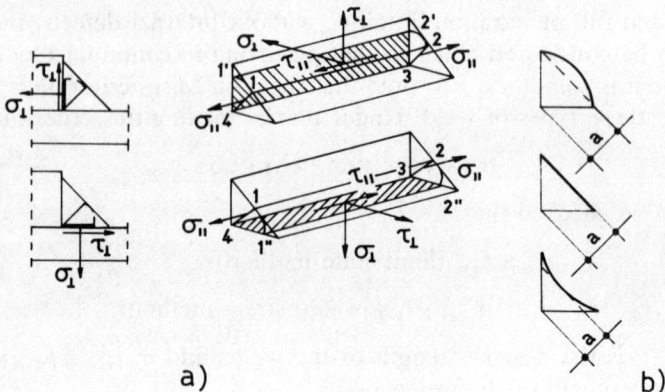

effective section of the fillet. This is usually assumed equal to the depth of the throat section of the fillet multiplied by the length of the fillet (Fig. 4.20). The depth a is the lesser height of the triangle inscribed in the cross-section of the fillet (Fig. 4.20b), and the effective length of the fillet is the total length provided that edges with defects are avoided.

The stress field in the fillet is usually referred to the throat section rotated to lie parallel to one side of the fillet (Fig. 4.20a). This stress field consists of the following components:

$\sigma_\perp$ tensile or compressive normal stress, which acts perpendicularly to the plane of one side of the fillet and the area of the throat section projected onto that plane

$\tau_\perp$ shear stress which acts perpendicularly to the axis of the fillet; it acts on a plane containing one side of the fillet and on the area of the throat section projected onto that plane

$\tau_\parallel$ is the shear stress, which acts in a direction parallel to the axis of the fillet; it acts on a plane containing one side of the fillet and on the area of the throat section projected onto that plane.

In the case of fillet welds there are two trends for considering or neglecting stresses $\sigma_\parallel$ (cf. butt-welded joints). It has to be noted that in this case it is more reasonable not to consider $\sigma_\parallel$, since the fillet is already loaded by the actions which tend to cause sliding of the two connected parts. Instead, $\sigma_\parallel$ acts on the overall cross-section which behaves as a built-up section due just to the efficiency of the welded connection itself. Therefore stresses have to be considered when checking the cross-section but not when checking the fillet welds.

It should be pointed out that there is a trend, even if it is not commonly adopted, to consider as an alternative the throat section in its actual position.

4.2.3.2 Simple stress state

When calculating a fillet-welded joint subjected to simple tension (or compression) perpendicular to the axis of the fillet, the effective section is assumed to be aL projected onto one side of the fillet (see Section 4.2.3.1). The tensile or compressive force N which acts on the joint perpendicularly to the axis of the fillet (Fig. 4.21) causes $\tau_\perp$ or $\sigma_\perp$ stresses, with respect to the direction of projection. They are given by

$$\left.\begin{array}{c}\tau_\perp \\ \sigma_\perp\end{array}\right\} = \frac{N}{aL} \qquad (4.14)$$

It has to be checked that

$\tau_\perp(\sigma_\perp) \leq f_{d,w}$ (limit state method)
$\tau_\perp(\sigma_\perp) \leq \tau_{adm,w}$ (allowable stress method)

where $f_{d,w}$ is the design strength of the fillet weld and $\tau_{adm,w} = f_{d,w}/\nu$ is the allowable shear stress in the fillet weld.

The tensile or compressive force N parallel to the fillets (Fig. 4.22) causes shear stresses parallel to the axis. They are given by

$$\tau_\parallel = \frac{N}{aL} \qquad (4.15)$$

It has to be checked that

$\tau_\parallel \leq \psi f_{d,w}$ (limit state method)

$\tau_\parallel \leq \psi \tau_{adm,w}$ (allowable stress method)

where ψ is an experimental coefficient which gives the ratio between the transverse and the longitudinal strength of a fillet weld.

Experimental research considered in the ECCS European recommendations gave the values ψ shown in Fig. 4.23.

Fig. 4.21

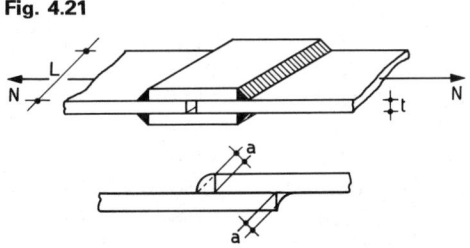

Aluminum alloy structures

Fig. 4.22

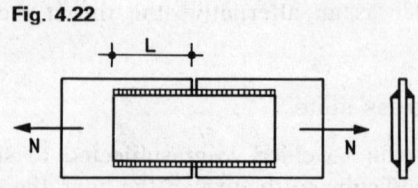

The design strength of a fillet weld $f_{d,w}$, if more precise data based upon experiments are not available, can be assumed to be as follows:

$$f_{d,w} = \eta_w \times \text{lesser of} \begin{cases} \sqrt{2}(f_{d,red}) \\ \gamma f_{d,0} \end{cases} \quad (4.16)$$

η_w coefficient of the joint, related to the quality of the joint itself
$f_{d,red}$ strength of the base metal in the reduced-strength zones (side of the fillet)
$f_{d,0}$ design strength of the weld metal
γ coefficient which depends upon the weld metal

Suggested values of γ are as follows:

Al99.8	0.75
Al99.5	0.75
Al99.5Ti	0.75
AlSi5	0.64
AlMg3Mn	0.75
AlMg3.5	0.75
AlMg4.5Mn	0.56
AlMg5	0.56

Fig. 4.23 ψ values

Parent metal	Condition	Weld metal	ψ
7020	T6	5356	0.69
7020	T6	4043	0.68
6082	T6	5356	0.64
6082	T6	4043	0.64
6061	T6	5356	0.64
6061	T6	4043	0.64
6061	T5	5356	0.67
5083	0	5356	0.71
5454	H24	5356	0.67

4.2.3.3 Combined stresses

Calculations of fillet welds in aluminum alloy joints make use of formulas for combining loadings which are derived from the development of the criteria adopted by the international organizations for welded steel structures (ISO, IIW).

For the shape of the failure curve, two tendencies are apparent: the ellipsoid and the sphere.

The ellipsoid is adopted in the ISO method, in which the reference stress is given by:

$$\sigma_{id} = \sqrt{[\sigma_\perp^2 + k_w(\tau_\perp^2 + \tau_\parallel^2)]} \qquad (4.17)$$

where the parameter k_w is not given. It has to be experimentally determined by different national specifications. It should be noted that if k_w is assumed to equal 3, the results are the same as those obtained using Von Mises criterion.

Equation 4.17 has to be used when the effective section is considered as the throat section of the fillet in its actual position.

In France (DTU régles Al) this formulation is adopted with $k_w = 2.7$. Thus we have

$$\sigma_{id} = \frac{1}{\alpha\beta\gamma}\sqrt{[\sigma_\perp^2 + 2.7(\tau_\perp^2 + \tau_\parallel^2)]} \qquad (4.18)$$

α a coefficient which characterizes the quality of execution ($\alpha = 1$ for radiographed welds and $\alpha = 0.8$ for welds in areas of difficult execution, not radiographed but correctly completed)

β a metallurgical efficiency coefficient, already defined in Section 4.2.1 (it can vary from 0.45 to 1 in the case of most common alloys)

γ a coefficient which takes into account the complex phenomena related to fillet welds (in the case of common alloys, it can vary from 0.65 to 1.0)

In other countries (UK, Germany) the failure curve has been assumed to be spherical. In this case it has to be checked that

$$\sigma_{id} = \sqrt{(\sigma_\perp^2 + \tau_\perp^2 + \tau_\parallel^2)} \le r \qquad (4.19)$$

r is the radius of the sphere. This is the parameter that has to be checked against the actual stress state in the weld. It is assumed to be equal to the design strength $f_{d,w}$ of the fillet weld (see Eq. 4.16) if the limit state theory is used, or equal to the corresponding allowable stress if the allowable stress method is used.

$\sigma_\perp$, $\tau_\perp$ and $\tau_\parallel$ are the components of the resultant stress on the throat section projected onto one side of the fillet (see Section 4.2.3.1).

In Italy, recent specifications for welded aluminum joints under static load (1978) [12] proposed the use of a criterion based upon the 'cut sphere', which had already been adopted in specifications for steel structures. Under this criterion it has to be checked that [6]

If all components of stresses are present:

$$\sqrt{(\sigma_\perp^2 + \tau_\perp^2 + \tau_\parallel^2)} \leq r \quad (4.20)$$

$$|\tau_\perp| + |\sigma_\perp| \leq 1.2r \quad (4.21)$$

$$|\tau_\parallel| \leq 0.85r \quad (4.22)$$

In the case in which only $\sigma_\perp$ and $\tau_\perp$ are present, it is sufficient to check:

$$\left.\begin{array}{l}|\tau_\perp|\\|\sigma_\perp|\end{array}\right\} \leq r \quad (4.23)$$

In the case in which only $\tau_\perp$ and $\tau_\parallel$ are present (or $\sigma_\perp$ and $\tau_\parallel$) it is sufficient to check Eqs 4.20 and 4.22.

It has been observed by the ECCS committee on aluminum alloy structures that the case of combined loading in which all the stress components $\sigma_\perp$, $\tau_\perp$ and $\tau_\parallel$ are present is not very common (Fig. 4.24). In fact this happens only in very unusual cases with the fillet welds inclined with respect to the direction of the external force (Fig. 4.24a) or in the case (Fig. 4.24b) in which the joint has been designed ignoring the fact that a shear force is best resisted by a fillet weld whose axis is in the same direction as that of the external force.

Usually only two stress components are present in well-designed cases of combined stresses $\tau_\perp$ and $\tau_\parallel$ (or $\sigma_\perp$ and $\tau_\parallel$ with respect to the direction of the section). In this case it has to be checked that

$$\left(\frac{\tau_\perp}{f_{d,w}}\right)^2 + \left(\frac{\tau_\parallel}{\psi f_{d,w}}\right)^2 \leq 1 \quad (4.24)$$

when the limit state method is used, or

$$\left(\frac{\tau_\perp}{\tau_{adm,w}}\right)^2 + \left(\frac{\tau_\parallel}{\psi \tau_{adm,w}}\right)^2 \leq 1 \quad (4.25)$$

Fig. 4.24

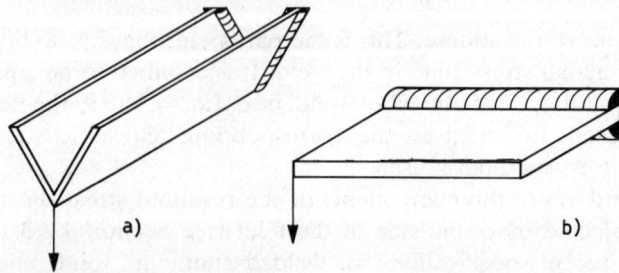

when the allowable stress method is used. Here the meaning of $f_{d,w}$, $\tau_{adm,w}$ and ψ is the same as in Section 4.2.3.2. The ECCS European recommendations introduced this formulation in the chapter on filletwelded joints [13].

References

1. Guiaux, P., Essais statique sur noeuds soudes en alliage d'aluminium (Statical tests on aluminum alloy welded joints), University of Liège, Int. Rep., 1965.
2. Brillant and Charleux, *Alliages d'aluminium corroyés soudés. Essais de rupture de soudures bout à bout, soudures d'angle longitudinales et transversales, détermination de la largeur de la zone à résistance réduite*, Report Nos. 860–865, Centre Téchnique de l'Aluminium, Paris, 1968.
3. Brillant and Charleux, *Alliages d'aluminium coulés soudés. Essais de résistance de soudures bout à bout, soudures d'angle longitudinales et transversales*, Report Nos. 873–988, Centre Téchnique de l'Aluminium, Paris, 1968.
4. Brillant and de Bony, *Alliages d'aluminium corroyés soudés. Compléments des Rapports 860–865*, Report Nos. 988–1212, Centre Téchnique de l'Aluminium, Paris, 1970–1972.
5. Molina, C., Bompard, S. and Goliard, F., *Calcul des assemblages soudés (Design of welded connections)*, Aluminum Rechiney, Rep. No. 1465, 1976.
6. Costa, G., Daddi, I. and Mazzolani, F. M., Collegamenti Saldati (*Welded Connections*), Ed. CISIA, Milan, 1977.
7. Kosteas, D., Steidl, G. and Strippelmann, W. D. *Geschweisste Aluminion Konstruktionen (Aluminum Welded Constructions)*, Vieweg, 1978.
8. IIW, *Colloquium on Aluminum and its Alloys in Welded Construction*, Porto, 1981.
9. Valtinat, G., Welded joints. Chapter 6 in: *European Recommendations for Aluminum Alloy Structures, IIW Colloquium*, Porto, 1981.
10. Werner, G., The calculation of welded joints in statically loaded aluminum structures, *IIW Colloquium*, Porto, 1981.
11. Brenner, P., *Alliages d'Aluminium Durcissables comme Matériaux pour les Constructions Soudées*, VDI, Vol. 103, No. 18, 1961.
12. UNI 8209, *Giunti saldati di alluminio e leghe di alluminio sollecitati staticamente: Istruzioni per il calcolo*, 1978.
13. ECCS, *European Recommendations for Aluminum Alloy Structures*, First Edn, 1978.

5 Mechanical joints

5.1 Technology of joints

5.1.1 *Materials, Types and Shapes*

5.1.1.1 Riveted connections

Riveted joints provide irremovable connections and can be considered as the oldest way to join metallic materials. Although this type of connection is now practically abandoned in steel structures, it is still in use when connecting light structures. For these structures, in contrast to steel structures, cold-driven rivets are used.

The most commonly used rivets for aluminum alloy joints are shown in Fig. 5.1. Their heads are shaped according to use (round, elliptical, conical, countersunk etc.).

Other types used in special applications include those of Fig. 5.2, which are:

(1) Countersunk head with semi-hollow shank
(2) Tubular type
(3) Tubular with threads and plane head
(4) Tubular with threads and countersunk head
(5) Explosive

In all these cases the rivet shank has to be long enough to allow riveting of the head.

The elements to be joined are first thoroughly cleaned and then fixed in their correct relative positions by using assembling bolts; sometimes they have to be clamped.

Another method, commonly used in the aeronautical industry, and in general when the head to be riveted is not easily accessible by the riveting machine, is the one which makes use of explosive rivets (see Fig. 5.2 case 5). These rivets have a semi-hollow shank in which an explosive charge is placed. The explosion is caused by the electric heating of the other head of the rivet previously introduced into the hole.

Mechanical joints

Fig. 5.1

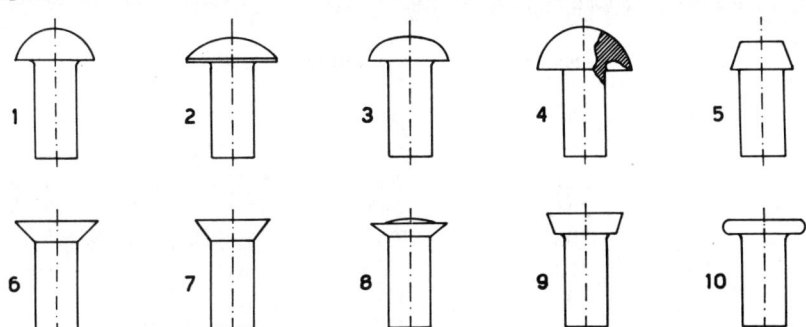

The materials which are commonly used in these types of joint are given in Fig. 5.3. Stainless or standard steel rivets to be hot riveted can also be used, but in these cases contact corrosion has to be avoided (see Section 1.6.3).

Riveted joints essentially resist forces by shear (Section 5.5) and they are not advisable in the case of significant tensile forces.

5.1.1.2 Bolted connections

Bolts are formed by a screw with a hexagonal head and a threaded shank which is then connected to a hexagonal nut. A circular washer is usually put under the nut (Fig. 5.4).

The optimum length of the unthreaded part of the shank is equal to the thickness of the plate to be joined, so that the contact between the plates and the bolts is along the unthreaded part of the shank and the thread begins at the washer.

Bolted connections between aluminum alloy elements can be made by:

Aluminum alloy bolts
Steel bolts (mild and high-strength steels)
Stainless steel bolts

Aluminum alloy bolts are usually made of the materials given in Fig. 5.5.

In the case of highly stressed connections, standard steel bolts can be

Fig. 5.2

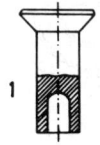 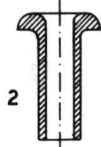 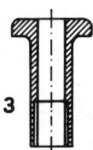 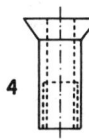 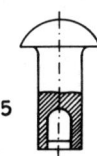

Aluminum alloy structures

Fig. 5.3 Materials for aluminum alloy rivets

Designation	Condition
1200	F
1050A	H14
2117	T4
2017A	T4
2024	T4
3103	H14
3003	H14
5083	H32
5056A	H32
6082	T4 or T6
6061	T4 or T6

used in shear joints (Section 5.3) and high-strength bolts (HS) in friction joints. The corresponding classes used in steel structures are those given in Fig. 5.6, with respect to the bolt–nut connection.

The use of steel bolts is only allowed if adequate superficial protection is provided in order to prevent contact corrosion (see Section 1.6.3). Therefore the bolts have to be galvanized or painted with a protective paint.

Alternatively, stainless steel bolts can be used. They are described in Fig. 5.6 (nickel-chrome-molybdenum, nickel-chrome-niobium etc.).

Tightening is of particular importance in bolted joints. When using high-strength bolts in friction joints, it is necessary to use a calibrated wrench and to check the effect of the applied torque by specific tests in the shop or during fabrication.

In bearing joints it is better to apply an adequate tightening since, even if it does not change the ultimate capacity of the joint, it provides more rigidity to the joint. This improves its performance with respect to the

Fig. 5.4

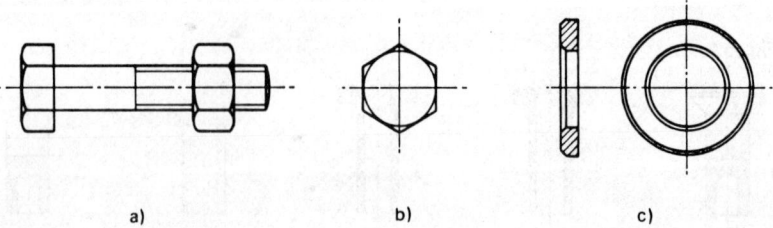

a)   b)   c)

Fig. 5.5 Materials for aluminum alloy bolts

Designation	Condition
2024	T4
5083	0
5056A	0
6082	T6
6061	T6
7075	T73

serviceability limit states, which are:

In the case of lap joints, the load exceeds the frictional resistance and the joint slips into bearing.

In the case of end plate joints, the load causes the uplift of the plates.

Tightening causes a preload N_s in the shank of the screw, the value usually being defined by

$$N_s = 0.8 f_y A_{res} \quad (5.1)$$

where f_y is the yield limit of the material of the bolt, and A_{res} is the resisting area of the bolt. A_{res} is conventionally assumed to be given by

$$A_{res} = \frac{\pi}{2}\left(\frac{3d_1 + d}{4}\right)^2 \quad (5.2)$$

where d and d_1 are respectively the diameter of the bolt and the diameter of the threaded part. In the case of threads with large grips, the values of A_{res} are given in Fig. 5.7. p is the grip of the threads.

Because of the preload N_s, the connected plates are in compression and

Fig. 5.6 Materials for steel bolts

Steel bolts	Materials	
	Screw	Nut
Standard	4.6	A4
	5.6	5D
	6.6	5S
High strength	8.8	6S
	10.9	8G
Stainless	×8 CND 1712	
	×8 CNNb 1811	
	×15 CN 1808	

Aluminum alloy structures

Fig. 5.7 A_{res} for threads with large grips

d	p	A_{res}	A	A_{res}/A
8	1.25	38.6	50.3	0.77
10	1.50	58.0	78.5	0.74
12	1.75	84.3	113	0.75
14	2.00	115	154	0.75
16	2.00	157	201	0.78
18	2.50	192	254	0.75
20	2.50	245	314	0.78
22	2.50	303	380	0.80
24	3.00	353	452	0.78
27	3.00	459	573	0.80
30	3.50	581	707	0.82

have the capacity to resist by friction. The torque T_s to be applied in order to achieve the preload N_s (Eq. 5.1) in the shank is given by

$$T_s = \chi d N_s \tag{5.3}$$

where χ is a coefficient which depends upon the material and the roughness of the surface, and can usually be assumed to be 0.20.

It is necessary to tighten the bolts with a calibrated wrench and to apply in the first stage only 60 percent of the total torque required. It is also advisable to begin from the interior bolts and successively tighten the outer ones; the operation is then repeated until all the bolts are completely tightened.

The efficiency of the applied torque can be checked:

The torque necessary to rotate the nut 10 percent is measured by a calibrating wrench.

The bolt and the nut are marked to identify their relative position, and the nut is loosened by at least 60°. The nut is now retightened with the required torque and is checked to see if the new position corresponds to the old.

If there is one bolt in the joint which fails this check, all the bolts have to be verified.

5.1.2 Dimensional Ratios and Tolerances

The diameter of the rivet or of the bolt is determined by the thickness of the smallest ply connected. Usually it has to be checked that

$$d = (1.5 \text{ to } 2.5) t_1 \tag{5.4}$$

Fig. 5.8

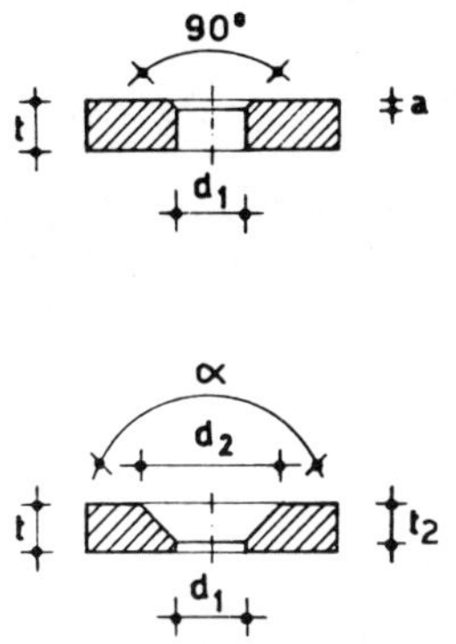

Diameter d (mm)	Dimensions of holes			
	a (mm)	d_2 (mm)	t_2 (mm)	α
4	0.4	10.7	1.9	120°
5	0.4	9.8	2.4	90°
6	0.4	11.7	2.8	90°
8	0.6	15.6	3.8	90°
10	0.6	15.5	4.7	60°
12	0.6	18.6	5.7	60°
14	0.8	21.7	6.6	60°
16	0.8	24.8	7.6	60°
18	1	28	8.6	60°
20	1	31.1	9.5	60°
22	1	34.2	10.5	60°

In the case of aluminum plates the holes are driven by screws (helix drill), and punch machines should be avoided. It is good practice to drill the hole in two separate stages:

Drill the hole using a screw smaller than the nominal diameter by 1 to 2 mm.

Finish the surfaces by reamer after having joined the plates.

In the case of rivets, the holes have to be made by a conical tool having an angle at the end of the tool with a depth from 0.05 to $0.1d$. The corresponding dimensions are given in Fig. 5.8 with respect to cylindrical countersunk holes.

The shank of the rivet must have the following tolerances on the nominal diameter:

$$\left.\begin{array}{c}+0\\-0.1\text{ mm}\end{array}\right\} \text{ for } d<8\text{ mm}$$

$$\left.\begin{array}{c}+0\\-0.2\text{ mm}\end{array}\right\} \text{ for } d\geq 8\text{ mm}$$

These values, referred to the most common diameter, are given in Fig. 5.9. The ranges of minimum thicknesses of plates t_1 determined using Eq. 5.4 are also given.

Aluminum alloy structures

Fig. 5.9 Rivet tolerances

Diameter (mm)	Tolerances		Thickness of plates t_1 (mm)
	d_{min} (mm)	d_{max} (mm)	
4	3.9	4.0	1.5 to 2
5	4.9	5.0	2 to 2.5
6	5.9	6.0	2.5 to 3.5
8	7.8	8.0	3.5 to 4.5
10	9.8	10.0	4.5 to 6
12	11.8	12.0	6 to 7.5
14	13.8	14.0	7.5 to 9
16	15.8	16.0	9 to 10.5
18	17.8	18.0	10.5 to 12
20	19.8	20.0	12 to 14
22	21.8	22.0	14 to 16

In the case of bearing joints it is necessary to connect the plies by 'calibrated holes' in order to reduce the clearance between the hole and the fastener – thus producing a negligible slippage of the joint. This is only possible if difficulties do not arise when fitting the two holes. The calibrated holes correspond to the following tolerances:

$$d_1 = d \begin{cases} +0.1 \text{ mm} \\ +0.2 \text{ mm} \end{cases} \quad d \leq 18 \text{ mm}$$

$$d_1 = d \begin{cases} +0.2 \text{ mm} \\ +0.3 \text{ mm} \end{cases} \quad d > 18 \text{ mm}$$

d nominal diameter of the rivet or of the bolt
d_1 diameter of the punched hole, values of which are given in Fig. 5.10

In the case of bearing joints a tolerance is permitted of 1 to 2 mm between the diameter of the hole and the nominal diameter of the bolt.

The distribution of holes in a joint has to be within the dimensional limitations required by specifications in order to produce a solid joint without weakening the plates with too many holes (see Fig. 5.11). The distance p between the axis of the holes has to be equal to:

$$(2.5 \text{ to } 3)d < p \leq (3 \text{ to } 4)d$$

The minimum distance between the axis of the hole and the edges of the plates to be connected has to be equal to

$a \geq (1.5 \text{ to } 3)d$ perpendicular to line of force

$a_1 \geq 1.5d$ parallel to line of force

Fig. 5.10 Rivet and hole tolerances

Diameter d (mm)	Diameter of drilled hole d_1 (mm)	
	min	max
4	4.1	4.2
5	5.1	5.2
6	6.1	6.2
8	8.1	8.2
10	10.1	10.2
12	12.1	12.2
14	14.1	14.2
16	16.1	16.2
18	18.1	18.2
20	20.2	20.3
22	22.2	22.3

Experimental results showed that it is also necessary to relate these distances to the smallest thickness of the plates to be connected.

The European recommendations provide the following limitations with respect to the edge distances of the joined plates:

Parallel to the line of force:

$$\text{joints under tension} \qquad a \leq \begin{cases} 16t_1 \\ 200 \text{ mm} \end{cases}$$

$$\text{joints under compression} \qquad a \leq \begin{cases} 8t_1 \\ 200 \text{ mm} \end{cases}$$

Perpendicular to the line of force:

$$a_1 \leq \begin{cases} 8t_1 \\ 100 \text{ mm} \end{cases}$$

Fig. 5.11

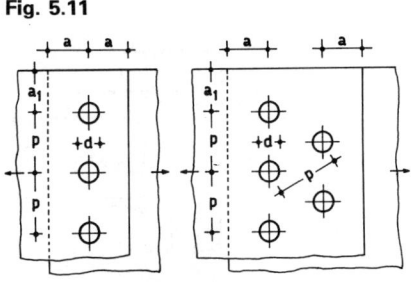

Aluminum alloy structures

The spacing between the centres of adjacent holes, whether staggered or not, must satisfy the following equations:

$$\text{joints under tension} \qquad p \leq \begin{cases} 32t_1 \\ 300 \text{ mm} \end{cases}$$

$$\text{joints under compression} \qquad p \leq \begin{cases} 20t_1 \\ 300 \text{ mm} \end{cases}$$

5.2 Strength of connections [1–8]

5.2.1 *Experimental Behavior*

In order to define the strength of a riveted or bolted connection and to specify necessary checks, it can be very useful to study the behavior of an elementary joint, as shown in Fig. 5.12, subjected to an increasing load.

The diagram (Fig. 5.13) provided by the testing machine, which relates the applied load F_v to the slippage ΔL between the points A and B of the plates, exhibits four distinct types of behavior (Fig. 5.13):

Phase 1 The slippage ΔL is practically equal to zero as the load increases: the load is therefore transmitted by friction of the plates. This phase ends when the load $F_{v,f}$ equals the friction between the plies.

Phase 2 the load exceeds the frictional resistance and the joint slips into bearing.

Phase 3 the joint continues to deform elastically and consequently the load/deformation relationship remains linear.

Phase 4 yielding of plates, fasteners or both occurs and results in plate fracture or complete shearing of the fasteners.

The second phase, which corresponds to slippage of the joint owing to the frictional resistance being exceeded, is not present in the case of cold-

Fig. 5.12

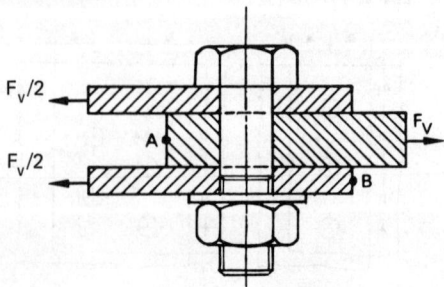

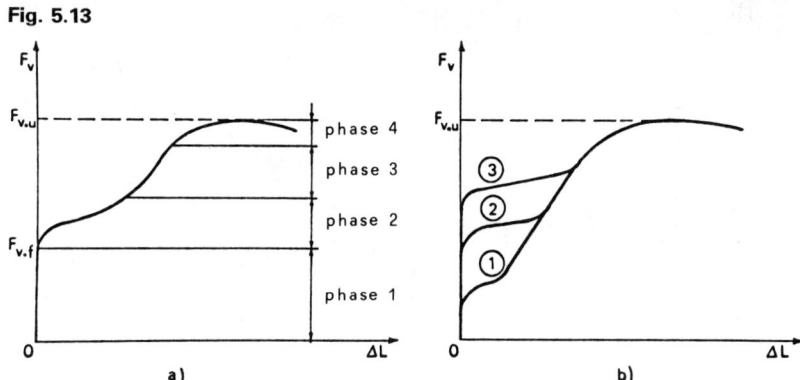

Fig. 5.13

driven rivets. This type of installation does not provide a tensile preload in the shank of the rivet, and so there is no precompression of the plates of the joint.

In the case of hot-driven rivets, as the rivet cools it shrinks and squeezes the connected plies together. A residual clamping force or internal tension results in the rivet. The residual clamping force contributes to the slip resistance of the joint just as in high-strength bolts. However, the clamping force in the rivet is difficult to control, is not as great as developed by high-strength bolts, and cannot be relied upon.

In the case of bolts, since it is possible to control the magnitude of the applied torque, the friction resistance of the joint can be computed exactly. The roughness of the surfaces as well as the clearance between the hole and the bolts influence the value of $F_{v,f}$ and therefore the extent of the region of the diagram $F_v - \Delta L$ that is being considered. In fact if we test the same specimen but change the preload of the bolt or the surface roughness, we can observe a behavior which is qualitatively the same but with a different extension of the second phase. The value of the load $F_{v,f}$ that corresponds to the slippage of the joint is variable, and the elastic region is decreased or increased. The plastic region as well as the value of the ultimate capacity of the joint do not vary because they are independent of the magnitude and type of tightening.

The ultimate capacity of the joint can be related to one of the following failure mechanisms (Fig. 5.14):

(a) Shear failure of the rivet or bolt
(b) Excessive ovalization of the hole
(c) Shear failure of the plate
(d) Tension failure of the plate

The weakest of these mechanisms will cause failure of the connection and is therefore assumed to be the design capacity of the connection itself.

Aluminum alloy structures

Fig. 5.14

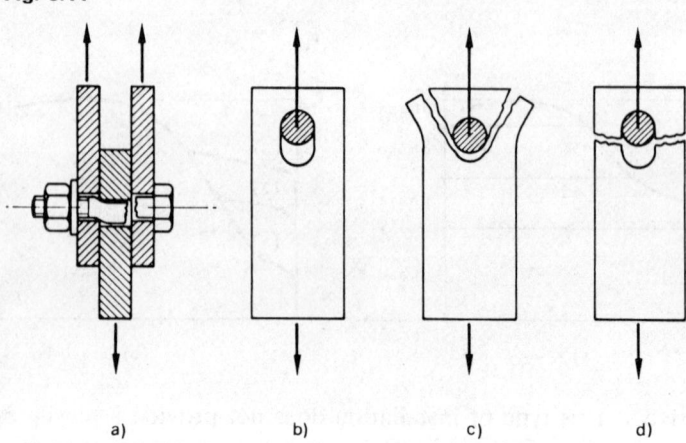

a) b) c) d)

5.2.2 Limit States

The riveted or bolted connection has to be checked against the loads it can carry. The following cases can be identified:

The rivet or the bolt are subjected to shear (Fig. 5.15a).
The bolt is subjected to tension (Fig. 5.15b).
The bolt is subjected to tension and shear (Fig. 5.15c).

The possibility of having rivets subjected to tension is therefore omitted.

The connections can also be classified with respect to the principal way that they are designed to resist the applied loads. *Shear connections* (bearing type) transmit the loads by shear in the shank of the fastener (rivet or bolt), and *friction connections* transmit the load by friction of the plates (in the case of high-strength steel bolts).

It is evident that in the first case it is important to take into account the bearing between the shank and the hole, whereas in the second case the operations concerned with tightening of the bolt are of extreme importance.

With respect to bolted joints, two main limit states can be identified:

(a) *Service limit state*, which corresponds to:
   (i) Slippage of the joint due to exceeding the friction resistance with consequent bearing of the bolt. This applies to lap joints of both the bearing and the friction type
   (ii) Decompression and consequent separation of the plates in the case of end plate joints, applying to bearing and friction types
(b) *Ultimate limit state*, which corresponds to failure of the connection

Fig. 5.15

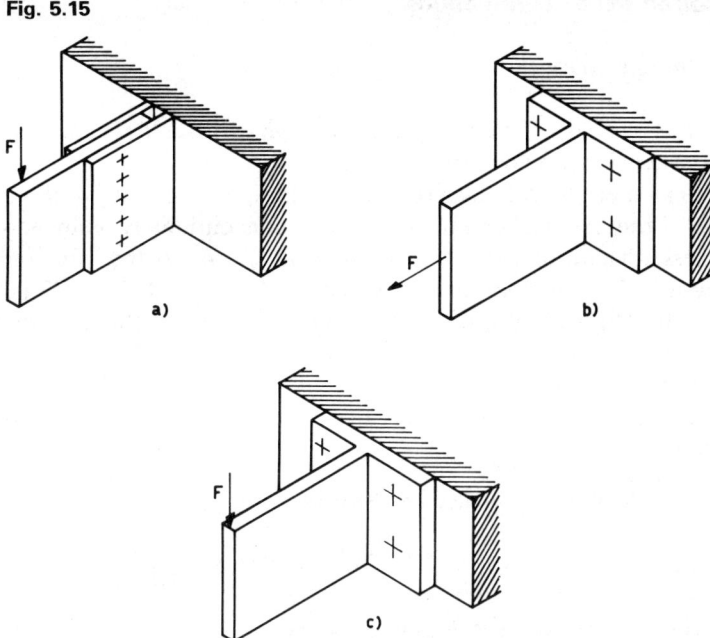

and requires the following to be checked:
  (i) Shear and/or tension stress in the fastener
  (ii) Shear and/or tension stress in the plates of the joint
  (iii) Bearing stress in the plates of the joint

The check of the service limit state (a)(i) is important only in the case where the connection must not be very deformable – in order to avoid inelastic displacement, particularly in the case of trusses, due to the bolt clearance. The major unfavorable consequences can be summarized as follows:

Large inelastic deformations additional to the elastic displacements of the structure that are due to live loads

Eccentricities, which may cause instability of the structure under compression or compression and bending

Check (a)(ii) is of particular importance because the separation of plates reduces the protection of the plies, in the region of contact, against atmospheric corrosive agents.

The mechanism of failure of the connection is not known *a priori*, and therefore checks (b) against the ultimate limit state have to be made in each case.

5.3 Bolted shear connections

5.3.1 Design Strength

In the case of bearing joints the transmission of loads depends upon the shear and tension capacity of the bolt itself.

It is convenient to define as the design strength of a bolt $f_{d,b}$ the design strength of the material of the bolt. This is conventionally assumed equal to the lesser value of either the yield limit $f_{y,b}$ (or to the stress corresponding to 0.2 percent strain) measured on the whole bolt or 70 percent of the ultimate strength f_t. If experimental results which have been statistically analyzed are considered, the design strength is equal to the characteristic strength.

The design strength to be used can be derived from the values of $f_{d,b}$. In particular:

The design shear strength $f_{d,V}$ is given by

$$f_{d,V} = \frac{f_{d,b}}{\gamma_V} \tag{5.5}$$

The design tensile strength $f_{d,N}$ is given by:

$$f_{d,N} = \frac{f_{d,b}}{\gamma_N} \tag{5.6}$$

The coefficients γ_V and γ_N have the same meaning as the coefficient γ_m defined in Section 3.4.2. In general they permit the design strength to be obtained from the characteristic strength by taking account of all the random processes which are present in a real structure.

In this specific case, the coefficient γ_V takes account of the relationship between the tension capacity of the specimen and the shear strength of the bolt in the joint. As it is not very significant to use a failure criterion, the values of γ_V are obtained from experimental tests. The European recommendations suggest:

$\gamma_V = 1/0.6$ aluminum alloy bolts

$\gamma_V = 1/0.7$ steel bolts

The coefficient γ_N depends upon two different phenomena. The first is that the head of the bolt, usually obtained by upsetting, can shear off the shank. This possibility is avoided by use of a controlled tightening which permits identification of defects in the bolt. The second phenomenon corresponds to the secondary stresses due to bending in the bolt (see Fig. 5.16).

The consequent tension combined with bending reduces the capacity of

Fig. 5.16

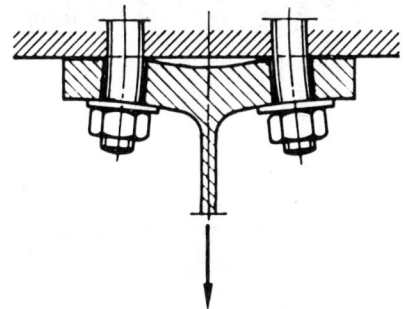

the bolt more significantly if the resistance of the bolt under simple tension is smaller. This effect can be negligible if the bolt is fastened so that the plates are unable to separate, producing less deformation capacity. Therefore the value of γ_N is governed by the magnitude of tightening because it affects both the phenomena which γ_N takes into account. The values suggested for γ_N are the following:

$\gamma_N = 1$ aluminum alloy and standard steel bolts with controlled tightening and where there is no secondary bending
$\gamma_N = 1.25$ high-strength steel bolts with controlled tightening
$\gamma_N = 1.33$ aluminum alloy and steel bolts without controlled tightening

When controlling tightening, one of the methods explained in Section 5.1.1.2 has to be used.

The values of the design strengths $f_{d,b}$, $f_{d,V}$ and $f_{d,N}$ for some types of bolt are given in Fig. 5.17.

If the semi-probabilistic method is adopted the reference stress computed with respect to the design loads has to be checked against the design strength.

If the allowable stress method is adopted, the design strength can be used to calculate the allowable stresses:

Allowable shear stress:

$$\tau_{adm,b} = \frac{f_{d,V}}{\nu} \tag{5.7}$$

Allowable tensile stress:

$$\sigma_{adm,b} = \frac{f_{d,N}}{\nu} \tag{5.8}$$

Fig. 5.17 Bolt design strengths

Type of bolt			Design strength ($N\,mm^{-2}$)				Allowable stresses ($N\,mm^{-2}$) Loading condition 1			Loading condition 2		
					$f_{d,N}$			$\sigma_{adm,b}$			$\sigma_{adm,b}$	
			$f_{d,b}$	$f_{d,V}$	Controlled tightening	Uncontrolled tightening	$\tau_{adm,b}$	Controlled tightening	Uncontrolled tightening	$\tau_{adm,b}$	Controlled tightening	Uncontrolled tightening
	AlCu1.5MgMn	TI	275	165	275	206	97	161	121	110	183	137
	AlMg1.5 AlMg5	R R	130	78	130	97	45	76	57	52	86	64
	AlSi1MgMn AlMg1SiCu	TA18 TA16	245	147	245	184	86	144	108	98	163	122
	AlZn5.8MgCu	TA	425	255	425	319	150	250	187	170	253	212
Steel	Standard screw	4.6	205	143	205	154	84	120	90	95	136	102
Steel	Standard screw	6.6	335	234	335	252	137	197	148	156	223	168
Steel	HY screw	8.8	550	385	440	413	226	258	242	256	293	275
Steel	HY screw	10.9	685	479	548	515	281	322	302	319	365	343
Stainless steel	× 8 CND	1712	260	182	260	195	107	152	114	121	173	130
Stainless steel	× 8 CNND	1811	285	201	285	216	118	159	127	134	192	144
Stainless steel	× 15 CN	1808	314	219	314	236	128	184	138	146	209	157

Mechanical joints

The safety factor ν is closely related to the type of structure and to the loading condition.

The corresponding values, given in Fig. 5.17, are referred to the following values of ν:

$\nu = 1.70$ principal loading condition

$\nu = 1.50$ secondary loading condition

5.3.2 Shear Failure of the Bolts

If a bolt is subjected to a shear force F_V (Fig. 5.12) then the average shear stress, assuming uniform distribution of stresses at the contact between the plate and the bolt and no secondary bending stresses, is equal to:

$\tau = \dfrac{F_V}{A}$ shank in contact with the plates of the joint

$\tau = \dfrac{F_V}{A_{res}}$ threaded part in contact with the plates of the joint

A nominal area of the shank
A_{res} resisting area defined by Eq. 5.2

It has to be checked that

$$\tau \leq f_{d,V} \qquad (5.9)$$

when the semi-probabilistic method is used ($f_{d,V}$ being defined by Eq. 5.5), or that

$$\tau \leq \tau_{adm,b} \qquad (5.10)$$

when the allowable stress method is used (and $\tau_{adm,b}$ is defined by Eq. 5.7).

5.3.3 Tension Failure of the Bolts

If a force F_N is applied to the shank of a bolt, it tends to lengthen the bolt and to separate the plates which are connected and are under compression due to the tightening force N_s (Fig. 5.18).

When the load F_N is applied the force acting on the shank of the bolt is increased by a magnitude X and the compression preload is decreased by a magnitude Y. The equilibrium condition is

$$X + Y = F_N \qquad (5.11)$$

Aluminum alloy structures

Fig. 5.18

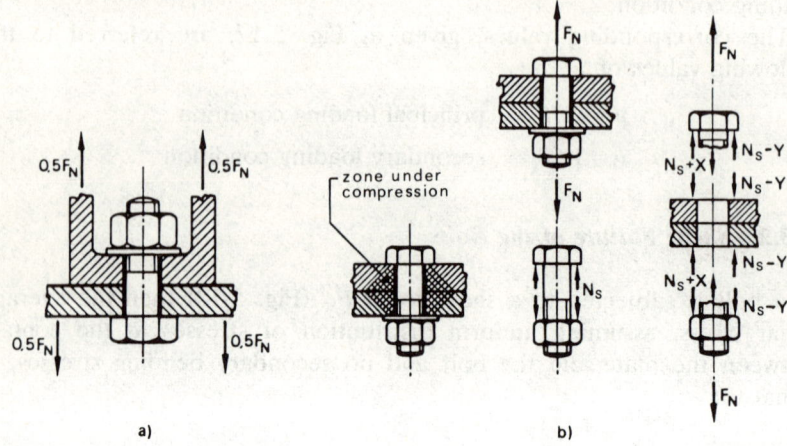

The plates do not separate if

$$Y \leq N_s \tag{5.12}$$

whereas in the case in which $Y > N_s$ only the bolt carries the external load F_N.

Some experimental results showed that usually the increment X of the tension force in the bolt is equal to 10 percent of F_N:

$$X = 0.1 F_N \qquad Y = 0.9 F_N \tag{5.13}$$

Therefore the plates begin to separate when

$$F_N \geq 1.1 N_s \tag{5.14}$$

This load can be assumed as a reference value for the serviceability limit state defined in Section 5.2.2 (a)(ii), and is strictly related to the magnitude of tightening.

The ultimate tensile capacity of the bolt is independent of the tightening and is checked with respect to the average nominal tension:

$$\sigma = \frac{F_N}{A_{res}} \tag{5.15}$$

It has to be checked that

$$\sigma \leq f_{d,N} \tag{5.16}$$

when the semi-probabilistic method is used, or that

$$\sigma \leq \sigma_{adm,b} \tag{5.17}$$

if the allowable stress method is adopted. $f_{d,N}$ and $\sigma_{adm,b}$ are defined by Eqs 5.6 and 5.8, respectively.

5.3.4 Bolts Subjected to Shear and Tension

When tension and shear act together on a bolt a failure criterion has to be chosen in order to verify the adequacy of the bolt. Several tests carried out on steel and aluminum alloy bolts determined a failure locus which can be defined by a quadratic curve (Fig. 5.19).

The relationship to be satisfied is therefore

$$\left(\frac{\tau}{f_{d,V}}\right)^2 + \left(\frac{\sigma}{f_{d,N}}\right)^2 \leq 1 \qquad (5.18)$$

if a semi-probabilistic method is used, or

$$\left(\frac{\tau}{\tau_{adm,b}}\right)^2 + \left(\frac{\sigma}{\sigma_{adm,b}}\right)^2 \leq 1 \qquad (5.19)$$

if the allowable stress method is used. The parameters are defined in Sections 5.3.2 and 5.3.3.

The European recommendations assume as an equivalent to the combined tension and shear load the value

$$F = \sqrt{(F_N^2 + \chi F_V^2)} \qquad (5.20)$$

$\chi = 2$ steel bolts

$\chi = 2.8$ aluminum bolts

The force F has to be checked against the design strength of the bolt F_d:

$$F \leq F_d \qquad (5.21)$$

where

$$F_d = f_{d,b} A_{res} \qquad (5.22)$$

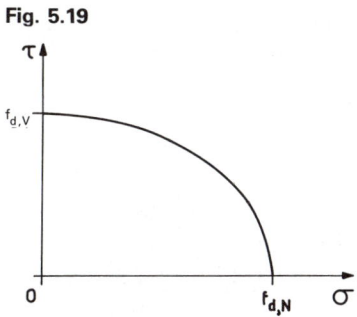

Fig. 5.19

Aluminum alloy structures

The criteria expressed by Eqs 5.18 and 5.21 are identical if we assume the values 1/0.6 and 1/0.7 for γ_V for aluminum alloy bolts and steel bolts, respectively, and $\gamma_N = 1$.

5.3.5 Bearing Stresses

The design strength with respect to bearing (Fig. 5.14b) and shear of the plate (Fig. 5.14c) depends on the distance of the bolt from the free edge of the plate in the direction of the force acting on the bolt. Therefore, the behavior of a joint in which the plates are subjected to compression (Fig. 5.20a) is different from that in which the plates are subjected to tension (Fig. 5.20b).

However, in both cases a conventional value of the bearing stress between the bolt and the plate is assumed. The actual distribution of the bearing stress is given in Fig. 5.20c in the elastic range, and Fig. 5.20d gives the distribution of bearing stresses in the elastoplastic range. However for computation a nominal average value is required (Fig. 5.20e), and this is the value that experimental results are referred to.

If t_{min} is the total thickness of the plates for one direction of contact pressure (referring to the symbols of Fig. 5.20, t_{min} is the lesser value of t_3 and $t_1 + t_2$), then the bearing design strength of the plates can be computed using the formula:

$$V_{d,b} = \alpha f_d d t_{min} \tag{5.23}$$

The value of the limiting bearing load is related to the design strength of the material f_d through a coefficient α (≥ 1), which is a function of the distance a of the holes from the free edge in the direction of the force applied to the bolt (see Fig. 5.11). This coefficient can be assumed equal to

$\alpha = 2$ to 2.5   plates under compression (Fig. 5.20a)

$\alpha = \alpha_1(a)$ plates are under tension (Fig. 5.20b): as in Fig. 5.21

$\alpha = \alpha_2(a)$ no bending in the bolts, and hole assumed to deform plastically: as in Fig. 5.21

It has to be checked that

$$F_V \leq V_{d,b} \tag{5.24}$$

where F_V is the force transmitted by the plate of thickness t_{min} corresponding to the design loading condition.

If the allowable stress method is used, it is sufficient to substitute σ_{adm} for f_d in Eq. 5.23.

Mechanical joints

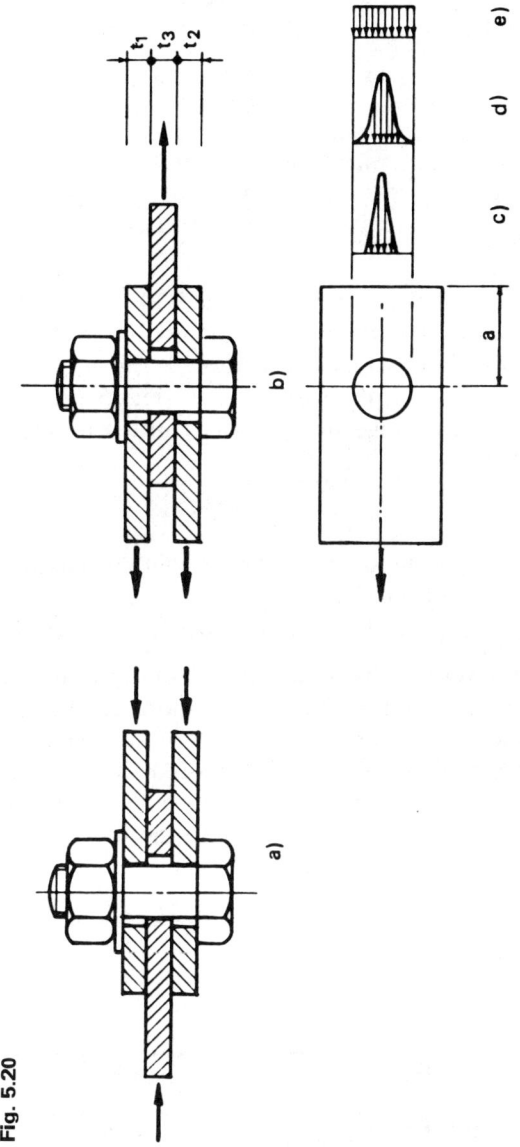

Fig. 5.20

Aluminum alloy structures

Fig. 5.21 α coefficient for bearing shear connections

	α_1	α_2
$a \geq 3d$	2.0	2.5
$3d > a \geq 2d$	1.5	2.0
$2d > a \geq 1.5d$	1.3	1.8
$1.5d > a \geq 1.2d$	1.0	1.7

5.3.6 Net Area

Failure of the plates of the connection (Fig. 5.14d) due to tension is also determined by means of a conventional method. The elastic distribution of stresses in a plate in proximity to the hole is shown in Fig. 5.22a. The redistribution of stresses in the plastic range (Fig. 5.22b) conventionally permits the use of an average strength value in tension equal to:

$$V_{d,t} = f_d A_{nom} \qquad (5.25)$$

where

f_d value of the design strength of the material of the plates

$A_{nom} = t_{min}(b - \phi)$ area of the critical section

When there are several bolts, definition of the critical section may be difficult. It is determined on the basis of the ultimate capacity of the plate subjected to tension and shear as a function of the possible lines of failure (Fig. 5.23a). An empirical rule, which has been shown as always safe, is that failure occurs along the minimum line going through several holes. For example the critical section of the plate shown in Fig. 5.23b is that produced by the minimum value of area from

$$2L_1 + 2L_2 \qquad 2L_1 + 2L_3 + L_4 \qquad 2L_1 + 2L_3 + 2L_5$$

Fig. 5.22

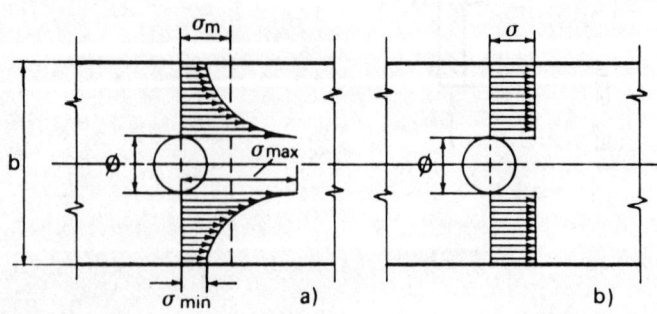

Mechanical joints

Fig. 5.23

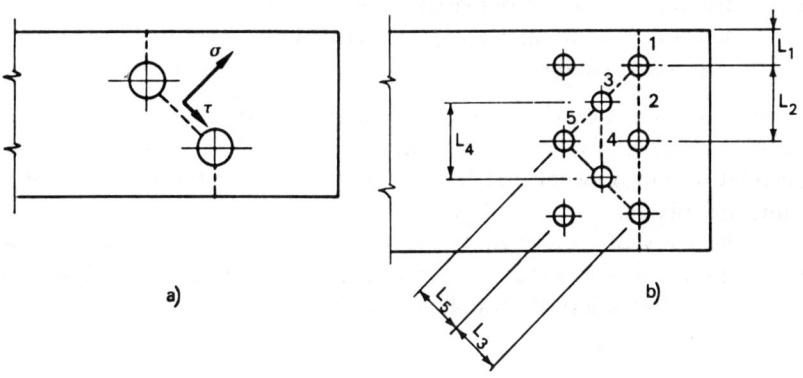

The value of $V_{d,t}$ (Eq. 5.25) must be checked against the tension load F_V transmitted by the joint.

5.4 Friction joints with high-strength steel bolts

5.4.1 Friction Joints

The friction resistance of a bolted connection can be considered as a serviceability limit state (see Section 5.2.2). It is an obligation to verify the connection against slippage in the case of friction joints.

Friction joints are permitted only when high-strength steel bolts are used, because it is important not to have a decrease of the tension preload in the bolt. The effect of tightening and its control is therefore of extreme importance, as explained in Section 5.1.1.2.

In the case of bearing joints it may be unnecessary to check the joint against friction (see Section 5.3.5) although if it is important to produce small deformations in the structure then this check can also be made.

The design value of the friction strength $V_{f,0}$ is defined as the value of the load which allows the joint to slip. This value depends upon the roughness of the surfaces to be joined, which affects the friction coefficient, and upon the value of the compression force N_s in the shank caused by fastening. Since it is difficult to statistically quantify this value, an average value in conjunction with a safety coefficient is usually assumed.

The European recommendations suggest the use of a value from 1.0 to 1.30. Therefore we have

$$V_{f,0} = \frac{\mu N_s n_f}{\gamma_f} \qquad (5.26)$$

Aluminum alloy structures

n_f number of surfaces in contact
μ friction coefficient between these surfaces
N_s tension preload subsequent to tightening
γ_f corrective coefficient

Some experiments carried out by Hacquart and Molina in France and by Valtinat in Germany on the specimens shown in Fig. 5.24 demonstrated the relationship between the friction coefficient and the thickness of the connected plates.

On the basis of these results the European recommendations suggest the friction coefficient values for treated surfaces given in Fig. 5.25. Treated surfaces are those roughened with particular mechanical (blasting

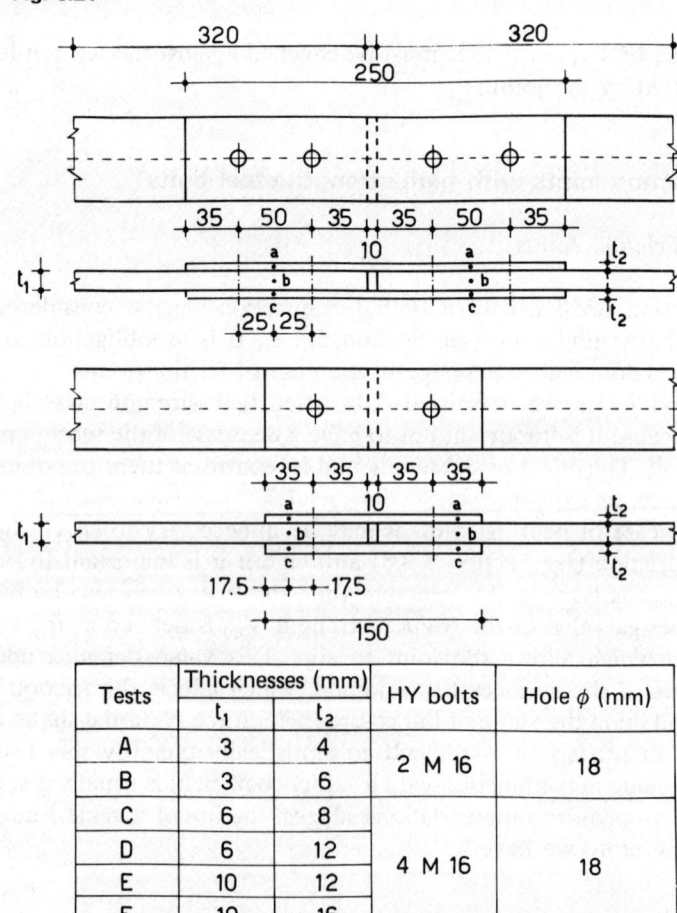

Fig. 5.24

Tests	Thicknesses (mm)		HY bolts	Hole ϕ (mm)
	t_1	t_2		
A	3	4	2 M 16	18
B	3	6		
C	6	8	4 M 16	18
D	6	12		
E	10	12		
F	10	16		

Fig. 5.25 Friction coefficient values

Thickness	Surfaces	
	Treated	Untreated
$t>10$ mm	0.45–0.55	0.30
$6<t<10$ mm	0.30–0.35	0.20
$3<t<6$ mm	0.20	0.15

of corindon, carborundum or sand) or chemical (zinc silicate or epoxy coating or resin) treatments. Values for untreated surfaces are also given; these values are also used in field connections.

It should be noted that the Poisson's ratio effect reduces the friction coefficient, especially in thin plates. In this case it is better not to use friction joints.

The values of $V_{f,0}$ for some types of high-strength steel bolts are given in Fig. 5.26 as a function of the diameter, with four different values of μ (0.15, 0.20, 0.30, 0.45) and with $\gamma_f = 1.10$ and 1.25.

The values of the friction coefficient can alternatively be computed experimentally following the procedure suggested in the European recommendations. This procedure consists of carrying out at least five tests, four of short duration and one of long duration, on specimens analogous to those shown in Fig. 5.27. The slippage of the plates a and c with respect to the upper plates b_{sup} and the lower plates b_{inf} has to be computed.

In the short-duration test, the load increases in increments equal to 10 to 20 kN/minute. The load corresponding to a relative slippage of 0.15 mm is calculated, and consequently the friction coefficient is given by

$$\mu = V_f/4N_s$$

where N_s is the tightening load.

The long-duration test is carried out with the same increment of load up to a value equal to 90 percent of the average of the loads calculated from the short-duration test to produce a slippage of 0.15 mm. The specimen is held at this value of load for a period of three hours.

If the slippage recorded in this test is smaller than 2×10^{-3} mm, the specimen is unloaded and incrementally loaded in the same way as the short-duration test. The load V_f which corresponds to a slippage of 0.15 mm is then computed (cf. the short-duration test).

The test is considered positive if the standard deviation, corresponding to the average of the two values of load which causes a slippage of 0.15 mm, is smaller than 8 percent.

Fig. 5.26 $V_{f,0}$ for high-strength steel bolts

	8.8 bolts								10.9 bolts							
	Loading condition 1				Loading condition 2				Loading condition 1				Loading condition 2			
	$\gamma_f = 1.25$				$\gamma_f = 1.10$				$\gamma_f = 1.25$				$\gamma_f = 1.10$			
d \ μ	0.15	0.20	0.30	0.45	0.15	0.20	0.30	0.45	0.15	0.20	0.30	0.45	0.15	0.20	0.30	0.45
12	0.46	0.62	0.93	1.40	0.52	0.70	1.06	1.59	0.63	0.84	1.27	1.90	0.72	0.96	1.44	2.16
14	0.63	0.84	1.27	1.90	0.71	0.95	1.44	2.16	0.87	1.16	1.75	2.62	0.99	1.32	1.99	2.98
16	0.87	1.16	1.75	2.62	0.98	1.31	1.99	2.98	1.18	1.58	2.37	3.56	1.35	1.80	2.70	4.05
18	1.06	1.42	2.13	3.20	1.20	1.61	2.42	3.64	1.45	1.93	2.90	4.35	1.65	2.20	3.30	4.95
20	1.36	1.82	2.73	4.10	1.54	2.06	3.10	4.66	1.86	2.48	3.72	5.58	2.11	2.81	4.22	6.34
22	1.68	2.24	3.36	5.04	1.90	2.54	3.81	5.72	2.29	3.05	4.58	6.87	2.60	3.47	5.20	7.81
24	1.96	2.62	3.93	5.90	2.22	2.97	4.47	6.70	2.67	3.56	5.35	8.02	3.04	4.05	6.08	9.12
27	2.55	3.40	5.11	7.66	2.89	3.86	5.80	8.71	3.48	4.64	6.46	10.44	3.95	5.27	7.90	11.86

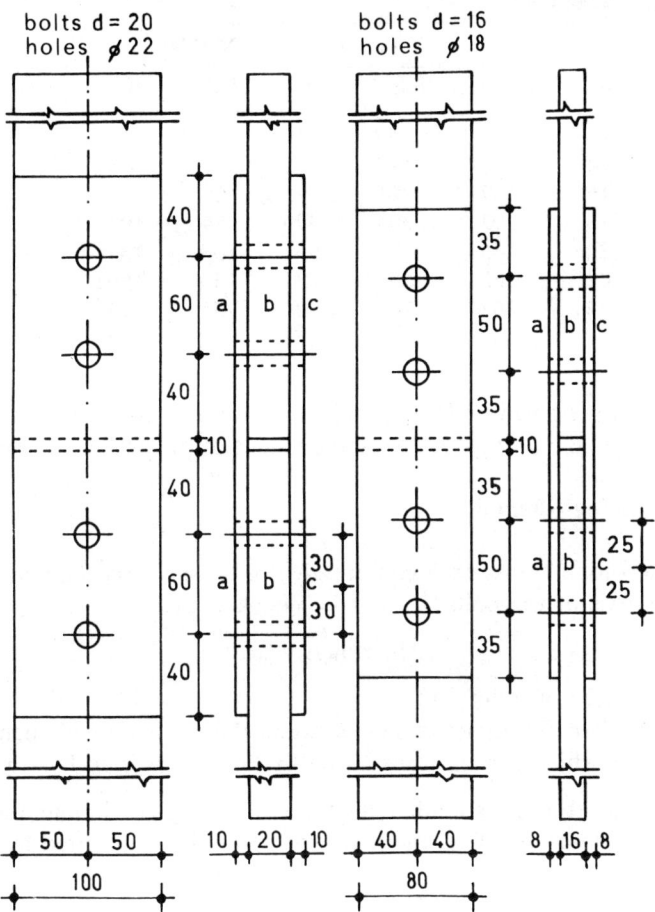

Fig. 5.27

If the result of the test is not positive, the test has to be repeated on a higher number of specimens, given by

$$n \geq \left(\frac{s}{2.5}\right)^2$$

where s is the standard deviation of the first five specimens.

If the long-duration test is not positive, it has to be proved, using at least three other specimens, that the load corresponding to the friction coefficient used in the calculations cannot cause slips higher than 0.3 mm during the whole life of the structure.

In the case of friction joints, tightening is to be undertaken to the

Fig. 5.28 Values of torque and preload for high-strength steel bolts

d	A_{res}	T_s (kNm) 8.8	10.9	N_s (kN) 8.8	10.9
12	84.3	89	110	37	46
14	115	142	176	51	63
16	157	221	275	69	86
18	192	304	379	84	105
20	245	431	537	108	134
22	303	587	731	133	166
24	353	746	929	155	193
27	459	1091	1358	202	252

values of torque given by Eq. 5.3. The values for high-strength steel bolts are given in Fig. 5.28 together with the values of the preload N_s (Eq. 5.1).

5.4.2 Slipping Strength

If a friction joint transmits a compressive or tensile force F_V perpendicular to the axis of the bolts, it must be checked that

$$F_V \leq n_b n_f V_{f,0} \qquad (5.27)$$

n_b number of bolts
n_f number of contact planes between the connected elements
$V_{f,0}$ load which can be transmitted by friction defined by Eq. 5.26.

When the joint is also loaded by forces causing tension in the bolts, without secondary bending, the load which can be transmitted by friction is equal to

$$V_{f,red} = V_{f,0}\left(1 - \frac{N}{N_s}\right) \qquad (5.28)$$

N axial load acting on the bolt
N_s tension preload

The load N must not exceed $0.8 N_s$.

5.5 Riveted shear connections

5.5.1 Design Strength

Riveted bearing connections are essentially used to resist loads which are perpendicular to the axis of these fasteners, and should not be used when

Fig. 5.29 Values of $f_{d,V}$ for some common materials

Rivets: designation	Condition	$f_{d,V}$ (N mm^{-2})	$\tau_{adm,r}$ (N mm^{-2}) Loading condition 1	Loading condition 2
Al99.0	F	38	22	25
Al99.5	H14	38	22	25
AlCu2.5MgSi	T4	104	61	69
AlCu4MgMn	T4	150	88	100
AlCu4.5MgMn	T4	167	98	111
AlMn1.2	H14	53	31	35
AlMn1.2Cu	H14	53	31	35
AlMg4.5	H32	116	68	77
AlMg5	H32	116	68	77
AlSi1MgMn	T4 or T6	95	56	63
AlMg1SiCu	T4 or T6	116	68	77
Steel Fe40		204	120	136
Stainless steel ×15 CN 1808		204	120	136

there are tension loads in the fasteners. Analogously to bolted bearing connections, the design strength of a rivet can be defined by

$$f_{d,V} = \frac{f_{d,r}}{\gamma_V} \quad (5.29)$$

where $f_{d,r}$ is the design strength of the material of the rivet, which can be defined as the lesser value of the yield stress and 70 percent of the ultimate strength, and γ_V is a safety coefficient.

The values of $f_{d,V}$ for some common materials are given in Fig. 5.29. From these values, the corresponding allowable shear stresses can be calculated:

$$\tau_{adm,r} = \frac{f_{d,V}}{\nu} \quad (5.30)$$

by introducing appropriate safety factors. If we use the values $\nu = 1.70$ (first loading condition) and $\nu = 1.50$ (second loading condition), the corresponding values of $\tau_{adm,r}$ are also given in Fig. 5.29.

5.5.2 Shear Failure of the Rivets

A rivet of area A loaded with a shear force F_V has an average shear stress equal to

$$\tau = \frac{F_V}{A} \quad (5.31)$$

Aluminum alloy structures

It has to be checked that

$$\tau \le f_{d,V} \quad \text{(semi-probabilistic method)}$$

or

$$\tau \le \tau_{adm,r} \quad \text{(allowable stress method)}$$

where $f_{d,V}$ and $\tau_{adm,r}$ are given by Eqs 5.29 and 5.30, respectively.

5.5.3 Bearing Stresses

The bearing stress between the shank of the rivet and the plate is considered in the same way as for bolted connections. Thus Eqs 5.23 and 5.24 have to be verified.

References

1. Steinhardt, O., Verbindungen bei Aluminium-Konstruktionen des Ingenieurbaus, *VDI-Zeitschrift*, **102,** No. 35, 1960, pp. 1729–42.
2. Valtinat, G., Untersuchungen zur Festlegung zulassiger Spannungen und Krafte bei Niet-Bolzen-und HV-Verbindungen aus Aluminiumlegierungen, *Aluminium*, **47,** H. 12, 1971, pp. 735–40.
3. Molina, C. and Chailleux, G., *Cornières et groussets en alliages d'aluminium. Détermination de la pression diamétrale limite en fonction des pinces d'extremités intermédiaires et latérales pour assemblages boulonnès en A-SGM-T66* (*boulons en acier*), Rep. No. 1159A, Centre Téchnique de l'Aluminium, Paris, 1972.
4. Molina, C. and Chailleux, G., *Cornières et groussets en alliages d'aluminium. Influence des boulons en 7075 T73 sur la pression diamétrale admissible dans les assemblages boulonnés en fonction des pinces et écartments déterminés entre trous et boulons,* Rep. No. 1209, Centre Téchnique de l'Aluminium, Paris, 1972.
5. Molina, C. and Chailleux, G., *Cornières et groussets en alliages d'aluminium. Determination de la pression diamétrale limite en fonction des pinces d'extremités et intermédiaires pour assemblages boulonnés en 7020 T6* (*boulons en acier*), Rep. No. 1214, Centre Téchnique de l'Aluminium, Paris, 1972.
6. Molina, C. and Chailleux, G., *Cornières et groussets en alliages d'aluminium. Détermination de la pression diamétrale limite en fonction des pinces d'extremités et intermédiaires pour assemblage boulonnés en 6005 A* (*boulons en acier*), Rep. No. 1329, Centre Téchnique de l'Aluminium, Paris, 1974.
7. Fisher, J. W. and Struik, J. H. A., *Guide to Design Criteria for Bolted and Riveted Joints,* John Wiley & Sons, New York, 1974.
8. Molina, C. and Hacquart, R., *Emploi des boulons à haute résistance à serrage controlé,* Rep. No. 1459, Centre Téchnique de l'Aluminium, Paris, 1976.

6 Strength of structural elements

6.1 Serviceability limit state

6.1.1 Influence of Deformation

The deformation of structures can be mainly grouped as axial, bending and shear.

Axial deformation is usually negligible in members subjected to pure or eccentric bending. On the other hand it cannot be neglected in the case of structures in which the members are primarily subjected to tension or compression.

Shear deformation is usually negligible in girders, although it must be taken into account in trusses.

Bending deformation is always greater than the other deformations and must always be taken into account in beams.

All deformations have to be limited to acceptable values in order to permit the use of the structures. A serviceability limit state check has to provide these bounds of utilization of the structure. This limit state is particularly severe for aluminum alloy structures because the material has a small value of Young's modulus together with relatively high values of strength.

When this serviceability limit state is not complied with, the functionality of the structure may be compromised by excessive values of displacement (equipment, floor, partitions, walls etc.). The reliability of the calculation method may also be affected because the structure may experience large deflections, requiring a second-order analysis.

However, the load-bearing capacity of the structure is independent of the attainment of this serviceability limit state. Therefore specifications do not usually impose these limits – they only suggest the bounds which vary with respect to the use of the structure.

These displacement limitations are usually given as percentages of the span of the beam L. The most significant values are:

$L/200$ for the total midspan deflection due to the dead and live loads on the floor beams

Aluminum alloy structures

$L/400$ for the midspan deflection due to the live loads on the girders
$L/500$ for the horizontal deflections of tall structures (multistorey frames, high-rise buildings, towers, masts, trusses) due to the wind action (in this case L is the total height of the structure)
$L/700$ for the deflection due to static loads on crane girders

In the case of cantilever beams, the same bounds are referred to an effective length L equal to twice the span of the cantilever.

6.1.2 Deformation of Plate Girders

In the case of girders it is of major concern to limit the effects of bending deformation, but the shear deformation can be neglected.

The maximum midspan deflections are computed using the usual formulas for elastic beams. In the case of a simply supported beam of span L and with a uniform load, the midspan deflection is given by

$$v_{max} = \frac{5}{384}\frac{qL^4}{EI} \tag{6.1}$$

It has to be checked that

$$v_{max} \leq \alpha L \tag{6.2}$$

where α is the limitation given in the form of a v/L ratio (1/200, 1/400, 1/500, 1/700 etc.).

If the allowable stress method is adopted, the same design loads are used for checking strength and deformation.

If the limit state method is used the enhancement factors γ, which define the design combination of loads and which are larger than 1 (1.3 or 1.5) when checking against the ultimate limit state, are made equal to 1 when checking against the serviceability limit state (see Section 3.4.4).

A qualitative evaluation, which shows the degree of influence of deformation checks on the ultimate capacity of a bent structural member, can be made by examining the elementary case of the simply supported beam subjected to uniform load. Under the allowable stress method, it is checked that the maximum stress

$$\sigma_{max} = \frac{M}{I}\frac{h}{2} = \frac{qL^2}{8}\frac{h}{2I} \tag{6.3}$$

is less than the allowable stress:

$$\sigma_{max} \leq \sigma_{adm} \tag{6.4}$$

If the q/I ratio is eliminated from Eqs 6.1 and 6.3, and it is assumed that

Fig. 6.1 L/h ratios

α \ σ_{adm}	100 N mm^{-2}	150 N mm^{-2}	200 N mm^{-2}	250 N mm^{-2}
1/200	16.70	11.20	8.35	6.70
1/400	8.35	5.55	4.15	3.35
1/500	6.70	4.50	3.35	2.70
1/700	4.75	3.15	2.35	1.90

$E = 70.000$ N mm^{-2}, then

$$\frac{v_{max}}{L} = 3 \times 10^{-4} \times \frac{L}{h}\left(\frac{\sigma_{max}}{100}\right) \tag{6.5}$$

where σ_{max} is given in N mm^{-2}.

If both deformation (Eq. 6.2) and strength (Eq. 6.4) requirements are satisfied then from Eq. 6.5:

$$\alpha = \frac{L}{h}\left(\frac{\sigma_{adm}}{100}\right) \times 3 \times 10^{-4} \tag{6.6}$$

Equation 6.6 shows that if deformation and strength requirements are satisfied together then no particular value of inertia is required. Instead a minimum value of the height is required with respect to the span of the beam. In the light of these considerations, and neglecting local and global buckling phenomena, slender profiles are to be preferred because they minimize the weight with respect to height.

The limiting values of L/h ratios, as functions of the strength given by the value of the allowable stress σ_{adm} and by the limitation on deformation α, can be calculated from Eq. 6.6, and the corresponding values are given in Fig. 6.1.

It should be noted that these values greatly penalize the structure because of the large σ_{adm}/E ratio, thus limiting the performance of the structure; it cannot fully utilize the load-bearing capacity.

In the case of bent large-span aluminum structures, large values of camber – larger than those of steel structures – should be used in order to compensate for dead load effects at least. In the case of continuous beams, with uniform load acting on all the spans, L/h values which satisfy the limitations valid for simply supported beams (Fig. 6.1) may be increased by 60 percent for the intermediate spans and by 30 percent for the end spans.

6.1.3 Deformation of Trusses

In the case of trusses (V, K, X diagonal types), or in the case of open-web expanded girders, shear deformations cannot be neglected because they affect the total deflection (by up to 30 percent). As a rule of thumb, an increase of deflection equal to one-third of that due to the bending moment may be assumed. If the classical methods of the theory of elasticity are used (e.g. strain energy methods) the correct elastic deformation of the truss can be obtained.

The approximate method of the equivalent web, suggested by French specifications (DTU, règles Al), can be considered as an intermediate one. It consists of defining a fictitious continuous web with cross-section A_w equivalent to the actual web from the point of view of shear deformation. The equivalent area is provided if the transverse deformation of a panel of length L_0 under the shear load V, that is

$$y_V = \frac{VL_0}{GA_w} \qquad (6.7)$$

is made equal to that of an equivalent girder. The corresponding midspan deflection

$$v_V = \frac{M_0}{GA_w} \qquad (6.8)$$

is added to the deflection v_M calculated for a girder with a moment of inertia equal to two concentrated masses which represent the chords of the truss. Hence, the total deflection will be given by

$$v = v_V + v_M \qquad (6.9)$$

In the most common cases of trusses and open-web girders, the French specifications provide expressions for the equivalent web A_w. The most significant are given here, with the notation of Fig. 6.2:

V symmetrical truss (Fig. 6.2a):

$$A_w = 1.3 A_d \frac{L_0 h^2}{L_d^3} \qquad (6.10)$$

N truss (Fig. 6.2b):

$$A_w = \frac{2.6 A_d \dfrac{L_0}{h}}{\dfrac{A_d}{A_t} + \dfrac{L_d^3}{h^3}} \qquad (6.11)$$

Open-web girder (Fig. 6.2c):

$$A_w = 31 \frac{I_t(I_1 + I_2)}{I_t L_0^2 + L_0 h (I_1 + I_2)} \qquad (6.12)$$

Strength of structural elements

Fig. 6.2

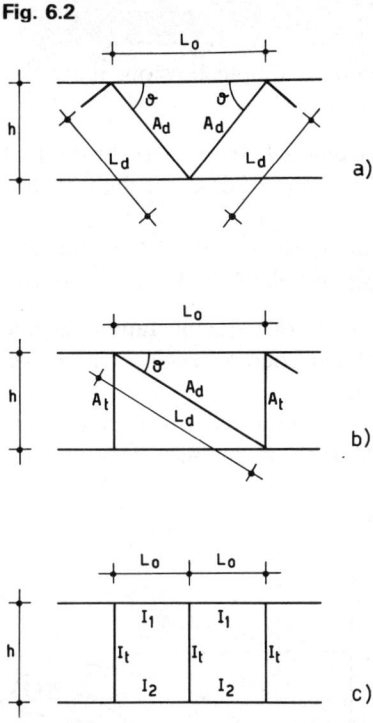

6.2 Ultimate limit state

Usually when a component reaches its load-bearing capacity 'rupture' is considered to have occurred. However, it is important whether brittle or ductile failure occurs.

Ductile failure can be described as due to a noncontrolled deformation which occurs before collapse and represents a warning sign.
Brittle failure occurs suddenly without prior plastic deformations.

The principal aim of limit state design is to identify each ultimate state of a structural element by characterizing the ductile or brittle behavior. The necessary checks are indicated with respect to the particular structural element and the way it is loaded and joined to other members.

A qualitative examination of the possible phenomena which correspond to an ultimate limit state for different structures will be given in the following sections with respect to the principal load acting in the bar. The ultimate state may be attained in a stable or an unstable way. The first case will be treated in this chapter, and unstable behavior will be extensively referred to in Chapter 7.

6.2.1 Tension

When a bar is subjected to simple tension, it reaches the ultimate limit state in the following cases:

(a) When the design strength f_d (usually coincident with the conventional elastic limit $f_{0.2}$ of the material) is reached in the cross-section of the bar of area A
(b) When rupture occurs in the net area A_n of the bolted connection or in the reduced strength zones of the welded connection A_{red}

The first case corresponds to a ductile failure because the material can elongate in the inelastic hardening region before reaching the ultimate stress f_t (case 1, Fig. 6.3).

The second case corresponds to failure of the connection, which can occur with different degrees of ductility depending upon the connection

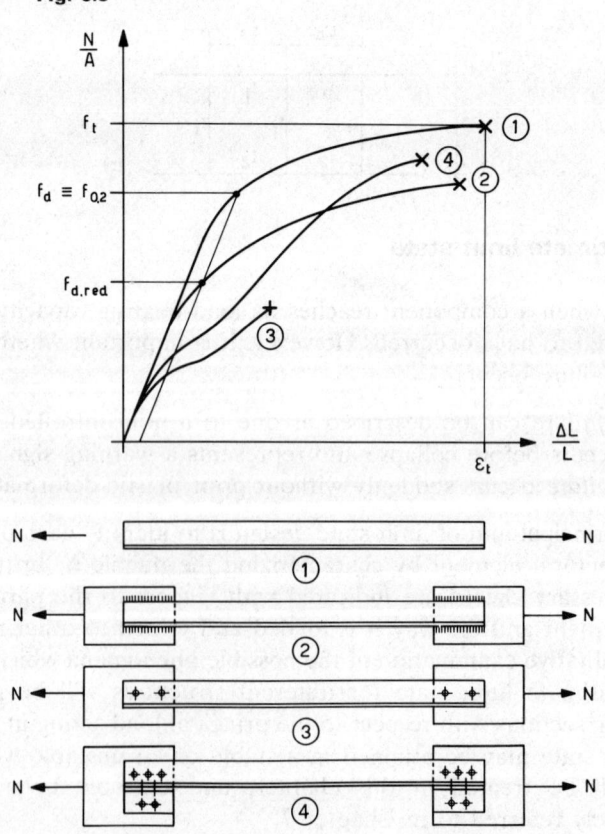

Fig. 6.3

itself (cases 2, 3 and 4 of Fig. 6.3). It is brittle in the case of bolted connections (case 3 of Fig. 6.3), in which rupture occurs when the bar is still in the elastic range: these connections are therefore defined as 'partial restored strength'. Those connections in which the failure load of the connection is equal to that of the elastic limit of the connected bar are termed 'complete restored strength'.

Welded connections can be considered as complete restored strength only in the case of non-heat-treated alloys, which are not affected by the heat input of the weld (no reduced-strength zones). Heat-treated alloys, in contrast, are to be considered as partial restored strength depending upon the extent of the reduced-strength zones. In both cases the ductility of the welded connection is sufficient. In the case of shear bolted joints, complete restoring of strength can be obtained, together with an elongation capacity, by connecting a sufficient number of bolts to all the parts of the cross-section (case 4 of Fig. 6.3).

The condition to be satisfied in order to have a complete restored joint in a bar subjected to tension is

$$R_b \geq N_d \qquad (6.13)$$

where R_b is the lesser of the values of the shear failure of the bolts and the bearing failure of the connected plate, and $N_d = f_{0.2} A$ is the design strength of the bar.

The following equation has to be satisfied together with Eq. 6.13:

$$R_t \geq N_d \qquad (6.14)$$

in order to guarantee failure in the section which is highly stressed by the bolted connection. R_t represents the collapse load of the net section and can be expressed as

$$R_t = \alpha f_t A_n \qquad (6.15)$$

f_t ultimate stress of the material
A_n net area of the cross-section
α a cocfficient less than 1 which takes into account the reduction in strength, with respect to the theoretical value, due to stress concentration along the holes and secondary stresses. This value has to be determined by tests.

We can assume $\alpha \simeq 1$ when the connection is designed, so that secondary stresses are eliminated (when the bar has double symmetry with respect to the gusset plate, thus eliminating eccentricities due to the shape of the cross-section between the axis of the bar and the center of the reaction of the support). In the case of asymmetrical connections the value of α can be considered to lie between 0.8 and 0.9.

Conditions 6.13 and 6.14, which guarantee complete restoration of the resistant cross-section, can be written in the form:

$$\begin{cases} R_b \geq f_{0.2} A \\ \alpha f_t A_n \geq f_{0.2} A \end{cases}$$

The first equation implies that rupture occurs in the bar, and the second provides plasticity in the bar before collapse of the connection. From the second equation the following is obtained:

$$\alpha \geq \frac{f_{0.2}}{f_t} \frac{A}{A_n} \qquad (6.16)$$

whereas the A_n/A ratio in a bolted joint can be considered to lie between 0.8 and 0.9, the $f_t/f_{0.2}$ ratio is dependent upon the type of alloy and varies between 1.1 and 2.5.

From Eq. 6.16 the following classification is derived:

(a) For $f_t/f_{0.2}$ ratios less than 1.25, it is advisable not to use any type of bolted connection because they are unable to provide the ductility requirements typical of complete restored joints.
(b) For $f_t/f_{0.2}$ ratios greater than 1.25 and less than 1.50, it is possible to obtain a complete restored joint provided that the secondary stresses are eliminated.
(c) For $f_t/f_{0.2}$ ratios greater than 1.50, the connection is ductile enough to be considered as complete restored even though it is not symmetrical.

From the examination of commercial aluminum alloys it can be observed that:

5000 series alloys fall in categories (b) and (c), allowing complete restored connections.

2000, 6000 and 7000 series alloys usually fall in category (b), even though in some cases they fall in category (a) and hence cannot be used in ductile structures with bolted connections. In fact, when $\alpha < 1$, tension members connected with bolts can reach the elastic limit of the material although failure occurs before the inelastic range is reached.

6.2.2 Bending

The ultimate limit state of a beam can occur in different circumstances depending upon the geometry of the beam (the span L, the L/h ratio of the individual parts etc.), the loading and support conditions and the type of connection [1–6].

In the case of beams with a compact cross-section, in which local buckling or flexural torsional buckling are not likely to occur, the beam experiences the inelastic range after reaching the limiting elastic moment

Fig. 6.4

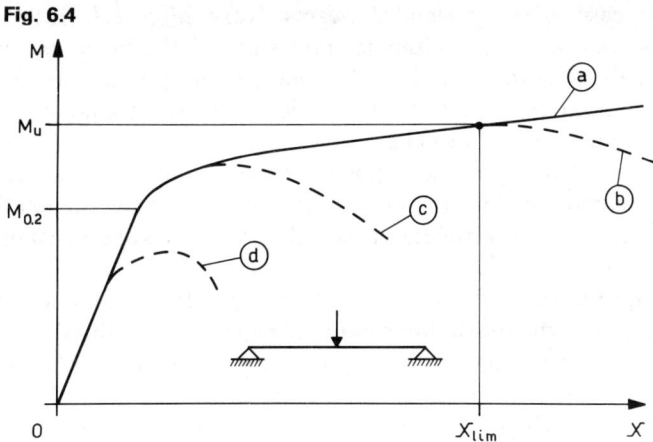

$M_{0.2}$ until the ultimate moment M_u is reached (Fig. 6.4a). This moment cannot be defined (as it is for steel structures) as the full plastic moment. In fact, due to the hardening behavior of the σ–ε law of aluminum alloys, a limiting curvature χ_{lim} has to be defined corresponding to the limit of large deformations in the inelastic range. The increase in strength, $M_u - M_{0.2}$, obtained in this phase can be quantified through a relation

$$M_u = \alpha M_{0.2}$$

which defines – in a generalized form – a new shape factor α which is not solely dependent upon the cross-sectional geometry, as is usual, but also depends upon the parameters of the σ–ε law ($f_{0.2}/E$ ratio, n exponent) and upon the definition of limiting curvature χ_{lim} (see Section 6.3.4.6).

In the case of open profiles, local buckling phenomena are most likely to occur (see Section 7.5) in the compressed regions of the cross-section and cause a decrease in the M–χ curve of the beam. This unstable behavior is dependent upon the b/t ratio (Fig. 6.4). If the decreasing portion of the curve occurs after the ultimate moment M_u is reached (Fig. 6.4b), the beam keeps the same maximum load-carrying capacity. In this case the rotational capacity of the cross-section, which characterizes the flexural ductility of the beam, allows redistribution of the internal actions, and it is therefore possible to carry out a limit analysis of the whole structure (see Section 6.4).

If the decreasing portion of the curve occurs before the ultimate moment M_u is reached (Fig. 6.4c), or even before the elastic moment $M_{0.2}$ (Fig. 6.4d), the load-carrying capacity of the beam is affected by local buckling phenomena, to a higher degree if the b/t ratio is large (e.g. thin profiles). Also ductility decreases to the extent that redistribution of internal actions cannot be considered.

Aluminum alloy structures

In the case of very slender beams (very high L/h ratio) without transverse restraints, the ultimate limit state of the beam can be determined by flexural–torsional buckling phenomena (see Section 7.3); these reduce the load-carrying capacity of the beam by an amount dependent upon the values of slenderness.

Failure of a beam can also happen because of failure of an intermediate or edge connection. This failure will be brittle or ductile depending upon the connection, with complete or partial restored flexural resistance of the beam.

The required ductility of a connection depends upon its position along the beam. Since the maximum rotational capacity is required at the end of a beam, the edge connections (beam–beam connections or beam–column connections) must be sufficiently ductile to guarantee the redistribution of internal actions when the stresses are in the inelastic range ($M > M_{0.2}$), in order to reach the maximum load-carrying capacity of the cross-section (M_u). Intermediate joints are used to connect the different sections of a beam constructed in more than one part. They usually correspond to the sections in which the bending moment is small (in close proximity to points of zero moment) and therefore do not need to exhibit complete restored flexural resistance.

6.2.3 Shear

Shear action usually occurs in conjunction with bending moment in beams and can cause collapse of the members, especially in the case of deep beams without stiffeners, in which the following unstable phenomena can occur:

(a) Local buckling of the web along the beam with consequent plane stress behavior and formation of diagonal stress fields (Fig. 6.5)
(b) Crippling of the web in positions corresponding to concentrated loads or support reactions (Fig. 6.6)

Fig. 6.5

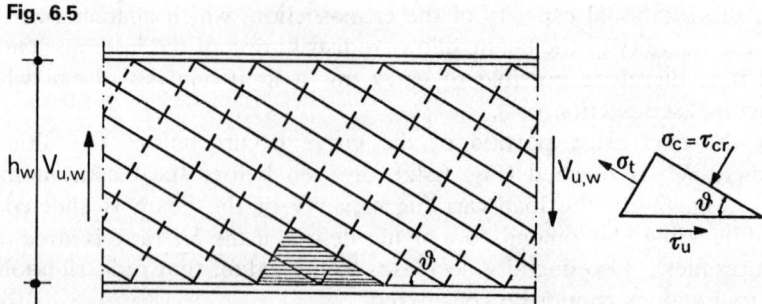

Strength of structural elements

Fig. 6.6

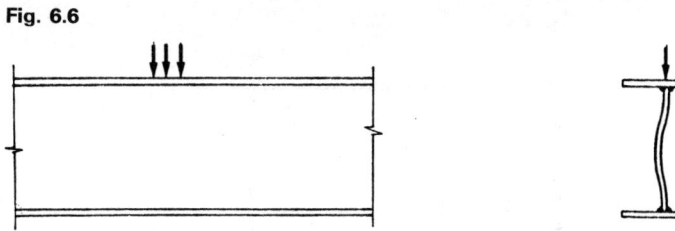

When stiffeners are present, the structure can exhibit additional resistance in the post-critical range.

From the constructional point of view, two systems can be identified:

(1) Webs with transverse and longitudinal stiffeners
(2) Webs without intermediate stiffeners along the span

In the first (traditional) system a critical state is characterized by local buckling of the web panel, with the flanges and the transverse stiffeners remaining rigid. This state can be identified provided that the appropriate values of extensional and flexural rigidities of the stiffeners are guaranteed (optimal relative rigidity: see Section 7.5.3.3). The critical state of the panel is idealized by diagonal bands of tensile stresses (Fig. 6.7a). Then, in the post-critical range, the beam can be approximated by a triangulated structure in which the flanges represent the chords, the verticals are given by the transverse stiffeners and the bands of tension stresses represent the diagonals (Fig. 6.7b).

In these beams two limit states can be identified:

A deformation limit state corresponding to local buckling of the panel
An ultimate limit state corresponding to collapse of the truss

Fig. 6.7

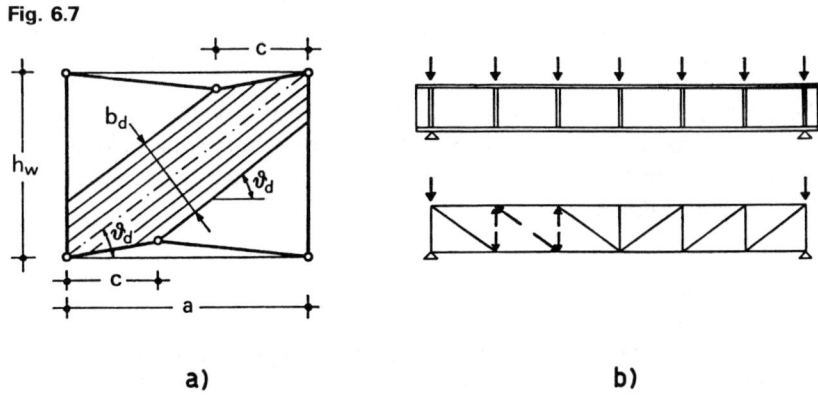

a) b)

Aluminum alloy structures

Fig. 6.8

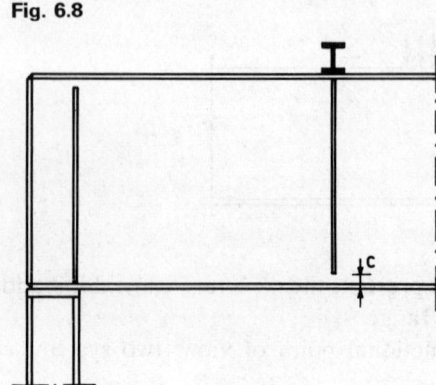

The first limit state, which represents the maximum load before large transverse deformations of the panel, can be considered a serviceability limit state because local buckling of the panels can cause psychological and esthetical inconvenience without compromising the load-carrying capacity of the beam.

The second system, which seems to be more convenient on the basis of research carried out in the USA and Sweden, requires stiffeners only at supports and at positions of concentrated loads. The design of these beams, on the basis of steel specifications (AISC, Swedish code), guarantees that local buckling will not occur at positions of concentrated loads (Fig. 6.8) even though it allows the web to buckle because of normal stresses due to shear and bending (Fig. 6.5). The ultimate limit state is reached when the shear value is given by

$$V_u = V_{u,w} + V_{u,f} \tag{6.17}$$

$V_{u,w}$ ultimate value of shear due to plane stress conditions in the web
$V_{u,f}$ ultimate value of shear which the flanges resist because of their flexural rigidity.

The post-critical behavior which provides the $V_{u,w}$ value can be considered as corresponding to a system of orthogonal bars whose slope varies with the load. In fact, since there are no stiffeners, the truss model is not applicable any more.

6.2.4 Compression

The load-carrying capacity of a compressed member is nearly always affected by buckling phenomena (see Section 7). The collapse modes

Fig. 6.9

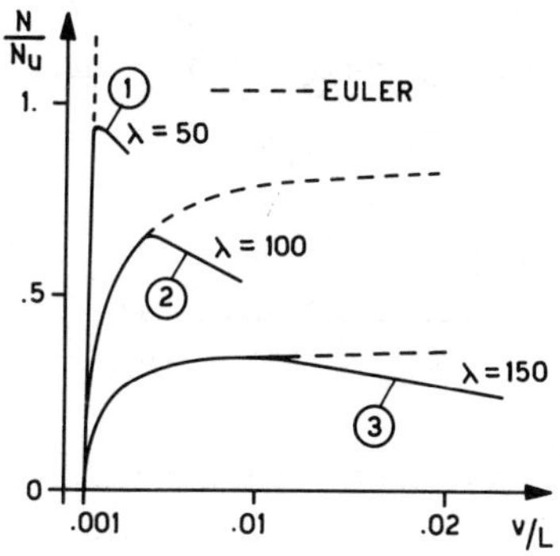

which correspond to the ultimate limit state can be of various types:

(a) Columns of medium–low slenderness ($\lambda \leq 50$) suddenly fail without going into the elastic/plastic and large-deflection range (curve 1 of Fig. 6.9).
(b) Columns with high values of slenderness ($\lambda \geq 150$), when close to the collapse load, experience an elastoplastic range with small increments of the load until the limiting stress is reached in the extreme fibers. At this point, the load decreases and the phenomenon follows an unstable curve in the post-critical range, with large deformations before complete failure (curve 3 of Fig. 6.9).
(c) Beam columns behave as in case (b), especially for large values of eccentricity (Fig. 6.10).
(d) Asymmetrical columns, which buckle with flexural–torsional modes, exhibit a sudden loss of load-carrying capacity when torsional rotations occur.

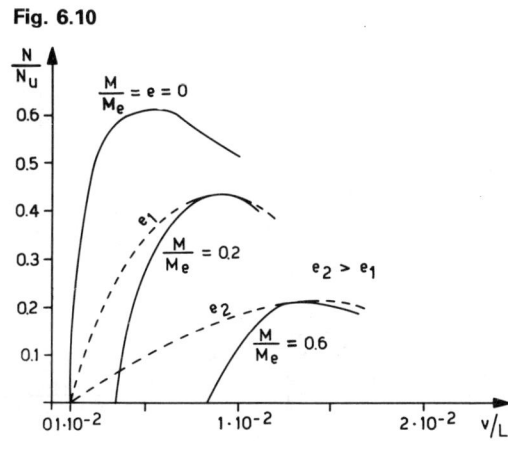

Fig. 6.10

Aluminum alloy structures

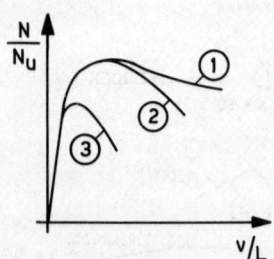

Fig. 6.11

(e) Columns in which local buckling of the cross-section occurs exhibit a sudden decrease of the load dependent upon the magnitude of local buckling phenomena (see Fig. 6.11).

Curve 1 of Fig. 6.11 corresponds to an absence of local buckling phenomena; curve 2 represents the case in which local buckling occurs in the large-deformation range. Thus load-carrying capacity of the bar is unaffected. Curve 3, in contrast, reaches maximum load-carrying capacity at values considerably less than curves 1 and 2 because of the severe interaction between global buckling of the bar and local buckling of its parts.

If these behaviors are examined it can be observed that a ductile ultimate state can only be obtained in the case of slender columns (case b), preferably with eccentric loads (case c).

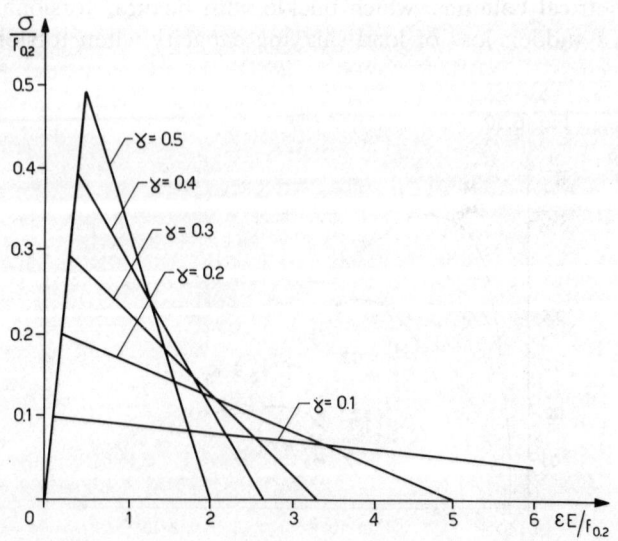

Fig. 6.12

The less is the slenderness, the more brittle is the behavior of stocky columns (case a). Brittle behavior can also be introduced by torsional (case d) and local buckling (case e) phenomena.

With respect to the interaction between slenderness and ductile behavior (cases a and b), Fig. 6.12 (proposed by Marsh [7]) schematically relates the normal stresses to the deformability of the bar ε as a function of the ratio $\gamma = \sigma_c/f_{0.2}$ which depends, among the other variables, upon the slenderness of the bar (σ_c is the failure stress of the bar, E is Young's modulus and $f_{0.2}$ is the elastic limit of the material).

6.3 Bending behavior of the cross-section

6.3.1 Definition of Limit States [8]

Calculation methods based upon the elastic analysis of structures are usually referred to a limit state of the cross-section called *conventionally elastic* (see Section 6.3.1).

One of the greatest limitations of this type of calculation is that it is not possible to calculate the inelastic capacities of the cross-sections which differ from each other. The importance of considering the inelastic behavior of the cross-section when checking a structure is evident if two cross-sections with different shape factors are considered. Under the same stress state (e.g. the allowable stress) they possess a different safety factor with respect to plastic failure.

This obvious consideration is taken into account by specifications which provide calculation methods for estimating the post-elastic behavior of the cross-section even though the more complex problem of plastic collapse is not dealt with (see Section 6.3.2). In fact these specifications introduce the plastic adaptation coefficients ψ. These coefficients amplify the section modulus of the cross-section by defining elastic limiting moments which are characterized by a permanent deformation of the cross-section instead of reaching the elastic limit of the material. The procedure is very similar to that used to define the conventional yield $f_{0.2}$ in aluminum alloys. In this way an elastic limit moment is defined for bent sections which can be associated with a new limit state, called *plastic adaptation limit state*. This is intermediate between the conventional elastic limit and the collapse limit state. Under this approach the plastic adaptation coefficient ψ can be considered as a reduced-shape factor ($1 < \psi < \alpha$), which represents a compromise between the elastic and the plastic analysis, in order to take more advantage of the strength of the materials even when using the 'elastic' design approach.

This methodology, which has long been used in steel specifications,

Aluminum alloy structures

cannot be used in aluminum structures without taking account of the actual inelastic behavior of the σ–ε law of aluminum alloys (see Section 6.3.2).

The *plastic collapse limit state*, in the case of aluminum alloy cross-sections, has a particular meaning because it does not depend solely upon the cross-section through the classical plastic modulus. Owing to the inelastic behavior of the material it requires, instead, a redefinition of the ultimate moment as a function of the alloy and the ductility requirements of the structure (see Section 6.3.3).

6.3.2 Conventional Elastic Limit State

6.3.2.1 Calculation method

As a consequence of the definition of the elastic limit of the material, conventionally identified as the value of the stress which corresponds to a deformation of 0.2 percent, a linear σ–ε law is assumed up to the $f_{0.2}$ value (Fig. 6.13). If this limit is reached in the highly stressed fibers of a section subjected to bending, a limit state (referred to as 'conventional elastic') is identified, which corresponds to the elastic moment

$$M_{0.2} = f_{0.2} W \qquad (6.18)$$

where W is the section modulus of the cross-section. Since perfectly elastic behavior is assumed below this value, the normal stresses can be computed with the relationships:

Symmetrical bending:

$$\sigma = \frac{M}{I} y \qquad (6.19)$$

Fig. 6.13

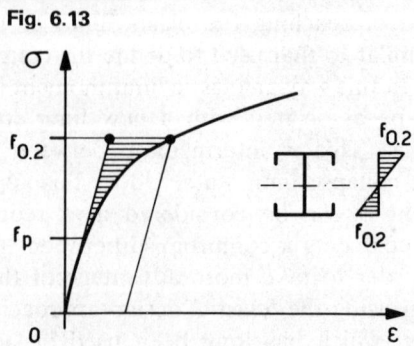

Asymmetrical bending:

$$\sigma = \frac{M_x}{I_x}y + \frac{M_y}{I_y}x \qquad (6.20)$$

When the moment of inertia is computed, the presence of holes has to be taken into account. As an approximation the equivalent moment of inertia (net) can be calculated by subtracting the moment of inertia of the areas of the holes with respect to the center of gravity of the cross-section.

When the allowable stress method is used, it has to be checked that

$$\sigma \leq \sigma_{adm} \qquad (6.21)$$

where

$$\sigma_{adm} = f_{0.2}/\nu$$

ν is an appropriate safety factor.

When the limit state method is used, it has to be checked that

$$\sigma \leq f_d \qquad (6.22)$$

f_d is usually assumed coincident with $f_{0.2}$.

In both cases the second terms of the inequalities are connected with the conventional elastic limit $f_{0.2}$.

It should also be noted that the difference between the two methods is that σ stresses are computed with service loads in Eq. 6.21, whereas in Eq. 6.22 the design combination of loads is used.

6.3.2.2 Residual deformation

The moment $M_{0.2}$ conventionally provides the limit of the elastic behavior. In fact, if we consider the real σ-ε law of the material, we observe that the limit of proportionality is smaller and that in the interval

$$f_p < \sigma < f_{0.2}$$

the stresses are in the inelastic range and cause residual deformations when the structure is unloaded (Fig. 6.14).

Since the stresses are in the inelastic range, a residual midspan deflection v_r can be observed whose magnitude depends upon numerous factors such as the type of structure, the loading and support conditions, the span length and cross-section of the beam and the type of alloy. Clearly the magnitude of this inelastic deformation depends upon the behavior of the structural element and upon the extent of the $f_{0.2}$-f_p range. If by convention it is assumed that

$$f_p = f_{0.01}$$

Aluminum alloy structures

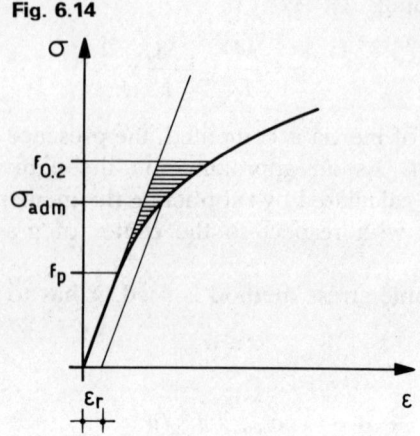

Fig. 6.14

then from the Ramberg–Osgood law

$$f_p = f_{0.2} \sqrt[n]{0.05}$$

This relationship allows the calculation of $f_{0.2}/f_p$ ratios as functions of n. By assuming the coincidence $n \equiv f_{0.2}$ (for units, see Eq. 2.33), curve (a) of Fig. 6.15 is obtained, showing a larger range for $f_{0.2}$–f_p in non-heat-

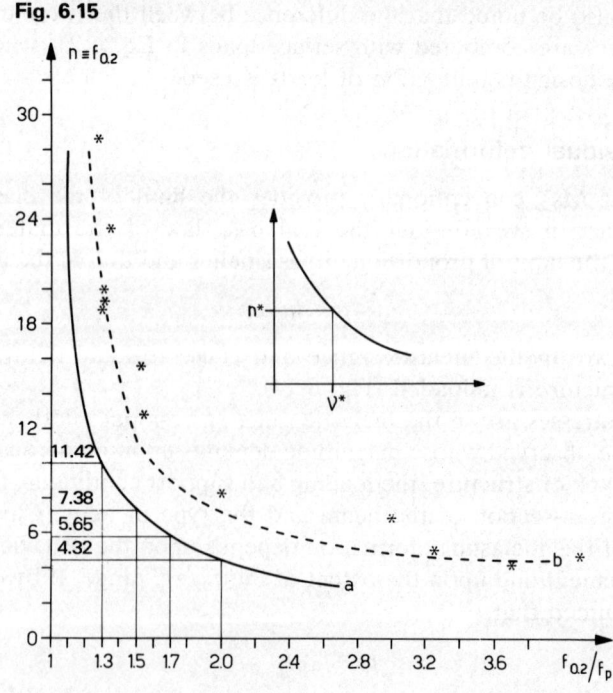

Fig. 6.15

Strength of structural elements

treated alloys (small values of n). This behavior is also confirmed by some experimental results on Swiss alloys, interpreted by curve (b).

If the allowable stress method is adopted, a relationship is obtained between the $f_{0.2}/f_p$ ratio, which characterizes the inelastic range, and the value of the safety factor ν, defined as $f_{0.2}/\sigma_{adm}$ ratio.

The value of the minimum safety factor needed to ensure that the stresses remain in the elastic range is obtained by taking $f_p = \sigma_{adm}$, and is equal to

$$\nu^* = f_{0.2}/f_p$$

Therefore we get:

for $\quad \nu < \nu^* \quad$ linear elastic behavior

for $\quad \nu > \nu^* \quad$ inelastic behavior

A value of n^* can be associated with ν^* through the Ramberg–Osgood law. In this way the behavior of alloys is different depending upon

$n > n^* \quad$ linear elastic behavior

$n < n^* \quad$ inelastic behavior

The values of n^* corresponding to the values of $\nu^* = 1.3$, 1.5, 1.7 and 2.0 are given in Fig. 6.15. They are respectively equal to 11.42, 7.38, 5.65 and 4.32.

Figure 6.16 gives the relationships which provide, for the same values

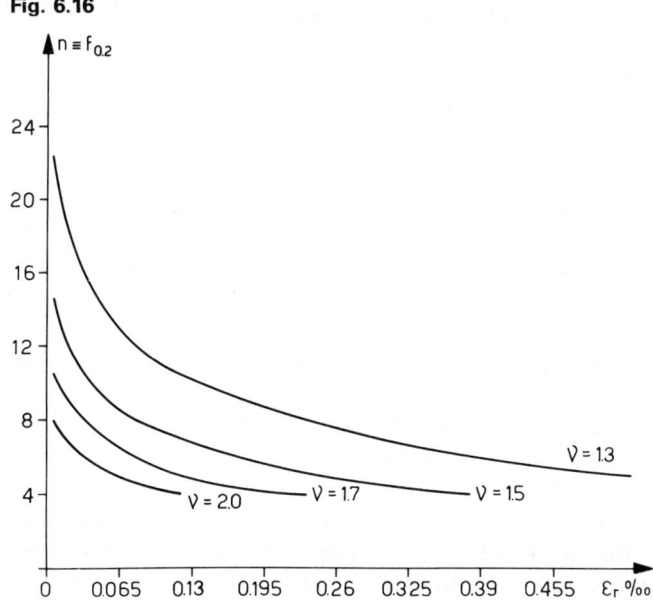

Fig. 6.16

Aluminum alloy structures

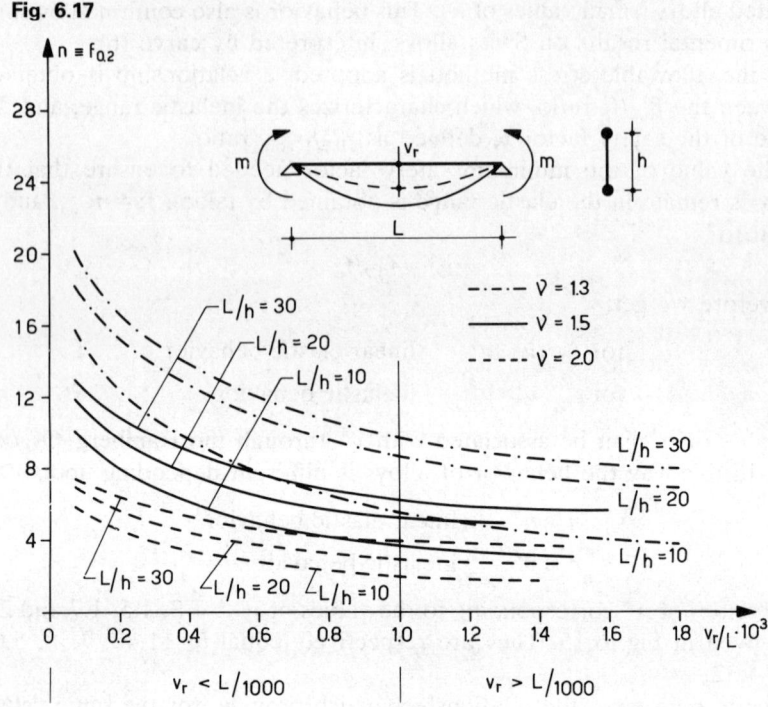

Fig. 6.17

of the safety factor, the residual deformation ε_r caused by the inelastic behavior.

An approximate evaluation of the consequences of these values of ε_r can be made by obtaining the corresponding residual midspan deflections. Consider a simply supported beam of length L under uniform bending, with a cross-section characterized by two concentrated masses at a distance h. Figure 6.17 shows the values of residual deflections (v_r/L) as functions of the type of alloy ($n \equiv f_{0.2}$), of slenderness (L/h) and of the safety factor (ν). If 1/1000 is assumed as the limiting value for v_r/L, which usually corresponds to the manufacturing tolerances, two different behaviors can be identified.

When $v_r/L < 1/1000$, even though there are some residual deformations in some fibers of the cross-section owing to the inelastic behavior of the σ–ε law, there is no significant inelastic behavior of the whole beam.

In contrast, when $v_r/L > 1/1000$, the inelastic behavior of the beam, when unloaded, leads to a residual curvature beyond tolerances. The bounding values of $n = f_{0.2}$, beyond which there are no residual deflection problems, are given in Fig. 6.18 as functions of L/h and ν. It should be noted that the commonly used alloys are characterized by $n \equiv f_{0.2}$ values

Fig. 6.18 Bounding values of $n \equiv f_{0.2}$

L/h ν	10	20	30
2.0	2	3	4
1.5	3.7	5.3	6.3
1.3	5.6	8.6	9.7

higher than the limiting values of Fig. 6.18. The most susceptible cases are those characterized by large values of slenderness, with small safety factors for non-heat-treated alloys.

In the case of heat-treated alloys it seems that, at least within the limits of the assumed approximations, there are no problems since the behavior of the material is not perfectly elastic. However, it should be noted that residual deflections larger than the tolerance limit represent a serviceability limit state, but do not compromise resistance of the structure.

6.3.3 Plastic Adaptation Limit State

6.3.3.1 Classical method

The plastic adaptation method was first used in French specifications on steel structures (1966). The same method is directly copied into French specifications for aluminum alloys [18], and in the first edition of the ECCS recommendations [19].

Although it seems logical to use this method in steel structures, it is not appropriate to use the same coefficient in steel and aluminum structures as has been done in French specifications (régles Al, 1976). This method is based upon the use of a plastic adaptation coefficient (which is referred to as η in the French code and ψ in the ECCS recommendations, even though the method is the same).

The ψ coefficient increases the section modulus in the elastic range, and the normal stress is therefore given by

$$\sigma = \frac{M}{\psi W} \qquad (6.23)$$

in the case of symmetrical bending, and by

$$\sigma = \frac{M_x}{\psi_x W_x} + \frac{M_y}{\psi_y W_y} \qquad (6.24)$$

in the case of asymmetrical bending.

Aluminum alloy structures

Fig. 6.19

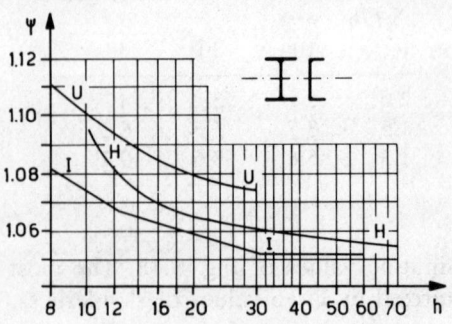

Fig. 6.20

POSITION	SHAPE	ψ
─┼─	I and H sections	1.20
─┤	T section	1.20
─>─	angles (t = a/10)	1.24
▮ ▨	rectangular section	1.185
─◆─	rhombic section	1.38
─●─	circular section	1.27
─○─	hollow circular section	1.093

ASYMMETRICAL BENDING

POSITION	SHAPE	ψ
─⊔─	channels	1.25
─⊥─	half HN, HE section half IPE, IPN section T section	1.20 1.21 1.23
─∨─	angles (t = a/10)	1.36
─⌐⌐─	square edged angles round cornered angles (t = a/10)	1.22 1.26

Strength of structural elements

In these specifications (refs. [18] and [19]) the numerical values of the coefficients were determined as follows:

The residual deformation ε_r in the highly stressed fiber when the component is unloaded, after experiencing a bending moment equal to $\psi W f_y$, must not exceed 7.5 percent of the elastic deformation ($\varepsilon_e = f_y/E$).

In the case of H, I, U sections for major axis bending, the values of ψ calculated using this criterion are given in the ECCS recommendations and in the régles Al in diagrammatic form as a function of the section depth (Fig. 6.19), and they vary from 1.05 to 1.11. In the case of other cross-sections subjected to symmetrical or asymmetrical bending, the ψ values are given in the same specifications in tabular form (Fig. 6.20).

These specifications justify this criterion, termed the 7.5 percent criterion, by arguing that it represents the limit of deformation to 'prevent serious damage to the structure'.

A more precise justification for this 7.5 percent limitation can be given if we relate the residual deformation of the cross-section to the midspan deflection of the whole member. In fact, in the most unfavorable case of uniform bending of a simply supported beam with a doubly symmetrical cross-section, the elastic midspan deflection can be expressed as

$$v = \frac{ML^2}{8EI} = \chi \frac{L^2}{8} \tag{6.25}$$

If the residual curvature is introduced as

$$\chi_r = \frac{2\varepsilon_r}{h} \tag{6.26}$$

the following residual midspan deflection is obtained:

$$v_r = \frac{\varepsilon_r}{h} \frac{L^2}{4} \tag{6.27}$$

h depth of the cross-section
L span of the beam
ε_r residual deformation, given by

$$\varepsilon_r = \frac{7.5}{100} \frac{f_y}{E} \tag{6.28}$$

If the value of ε_r is defined, the v_r/L ratio can be computed for different types of steels and for different values of L/h (10, 20, 30). The corresponding values are given in Fig. 6.21 for mild steels (Fe360, Fe430, Fe450).

From this figure we observe that in the most unfavorable case of those analyzed by the specifications (Fe510, $L/h = 30$) the residual deflection is

Aluminum alloy structures

Fig. 6.21 v_r/L ratios

	$\varepsilon_r = 7.5\% \dfrac{f_y}{E}$	L/h		
		10	20	30
Fe360	8.57×10^{-5}	1/4667	1/2334	1/1555
Fe430	10.00×10^{-5}	1/4000	1/2000	1/1333
Fe510	12.85×10^{-5}	1/3112	1/1556	1/1037

Fig. 6.22

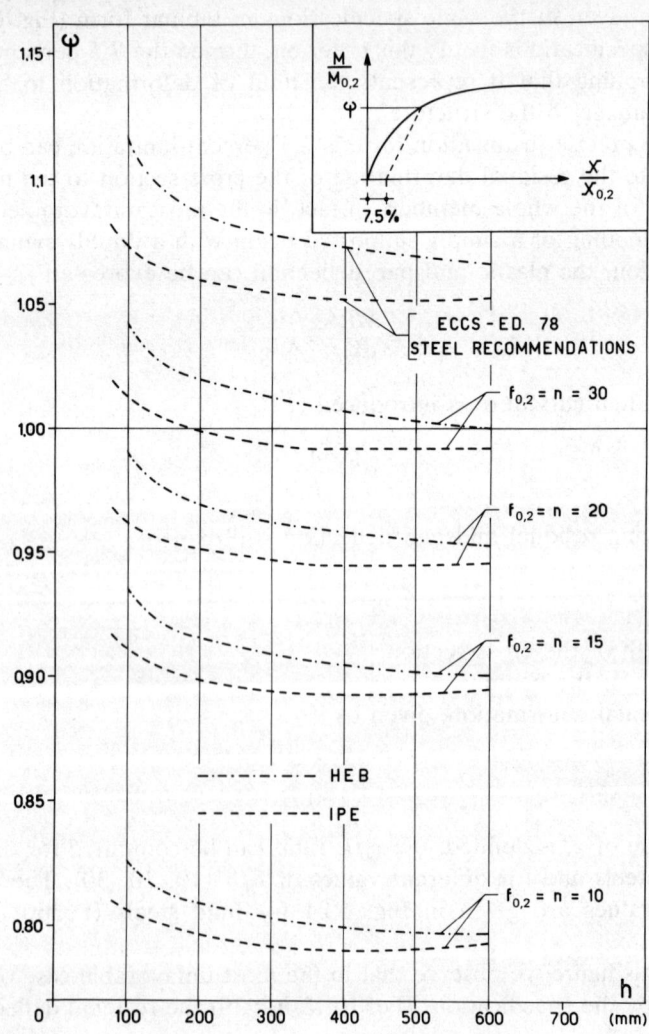

less than $L/1000$, which is universally accepted as an allowable limit for geometrical imperfections due to manufacturing tolerances in hot-rolled steel profiles.

This interpretation provides a quantitative explanation of the 7.5 percent deformation limit, but it also emphasizes that there is a lack of conformity in this method because it produces different limits of residual deflection in all beams, whatever their slenderness (L/h) and material. The 7.5 percent value seems to be conservative only in the case of mild steel, which is what these specifications refer to.

This method is not entirely satisfactory in the case of aluminum alloys. Figure 6.22 compares the behaviors of the ψ coefficients as a function of the section depth. The 7.5 percent criterion has been applied to double T sections for major axis bending, and curves are shown for I shapes (dashed line) and wide flange profiles (dashed and dotted line). As previously anticipated, ψ coefficients depend upon the type of alloy and are always less than those provided by the specifications.

The numerical analysis has been carried out by assuming the Ramberg-Osgood law with the values $f_{0.2} \equiv n = 10$, 15, 20 and 30; these more or less cover both heat-treated and non-heat-treated alloys.

By comparing the different curves it can be noted that values are smaller than 1 in all cases except for wide-flange heat-treated profiles ($f_{0.2} \equiv n = 30$). Hence the use of the same plastic adaptation coefficients for steel and aluminum alloys is not always conservative. On the other hand the use of different ψ coefficients, correctly calculated with the 7.5 percent method for the different aluminum alloys by taking account of the actual $\sigma-\varepsilon$ law, is not economical since the values are almost always less than 1.

6.3.3.2 Residual displacement method [9–11]

The plastic adaptation method described in the previous section leads to results that are oversimplified and often unconservative and illogical for aluminum alloys. Even though this method is correctly used, the criterion of limiting the plastic deformation to an assumed value (7.5 percent of the elastic deformation) does not take account of the geometry of the structure, which can be characterized in beams by the slenderness ratio L/h. On the basis of these considerations, and in order to eliminate the inconsistencies of the classical method, De Martino, Faella and Mazzolani [9] proposed a new plastic adaptation method based on the following criterion:

The plastic adaptation coefficient ψ can be calculated using the condition that the residual midspan deflection v_r due to bending is less than an assumed value αL, a function of the length L of the beam.

Aluminum alloy structures

Consequently the residual deformation becomes independent of the material and is only related to the geometry of the beam through the slenderness parameter L/h. In fact, the position is

$$v_r = \alpha L \qquad (6.29)$$

where, in the case of uniform bending,

$$\frac{v_r}{L} = \frac{\varepsilon_r}{4} \frac{L}{h} \qquad (6.30)$$

and hence

$$\varepsilon_r = 4\alpha \frac{h}{L} \qquad (6.31)$$

If α is taken as 1/1000, conventionally used to indicate initial values of the midspan deflection within tolerance limits, the following values of ε_r are obtained:

$$0.0004 \quad \text{for} \quad L/h = 10$$
$$0.0002 \quad \text{for} \quad L/h = 20$$
$$0.00013 \quad \text{for} \quad L/h = 30$$

This corresponds to an approach which is more consistent than the previous definition based upon a constant percentage deformation. The values of the computed deformation have the same meaning as the $\varepsilon_r = 0.002$ deformation conventionally used in defining the elastic limit of the material and do not require a plastic deformation proportional to the strength. This would be illogical because real materials in general become less ductile as strength increases.

The percentage value of residual deformation, which must be used instead of the classical value of 7.5 percent is given by

$$\frac{\varepsilon_r}{\varepsilon_{0.2}} = \frac{4E}{1000} \frac{h}{L} \frac{1}{f_{0.2}}$$

and varies with the material $(E, f_{0.2})$ and the slenderness of the beam (L/h).

This method seems to be particularly applicable to families of materials with a large range of mechanical properties, such as aluminum alloys. For aluminum alloys characterized by $f_{0.2} \equiv n = 10, 15, 20$ and 30, and with slenderness ratios $L/h = 10, 15, 20$ and 30, the $\varepsilon_r/\varepsilon_{0.2}$ values are given in Fig. 6.23. As can be seen, the deformation varies between 3.1 and 28 percent of the elastic deformation, depending upon the type of alloy. This variation is inversely proportional to the slenderness (L/h) and to the strength $(f_{0.2})$.

Fig. 6.23 $\varepsilon_r/\varepsilon_{0.2}$

$f_{0.2} \equiv n$ \ L/h	10	15	20	30
10	0.28	0.187	0.140	0.093
15	0.187	0.124	0.093	0.062
20	0.140	0.093	0.070	0.047
30	0.093	0.062	0.047	0.031

In order to use this method and to calculate the plastic adaptation coefficient ψ, the moment curvature relationships are needed for the different cross-sections and for the alloys commonly used. These relationships can be obtained through a numerical procedure which discretizes the cross-section and uses small steps of the curvature. This procedure, which will be described in Section 6.3.4.2, allows in this specific case a study of any type of cross-section, whether symmetrical, asymmetrical or more generally complicated shapes.

When the moment–curvature relationships are known, the values of the plastic adaptation coefficient ψ can be obtained by the intersection point between the moment–curvature diagrams with the unloading lines given by the following equation (Fig. 6.24):

$$\frac{M}{M_{0.2}} = \frac{\chi}{\chi_{0.2}} - \frac{\chi_r}{\chi_{0.2}} = \frac{\chi}{\chi_{0.2}} - \frac{\varepsilon_r}{\varepsilon_{0.2}} \qquad (6.32)$$

and assuming for $\varepsilon_r/\varepsilon_{0.2}$ the values computed in Fig. 6.23 according to the material and to the slenderness.

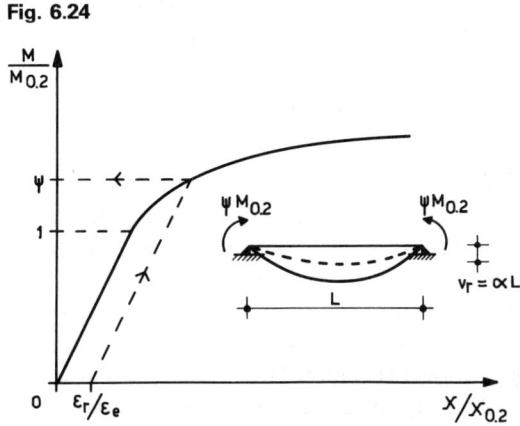

Fig. 6.24

Aluminum alloy structures

It should be noted that the position

$$\frac{\chi_r}{\chi_{0.2}} = \frac{\varepsilon_r}{\varepsilon_{0.2}}$$

is only valid in the case of profiles symmetrical with respect to the neutral axis. In the case of asymmetrical profiles the coefficient can be obtained by the relationship between the moment M and the maximum deformation ε_{max} of the highly stressed fiber. In this case Eq. 6.32 becomes:

$$\frac{M}{M_{0.2}} = \frac{\varepsilon_{max}}{\varepsilon_{0.2}} - \frac{\varepsilon_r}{\varepsilon_{0.2}} \qquad (6.33)$$

When this approach is used, the ψ value represents the ordinate of the intersection point between the M–χ curve (in the symmetrical case) and the M–ε_{max} curve (in the asymmetrical case) and the line represented by Eq. 6.32 or Eq. 6.33, respectively.

Some of the results are given in Figs 6.25 and 6.26 for the same values of $f_{0.2} \equiv n$ and of L/h considered in Fig. 6.23. Figures 6.25 and 6.26 are referred to two different procedures of building up the cross-section in order to change from two concentrated masses ($\alpha_p = 1$) to the rectangular cross-section ($\alpha_p = 1.5$), where α_p is the geometrical shape factor of the cross-section (see Section 6.3.4.4). The first procedure consists of adding the extreme areas towards the center. The second adds areas on both sides, thus increasing the thickness of the web which connects the two concentrated masses (flanges).

As might be expected, these two procedures lead to a lower and upper bound with respect to the real behavior of the cross-sections commonly used. From these results it can be observed that in the case of profiles having $\alpha_p = 1.1$–1.2 (I and H profiles) the ψ coefficients are always less than 1 for non-heat-treated alloys ($f_{0.2} \equiv n = 10$ to 15), whereas for heat-treated alloys they are greater than 1 only in some cases of beams with small slenderness ($L/h = 10$ to 15).

6.3.4 Plastic Ultimate State [12, 13]

6.3.4.1 σ–ε law at collapse [14]

As has been observed in Section 2.2, the σ–ε relationship in aluminum alloys cannot be expressed in a simplified form, as for steel structures, because there is continuity between the quasi-elastic and the inelastic hardening behavior. In the range of deformation which is of interest for buckling phenomena ($\varepsilon = 1$–2 percent), very accurate results can be obtained if the Ramberg–Osgood law in the classical form is adopted (see Section 2.2.5). If the ultimate capacity of the cross-section is of concern,

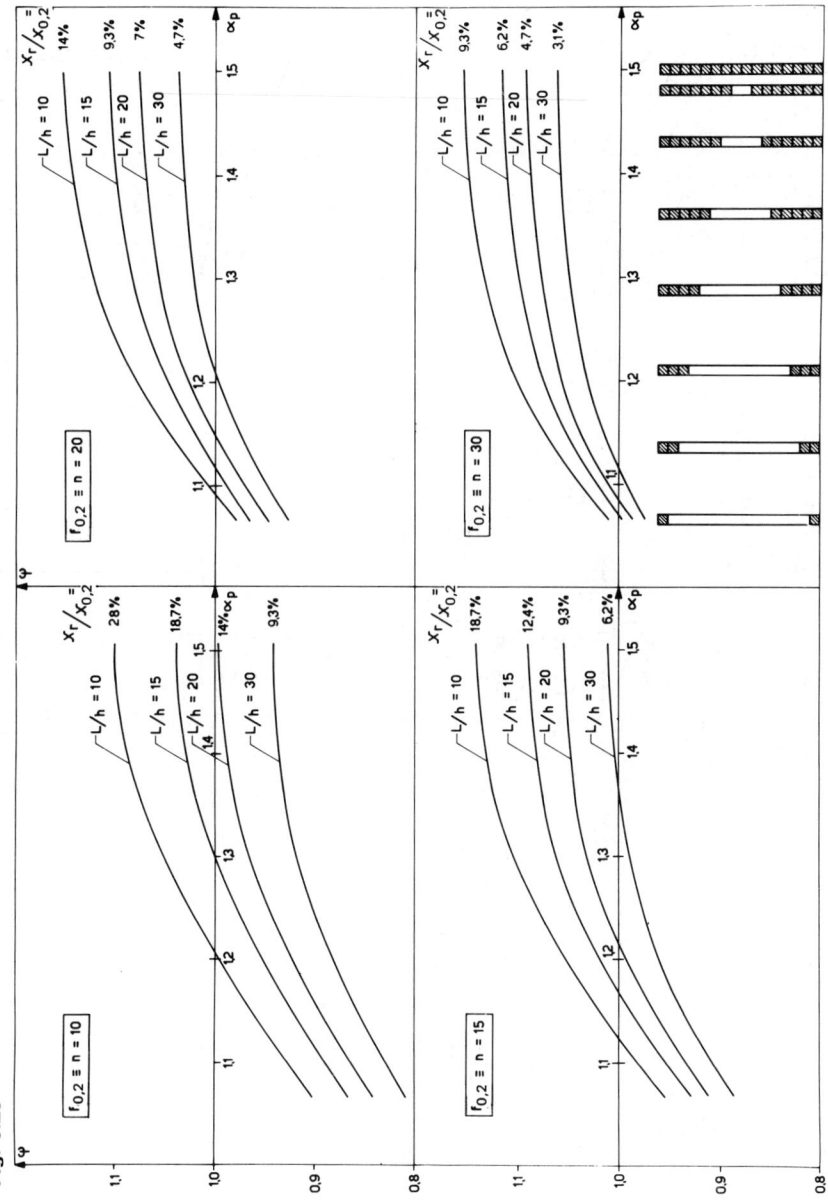

Fig. 6.25

Aluminum alloy structures

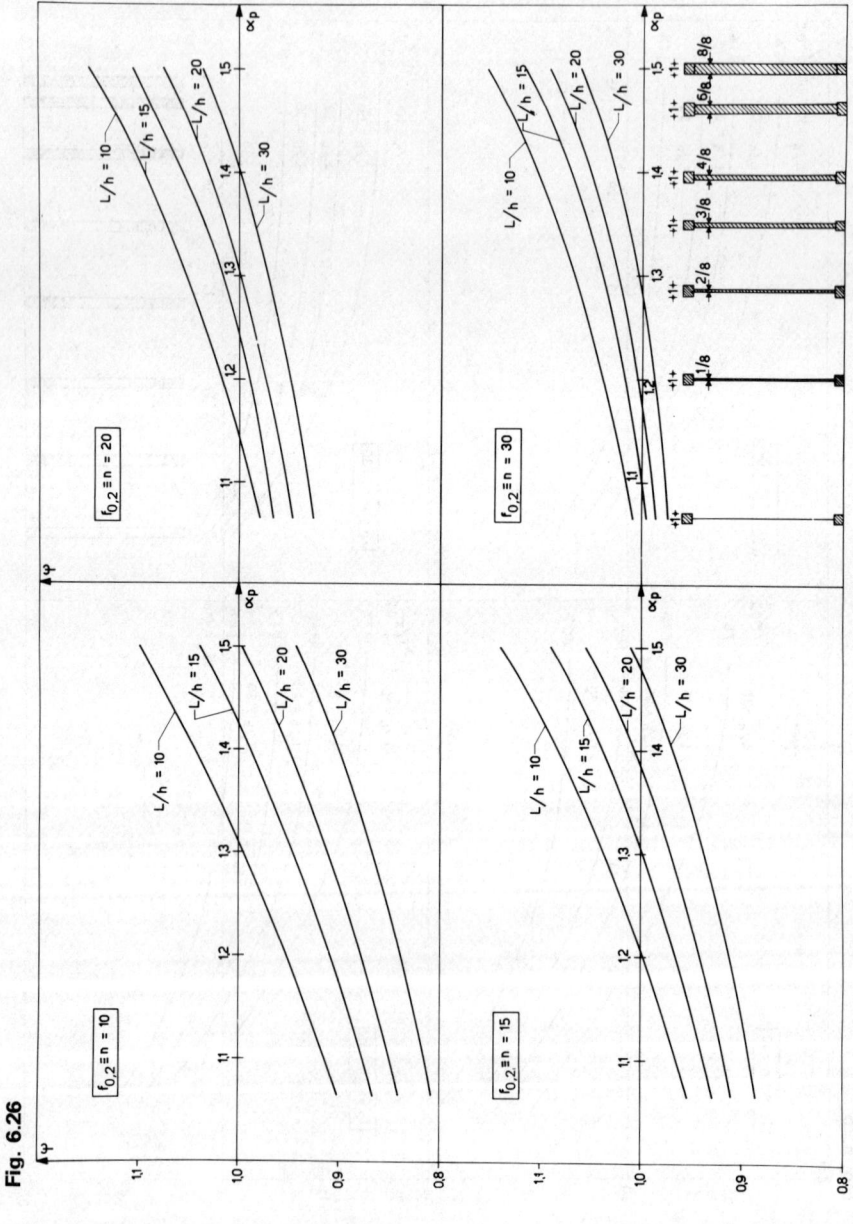

Fig. 6.26

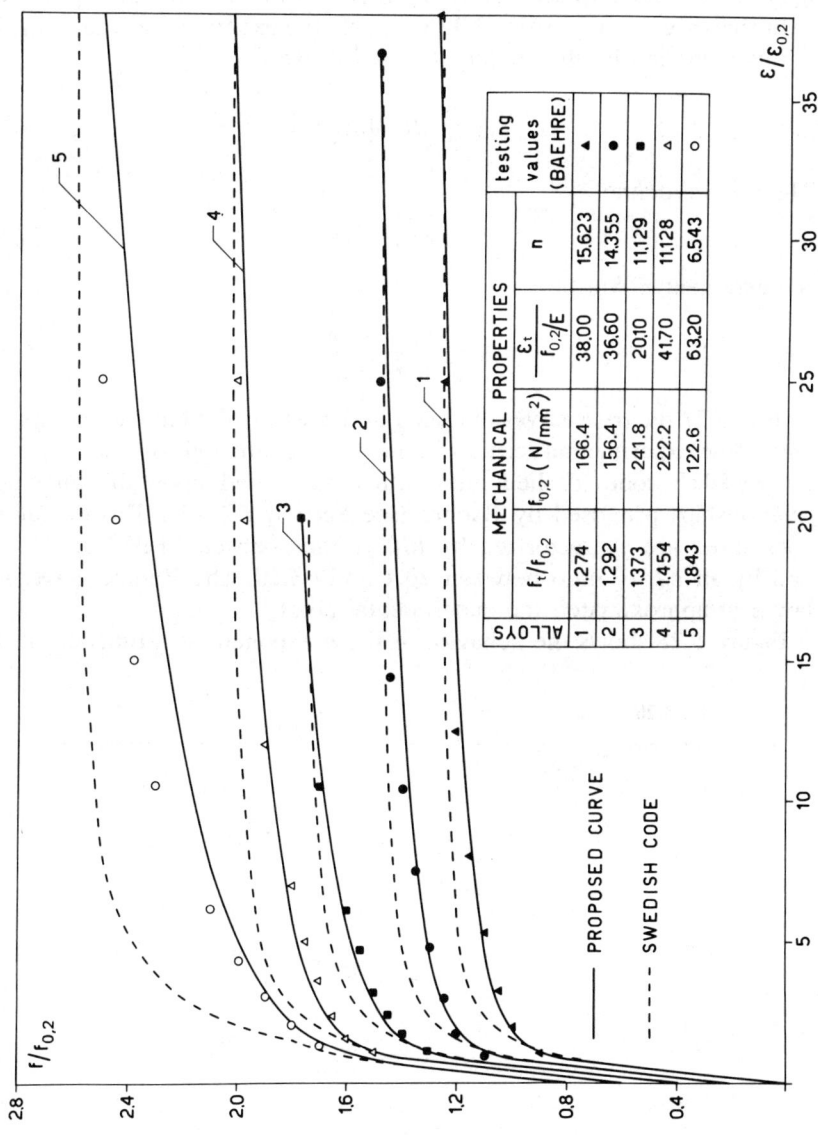

Fig. 6.27

Aluminum alloy structures

the Ramberg–Osgood law can still be used provided that the main parameters are redefined and calibrated as the point of rupture of the material, i.e. the rupture deformation ε_t and the ultimate strength f_t.

In this case, if we use Eq. 2.32 to satisfy the experimental values of the points identified by the stresses $f_{0.2}$ and f_t, we get

$$\varepsilon_t = \frac{f_t}{E} + 0.002 \left[\frac{f_t}{f_{0.2}}\right]^n \tag{6.34}$$

Then, if we define

$$\varepsilon_t' = \varepsilon_t - f_t/E \tag{6.35}$$

the exponent n will be

$$n = \frac{\log(\varepsilon_t'/0.002)}{\log(f_t/f_{0.2})} \tag{6.36}$$

In Fig. 6.27 the σ–ε curves obtained in this way (solid line) are compared with some experimental results obtained by Baehre and with the curves of the Swedish code (dashed line), which are based upon the analytical relationships proposed by Baehre (see Section 2.2.4.1). The five alloys considered are characterized by $f_t/f_{0.2}$ ratios between 1.274 and 1.843 and by an $\varepsilon_t/\varepsilon_{0.2}$ ratio between 20.1 and 63.20. The Ramberg–Osgood law best approximated the experimental points.

Figure 6.28 shows the behavior of the n exponent as a function of the

Fig. 6.28

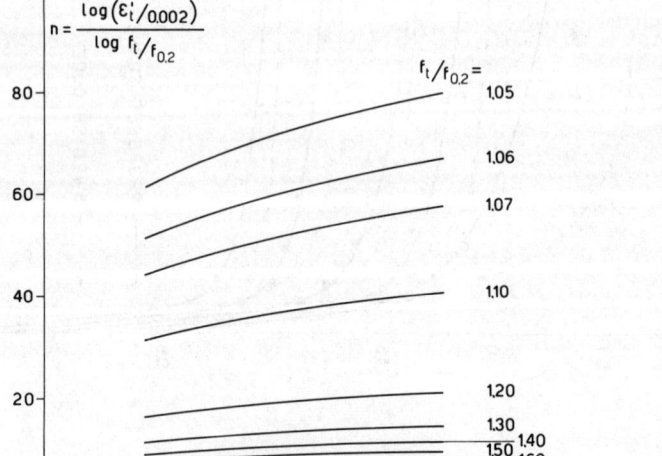

Fig. 6.29

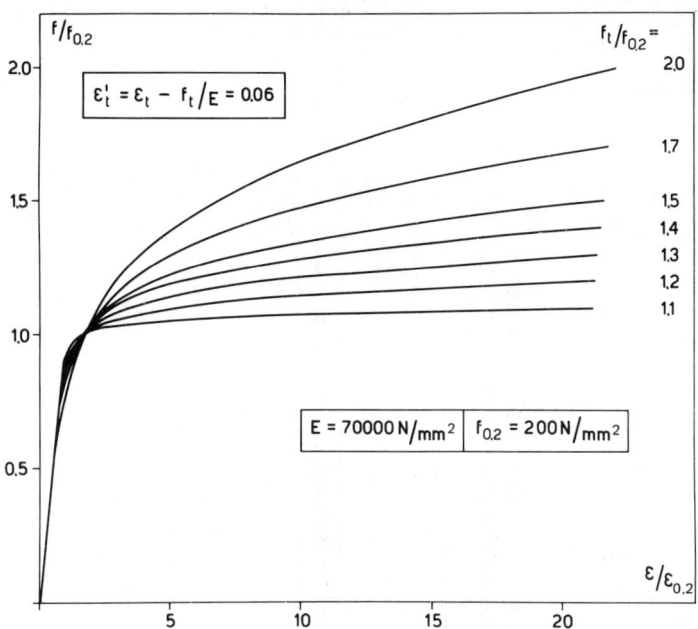

$f_t/f_{0.2}$ ratio and of ε'_t (Eq. 6.35). It should be noted that the influence of ε'_t is less than that of the $f_t/f_{0.2}$ ratio. As an example, the σ–ε behavior is shown for a given value of ε'_t as a function of the $f_t/f_{0.2}$ ratio (Fig. 6.29) and for a given value of the $f_t/f_{0.2}$ ratio as a function of ε'_t (Fig. 6.30).

It should be noted that the value to be used for the ultimate elongation should correspond to the maximum value given by the testing machine. The value of elongation ε_u – termed uniform elongation – identifies the beginning of necking. The ε_u value can be up to 30 percent smaller than the elongation given by standard tests on five diameters, which is measured by joining the two parts of the specimen. On the other hand, the measurement of ε_u is not standardized (Fig. 6.60).

Figure 6.30 shows that even if there are high variations of ε_t through Eq. 6.36, the σ–ε curve up to failure can be defined. We can therefore say that it is more conservative to use the value of ultimate elongation given by specifications instead of that corresponding to the uniform deformation ε_u.

6.3.4.2 Moment–curvature relationship [14]

In order to define the moment–curvature relationship of a general cross-section subjected to bending and compression, the following hypotheses

Aluminum alloy structures

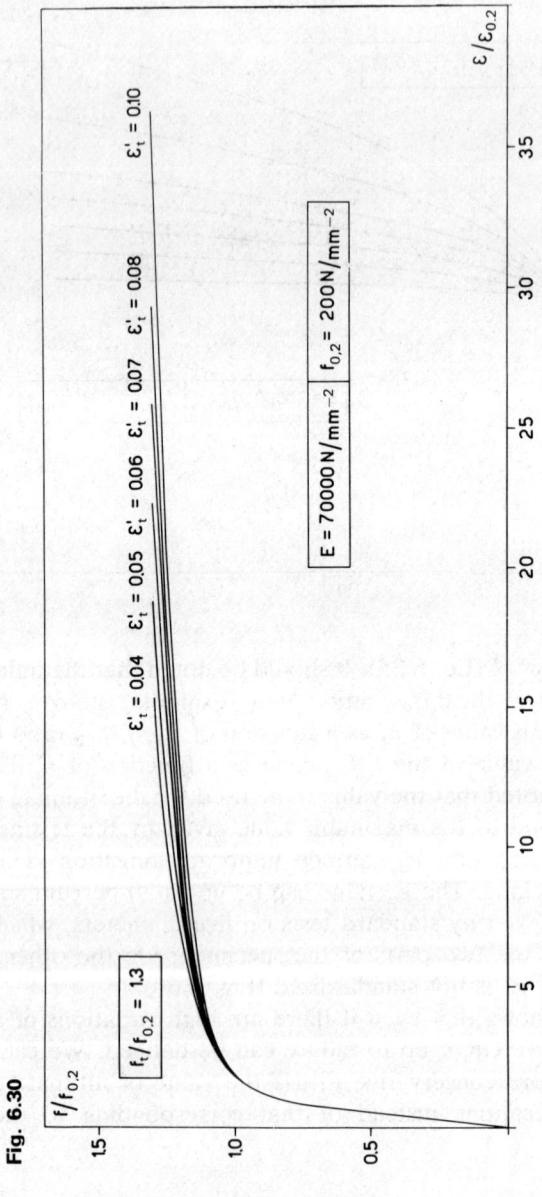

Fig. 6.30

are made:

(a) The cross-section remains plane.
(b) The constitutive law, in the form of Eq. 6.34, is irreversible.
(c) Local unloadings are elastic.

Hypothesis (a) leads to the following linear relation:

$$\varepsilon = \varepsilon_0 - \chi y \qquad (6.37)$$

where ε_0 is the deformation of a fiber at the center of gravity and χ is the curvature of the cross-section.

Equilibrium of the cross-sections is guaranteed by the relationships:

$$\int \sigma(y) b(y) \, dy = N \qquad (6.38)$$

$$\int \sigma(y) b(y) y \, dy = M \qquad (6.39)$$

where N and M are the internal actions on the considered cross-section. Equations 6.34, 6.35, 6.38 and 6.39 relate the variables N, M, ε_0 and χ through a system of nonlinear equations. The solution of this system can be obtained in a closed-form fashion only in the case of symmetrical bending without normal stresses. A more general solution to the problem can be obtained through a numerical procedure by discretizing the cross-section into small areas which are considered as concentrated at their centers of gravity (Fig. 6.31). In this case Eqs 6.37, 6.38 and 6.39 become:

$$\varepsilon_i = \varepsilon_0 - \chi y_i \qquad (6.40)$$

$$\sum_i \sigma_i A_i = N \qquad (6.41)$$

$$\sum_i \sigma_i A_i y_i = M \qquad (6.42)$$

Fig. 6.31

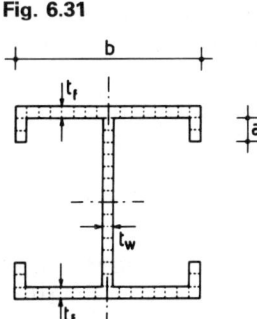
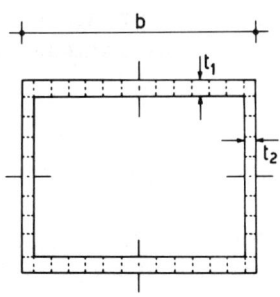

Aluminum alloy structures

As the σ–ε law (Eq. 6.34) is not reversible, it is valid only in the case of deformation increasing at each point of the cross-section without changing the sign of variation.

Since this hypothesis is not always satisfied, owing to the asymmetry of the cross-section with respect to the center of gravity or to the presence of normal stresses, the determination of the moment–curvature relationship is not uniquely defined. It depends instead upon the loading process, which has to use stepped increments.

In fact, if we indicate with subscripts i, j the properties related to the jth step in the area i, with subscripts i, $j+1$ the properties related to the next step, and with $\varepsilon_{i,\max}$ the maximum deformation reached at that point, we have the following possibilities for the σ–ε law in the $j \to (j+1)$ step:

$$\frac{\dot{\varepsilon}_{i,j}}{\varepsilon_{i,j}} > 0; \quad \varepsilon_i = \varepsilon_{i,\max}; \quad \varepsilon_{i,j+1} = \sigma_{i,j+1} + 0.002\left[\frac{\sigma_{i,j+1}}{f_{0.2}}\right]^n \quad (6.43)$$

$$\frac{\dot{\varepsilon}_{i,j}}{\varepsilon_{i,j}} > 0; \quad |\varepsilon_i| < |\varepsilon_{i,\max}|; \quad \sigma_{i,j+1} = \sigma_{i,j} + (\varepsilon_{i,j+1} - \varepsilon_{i,j}) \quad (6.44)$$

$$\frac{\dot{\varepsilon}_{i,j}}{\varepsilon_{i,j}} < 0; \quad \sigma_{i,j+1} = \sigma_{i,j} + (\varepsilon_{i,j+1} - \varepsilon_{i,j}) \quad (6.45)$$

The numerical evaluation of the moment–curvature relationship is carried out by a computer routine which takes the following steps:

(a) The properties of the cross-section of the material (E, $f_{0.2}$, n) and the axial load are set. The axial load is considered constant in each step.

(b) The increment of curvature $\Delta\chi$ is computed as a percentage of the conventional elastic curvature $\chi_{0.2} = 2\varepsilon_{0.2}/h$.

(c) The actual values of ε_0 and χ (initially $\varepsilon_0 = \chi = 0$) and the previous values of $\varepsilon_{i,j}$ and $\varepsilon_{i,\max}$ (initially equal to zero) are assumed. The maximum local deformation is then computed through Eqs 6.43, 6.44 and 6.45, and the subsequent stresses by taking into account Eqs 6.41 and 6.42.

(d) The axial load computed in this way is compared with the assumed one, and it is corrected if the deviation is larger than the assumed value by an elastic variation

$$\Delta\varepsilon_0 = N - \frac{\sum \sigma_i A_i}{E \sum A_i} \quad (6.46)$$

In this case steps (c) and (d) are repeated, otherwise step (e) is carried out.

(e) The assumed value of $\varepsilon_{i,j+1}$ and $\varepsilon_{i,j+1,\max}$ are memorized, the curvature is incremented $\chi_{j+1} = \chi_j + \Delta\chi$, and then the process returns to step (c).

Strength of structural elements

The procedure stops when a generic ε_i is larger than the assumed value of ε_t.

It should be noted that $\varepsilon_{i,j}$ and $\varepsilon_{i,j+1,max}$ are memorized only when these values correspond to an equilibrium situation (step (e)). Otherwise they do not change.

A theoretical closed form expression of the M–χ relationship can be obtained in the simple case of a double T section made of two concentrated areas A at a distance h. In this case the following relationships hold:

$$\varepsilon = \frac{\sigma}{E} + 0.002 \left[\frac{\sigma}{f_{0.2}}\right]^n \tag{6.47}$$

$$\sigma = \frac{M}{Ah} \tag{6.48}$$

$$M_{0.2} = f_{0.2} Ah \tag{6.49}$$

$$\chi = \frac{2\varepsilon}{h} = \frac{2}{h}\left\{\frac{\sigma}{E} + 0.002\left[\frac{\sigma}{f_{0.2}}\right]^n\right\} \tag{6.50}$$

If Eq. 6.48 is substituted in Eq. 6.50 we get

$$\chi = \frac{2}{h}\left\{\frac{M}{EAh} + 0.002\left[\frac{M}{Ahf_{0.2}}\right]^n\right\} \tag{6.51}$$

or through Eq. 6.49

$$\chi = \frac{M}{EI} + \frac{0.004}{h}\left[\frac{M}{M_{0.2}}\right]^n \tag{6.52}$$

If we assume:

$$\chi_{0.2} = \frac{2\varepsilon_{0.2}}{h} \tag{6.53}$$

then Eq. 6.51 can be written in the form:

$$\frac{\chi}{\chi_{0.2}} = \frac{M}{M_{0.2}} + \frac{0.002}{\varepsilon_{0.2}}\left(\frac{M}{M_{0.2}}\right)^n \tag{6.54}$$

6.3.4.3 Analysis of influence parameters

The numerical analysis method described in Section 6.3.4.2 has various uses according to the variability of the material, to mechanical imperfections of the cross-section and to the presence of axial loads.

As a first application, Fig. 6.32 shows the comparison of the non-dimensionalized M–χ curves obtained by the numerical analysis (solid line) and the curves obtained by Baehre (dashed line) for three different

Aluminum alloy structures

Fig. 6.32

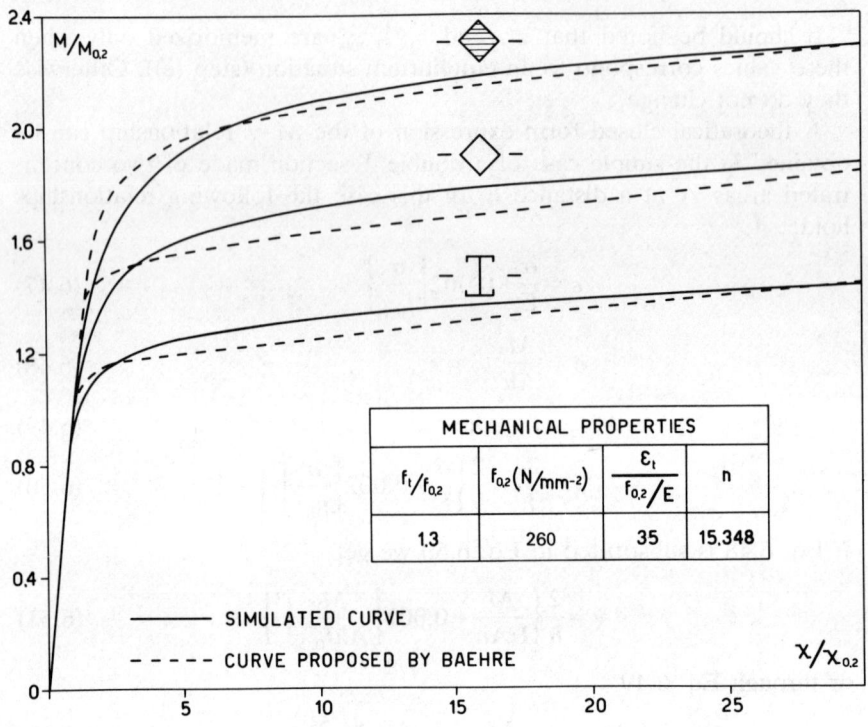

sections with shape factors varying from 1.14 to 2.0. The law has been based on the data provided by the same author. In fact, since:

$$f_{0.2} = 260 \text{ N mm}^{-2} \quad E = 70\,000 \text{ N mm}^{-2}$$

$$\frac{f_t}{f_{0.2}} = 1.308 \quad \varepsilon_t = 12.81 \text{ percent}$$

then from Eq. 6.36:

$$n = \frac{\log\left[(\varepsilon_t - f_t/E)/0.002\right]}{\log(f_t/f_{0.2})} = 15.348$$

Agreement between the curves, usually acceptable along the entire curve, is excellent at the maximum values of bending moment.

The next aspect to be examined deals with the influence of mechanical imperfections of the cross-section, such as residual stresses and nonuniform distribution of the elastic limit along the cross-section which particularly affects the buckling behavior of structures (see Chapter 7). Figures 6.33 and 6.34 show the M–χ curves for two welded cross-sections

Strength of structural elements

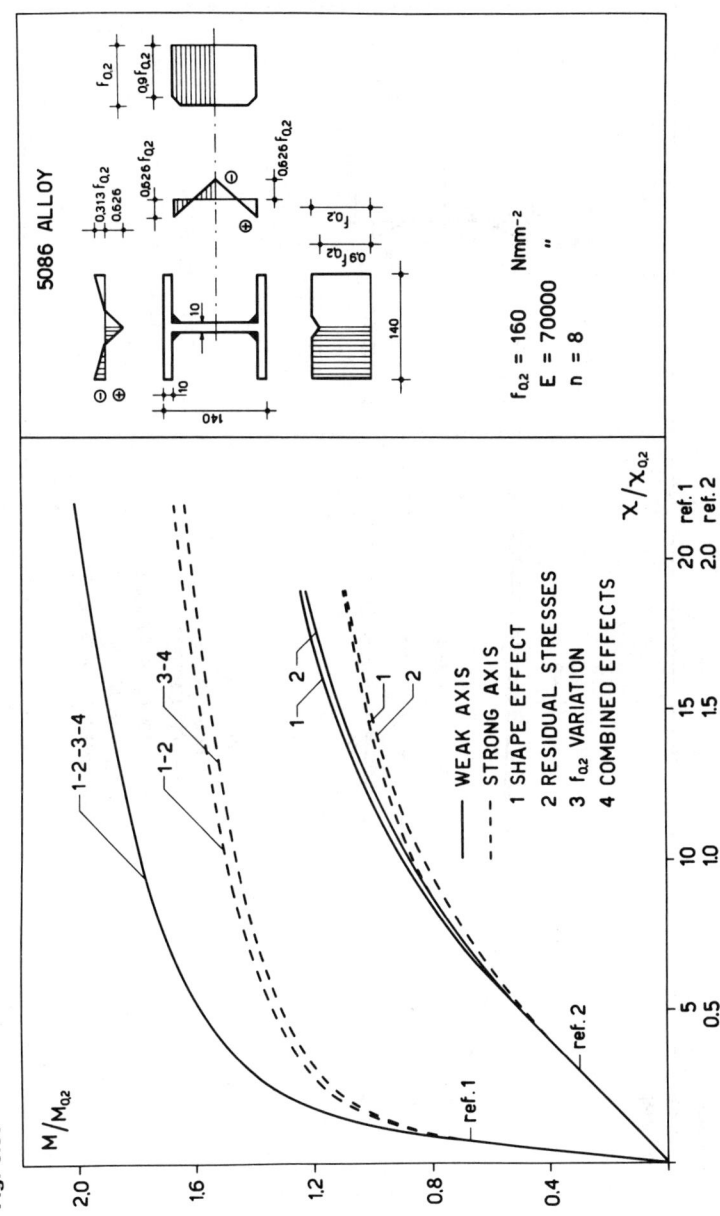

Fig. 6.33

Aluminum alloy structures

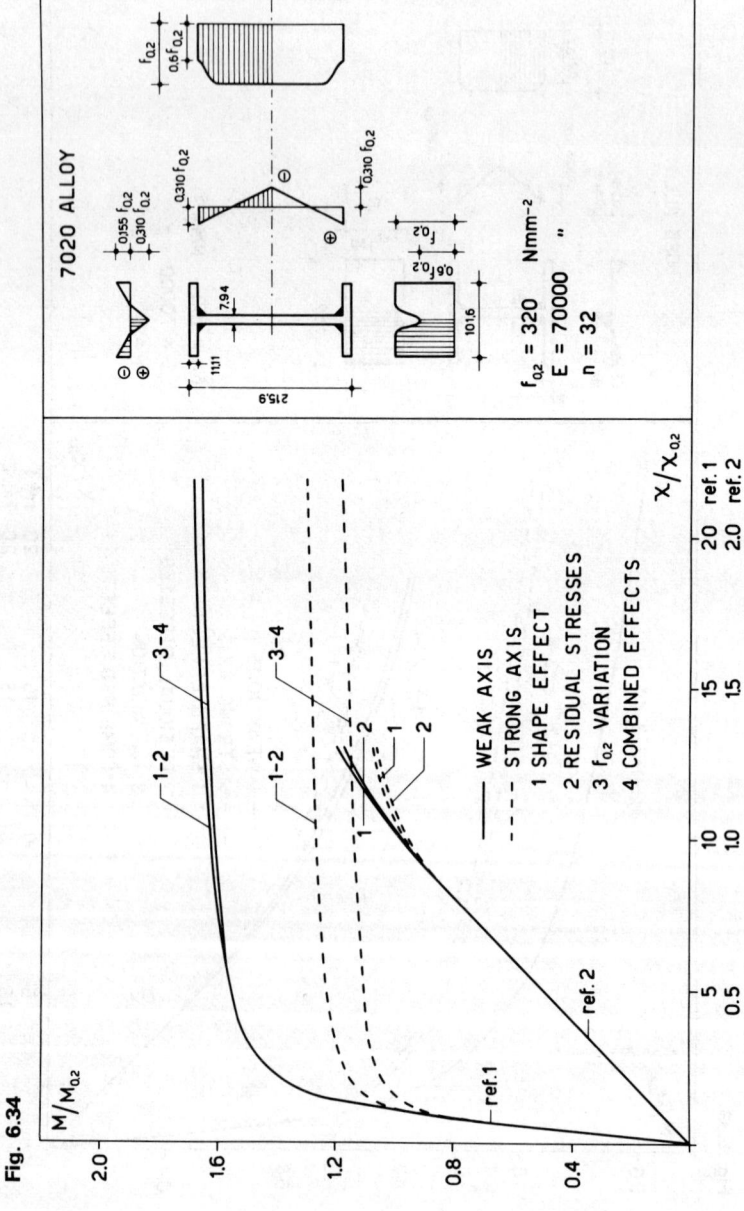

Fig. 6.34

in which there are both residual stresses and variations of elastic limit along the cross-section. Four different cases are examined:

(1) Cross-section without imperfections
(2) With residual stresses
(3) With variations of elastic limit in the cross-section
(4) With residual stresses and variations of elastic limit in the cross-section

The influence of residual stresses on the behavior of beams is evident in the first phase of plasticity of the cross-section (lower curve, ref. 2 on the x axis), whereas for increasing values of the curvature the only mechanical imperfections which produce considerable effects (50–60 percent) are reduced-strength zones, mainly in heat-treated profiles (Fig. 6.34).

Figure 6.35 shows the influence of a constant axial load N in the loading process whose values are $N/N_{0.2} = 0.2$, 0.4, 0.6 and 0.8, where $N_{0.2} = f_{0.2} A$.

In order to emphasize the spectrum of characteristic behavior of moment–curvature relationships in aluminum alloy profiles – as a function of the different cross-sections and of the mechanical properties of the material (exponent n of the σ–ε law and $f_{0.2}/E$ ratio) – the results of several applications on different double T and hollow sections of Fig. 6.36 are given in the following paragraphs.

Double T cross-sections can be defined by the geometrical parameters

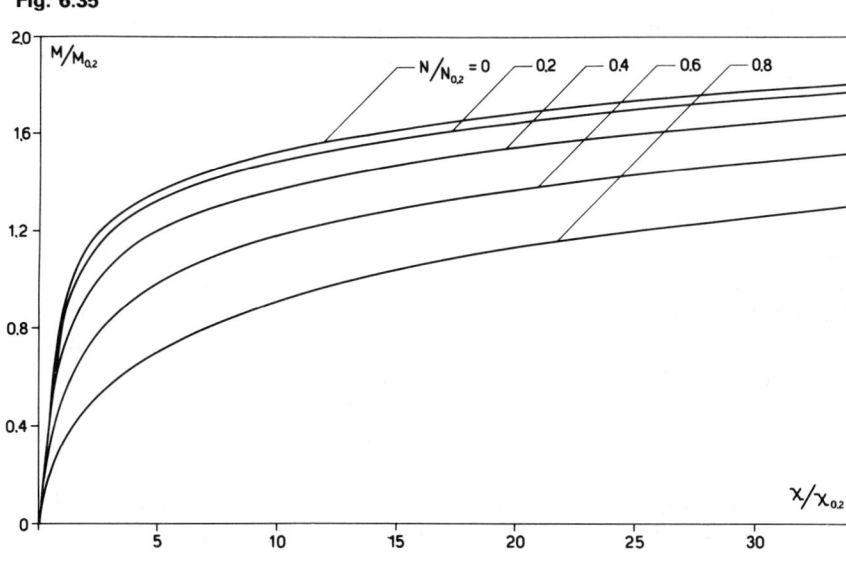

Fig. 6.35

Aluminum alloy structures

Fig. 6.36

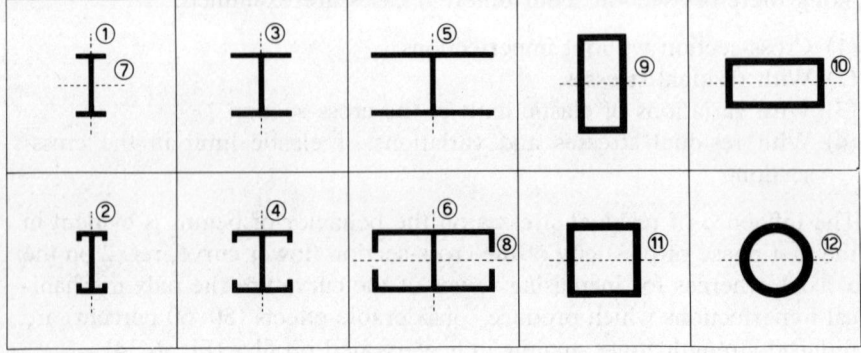

(Fig. 6.31) b/h, b/t_f, t_f/t_w, a/b with the following values:

$$b/h = 0.5 \text{ to } 2 \qquad t_f/t_w = 1$$
$$h/t_f = 15 \qquad a/b = 0, 0.15, 0.30$$

In the same way hollow sections can be defined by b/h, h/t_w, t_1/t_2 with the following values:

$$b/h = 0.5 \text{ to } 1.2$$
$$h/t_1 = 20$$
$$t_1/t_2 = 1$$

Cylindrical tubes have a d/t ratio equal to 20.

The values of the exponent n of the σ–ε law of the chosen material were equal to 8, 16, 24 or 32, which cover most of the structural alloys. The $f_{0.2}$ values used were, for each n case, equal to 80 and 320 N mm^{-2}. Thus the total range of aluminum alloys was covered, though some extreme combinations ($n = 8$, $f_{0.2} = 320$) have only a theoretical meaning (Fig. 6.37).

Figures 6.38–6.41 show the moment–curvature diagrams obtained for the sections under consideration: curve numbers correspond to the cross-sections and bending axes given in Fig. 6.36. These diagrams demonstrate the influence of the geometrical properties of the cross-section, which cause large differences in the moment–curvature diagram even with the same material. They suggest for limit behavior the ideal cross-section made of two concentrated masses (curves identified by the index 0).

The influence of the σ–ε law can be seen by comparing the diagrams for different materials. From these diagrams it can be seen that ductile behavior varies greatly for materials having the same ε_t but different $f_{0.2}$. We define the ductility parameter as the ratio between the ultimate

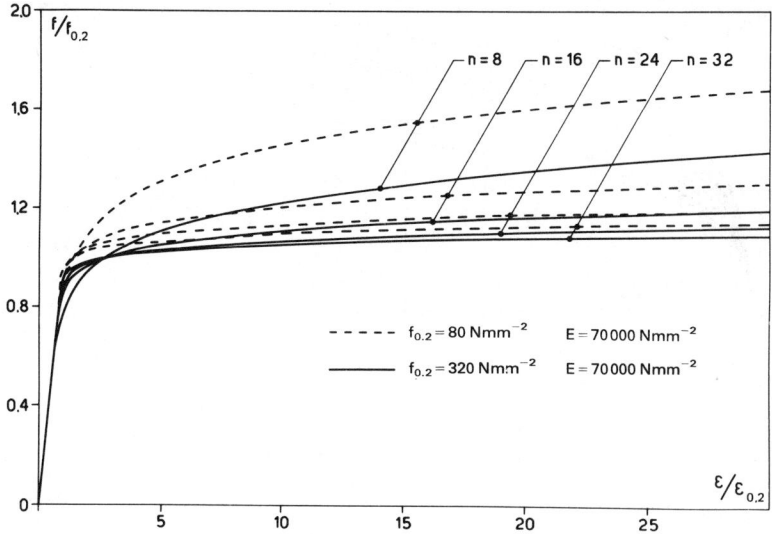

Fig. 6.37

curvature and the conventional limit elastic curvature, as follows:

$$\frac{\chi_t}{\chi_{0.2}} = \frac{\varepsilon_t}{f_{0.2}/E}$$

Then, assuming ε_t to vary between 4 and 20 percent and the $f_{0.2}$ stress to vary between 80 and 320 N mm^{-2}, we obtain extreme values very different from each other. They vary from a minimum of less than 10 to a maximum of greater than 100.

6.3.4.4 Shape factor of the cross-section

In the case of a cross-section of an elastic/perfectly plastic material, such as mild steel, the ratio between the fully plastic moment $M_{pl} = f_y Z$ and the maximum elastic moment $M_e = f_y W$, or between the plastic modulus of the cross-section Z and the elastic one W, i.e.

$$\alpha_p = \frac{M_{pl}}{M_e} = \frac{Z}{W} \qquad (6.55)$$

is called the *shape factor* (geometrical) of the cross-section. It reflects the gain in resistance between reaching the elastic limit and reaching complete plasticity of all the fibers of the cross-section.

The ratio 6.55 only depends upon the geometry of the cross-section. The following values are typical of the most common cases found in steel

Aluminum alloy structures

Fig. 6.38

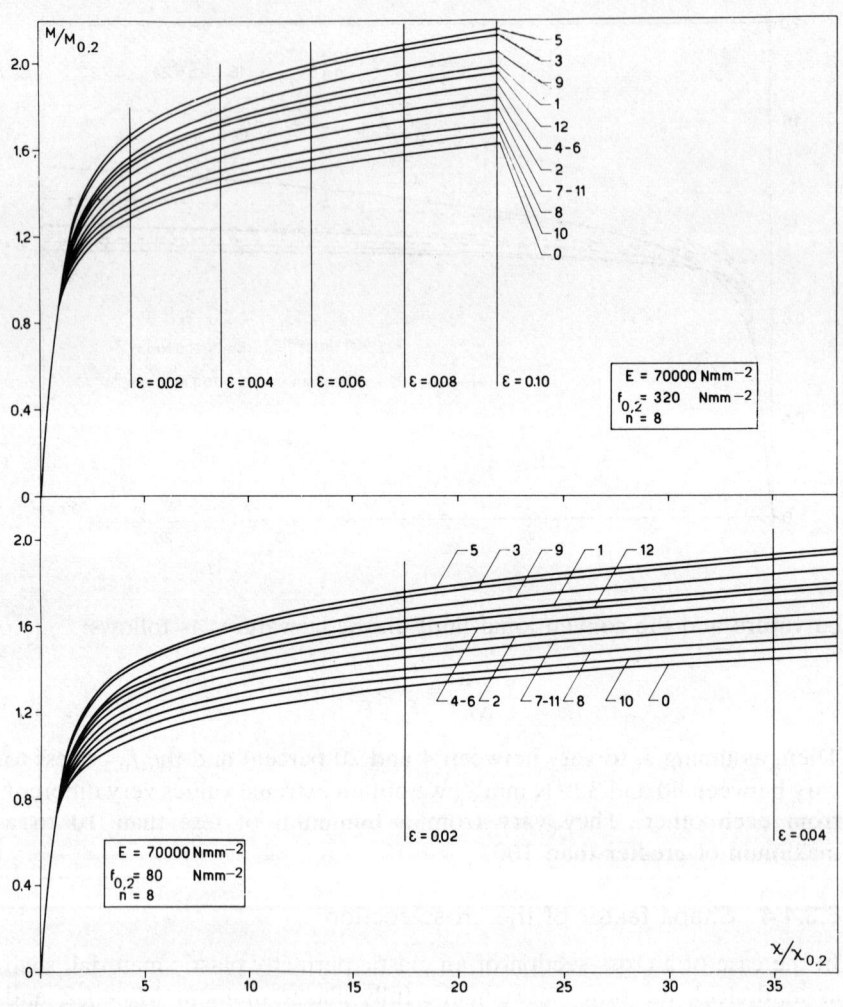

structures:

1.10 to 1.20	double T and U profiles
1.27	thin tubes
1.50	rectangular sections
1.70	solid circular sections
2.00	solid rhomboidal sections
2.37	solid triangular sections

In aluminum alloy sections, owing to the hardening behavior of the σ-ε

Fig. 6.39

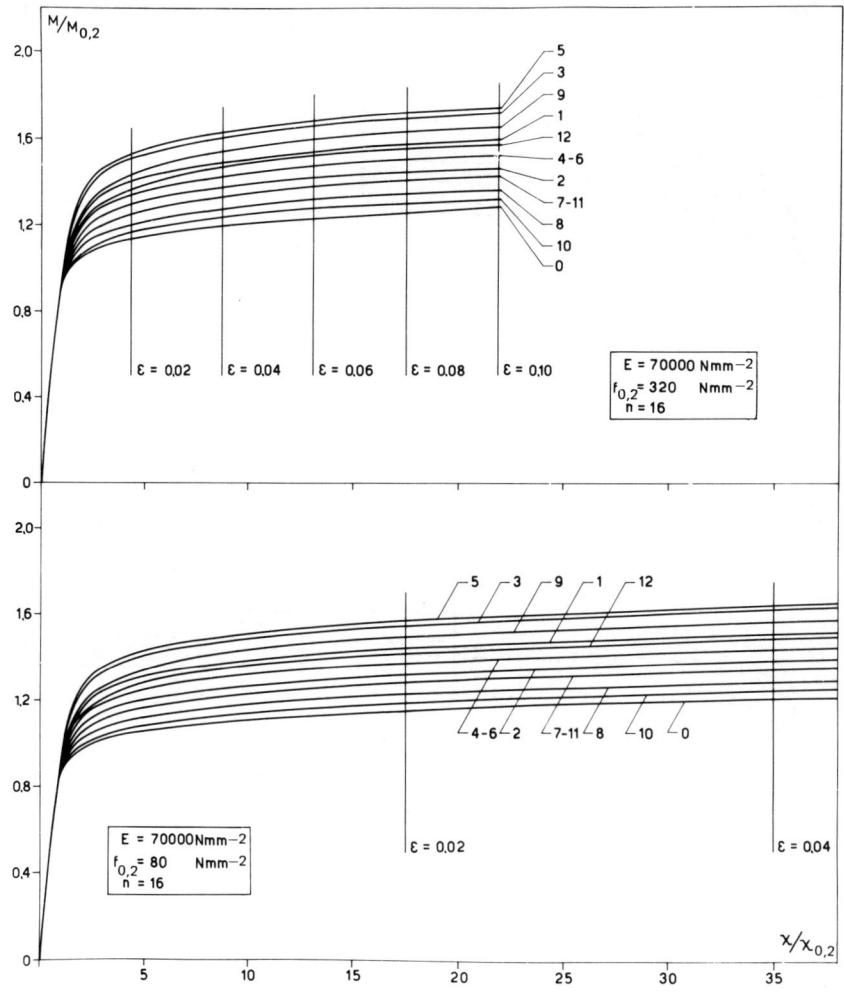

law, the shape factor given by Eq. 6.55 is no longer valid. In fact, whereas $M_{0.2}$ can be conventionally assumed to be the elastic moment M_e (Eq. 6.18), it is not logical to define a fully plastic moment because the fibers of the cross-section never reach a bound characterized by yielding. They experience, instead, increasing stresses for higher values of deformation until failure is reached in the highly stressed fibers. This shape factor (Eq. 6.55) loses its mechanical meaning in aluminum alloys while keeping its geometrical meaning.

Fig. 6.40

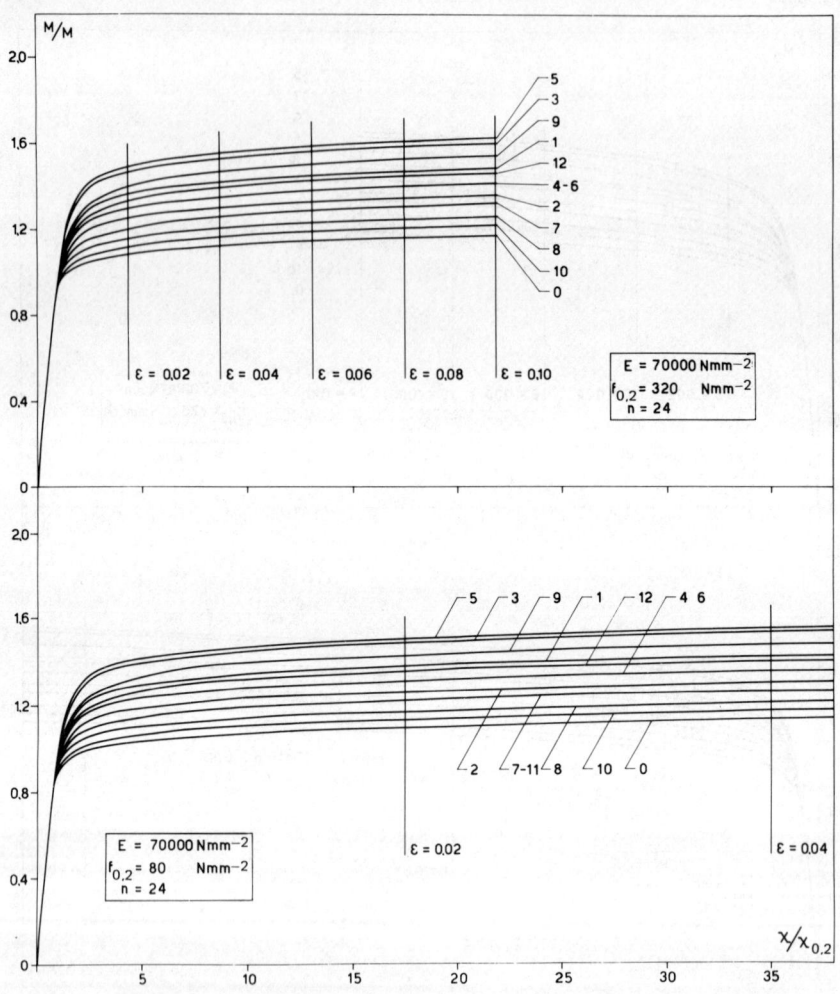

6.3.4.5 Definition of ultimate moment

The moment–curvature relationship in elastic/perfectly plastic materials – such as steel – shows an asymptotic behavior. The asymptote represents the fully plastic moment for increasing values of curvature.

In the case of elastic-hardening materials – such as aluminum alloys – this relationship is continuously increasing with a slope which is proportional to the degree of hardening of the alloy (Fig. 6.42).

The value of the ultimate moment M_u, which gives the highest capacity of the cross-section, has to be related to the capacity of the material to

Fig. 6.41

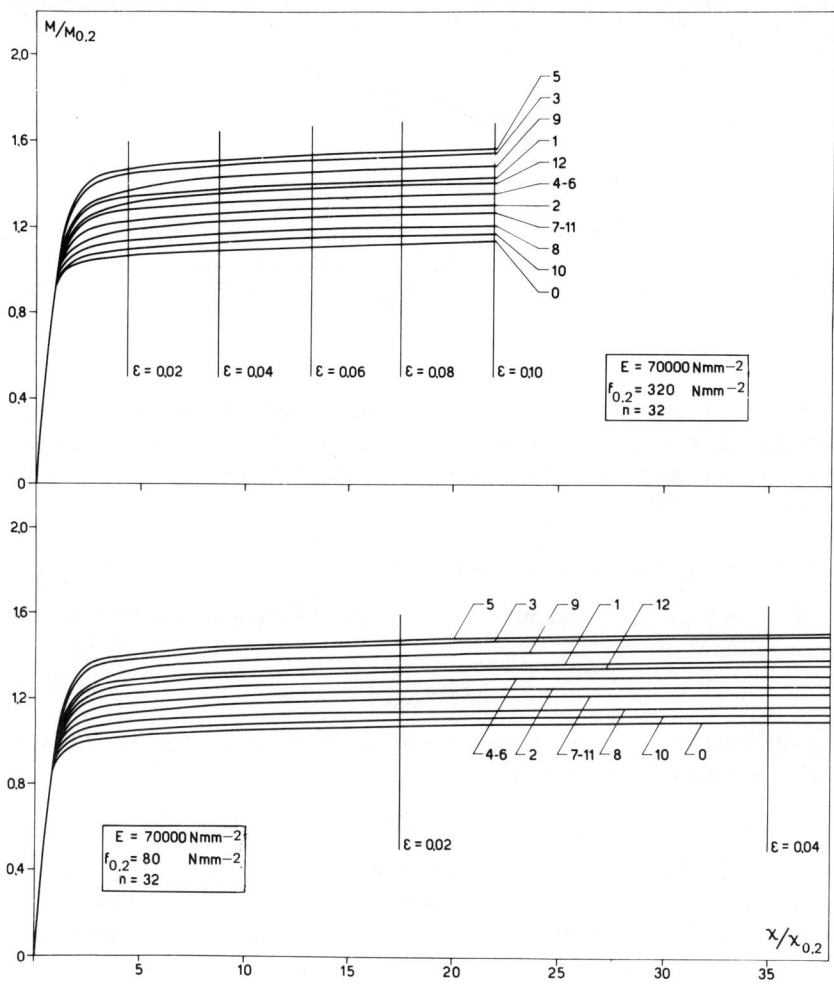

deform without danger of early brittle failure. Therefore, the curvature has to be identified as the controlling ductility parameter by setting a limiting value of $\chi_{\lim}$ for it. The conclusions of Section 6.3.4.3 lead to the conservative value

$$\chi_{\lim} = 5 \text{ to } 10 \chi_{0.2} \qquad (6.56)$$

because the most unfavorable failure conditions for the less ductile materials can fall within this range. The ultimate moment is therefore:

$$M_u = M(\chi_{\lim}) \qquad (6.57)$$

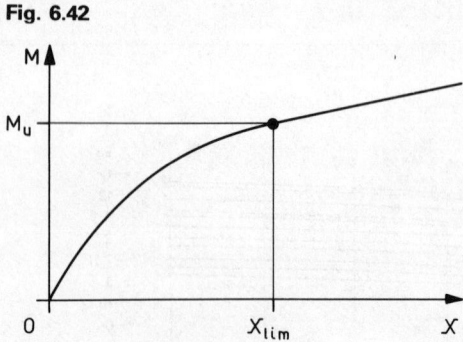

Fig. 6.42

This relationship is not in a closed form. Instead of the oversimplified assumption of Eq. 6.56, different χ_{lim} values can be taken for each alloy, and also the ductility requirements of each particular structure can be taken account of.

6.3.4.6 Generalized shape factor [15]

Analogously to steel structures, in which the fully plastic moment can be assumed as

$$M_{pl} = \alpha_p M_e$$

the ultimate moment in aluminum alloys can be related to the limiting elastic moment by the relation

$$M_u = \alpha M_{0.2} \qquad (6.58)$$

where α is the generalized shape factor. It depends upon the cross-sectional geometry, the σ–ε law of the material (if the Ramberg–Osgood law is used the exponent n and $f_{0.2}$ are the main parameters) and upon the assumed χ_{lim} value.

The two shape factors, geometrical α_p and generalized α, with $\alpha > \alpha_p$, can be related through nomograms (Fig. 6.43). On the horizontal axis are the α_p values which identify the geometry of the cross-section, and on the vertical axis are the values defined by Eq. 6.58 as ratios between the ultimate moment M_u (computed for a given limit curvature χ_{lim}) and the elastic limit moment $M_{0.2}$. Each line characterized by a value of n (exponent of the Ramberg–Osgood law) and of χ_{lim} represents the points obtained by numerical simulation of the moment–curvature diagram and allows the actual shape factor α to be related to the conventional one α_p.

In the case of different aluminum alloys having a conventional elastic limit $f_{0.2}$ varying from 100 to 300 N mm^{-2} and the exponent n varying

Fig. 6.43

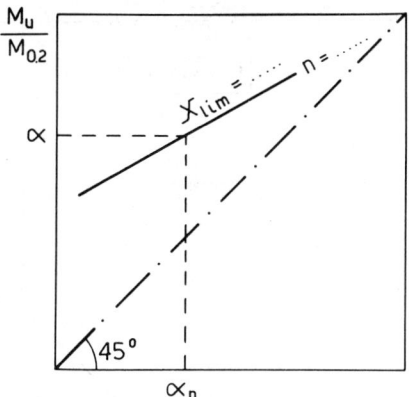

from 10 to 30, Figs 6.44–6.48 give the α_p–α relationships corresponding to the two limiting values of χ_{lim} proposed in Eq. 6.56.

It should be noted that the line corresponding to $n = \infty$ starts at the origin and represents all the points where $\alpha_p = \alpha$ which is valid for elastic/perfectly plastic materials such as mild steels. As n reduces, that is as the hardening behavior of the materials increases, the line rises and gives $\alpha > \alpha_p$, which shows the greater resistance in the inelastic range of aluminum alloys with respect to steel.

As an example, consider two double T sections with $\alpha_p = 1,2$, one made of mild steel and the other of aluminum alloy. For steel, since $n = \infty$ for whatever type of mild steel, $\alpha = \alpha_p = 1.2$. For aluminum alloy, the shape factor depends upon the type of alloy. In the case of non-heat-treated alloys, such as a 5000 series alloy with $f_{0.2} = 150 \text{ N mm}^{-2}$, $n = 10$, the following is obtained from Fig. 6.45:

$$\alpha = 1.34 \quad \text{for} \quad \chi_{lim} = 5\chi_{0.2}$$
$$\alpha = 1.46 \quad \text{for} \quad \chi_{lim} = 10\chi_{0.2}$$

In the case of a heat-treated alloy, such as a 6000 series alloy with $f_{0.2} = 250 \text{ N mm}^{-2}$, $n = 30$, from Fig. 6.47 is obtained:

$$\alpha = 1.26 \quad \text{for} \quad \chi_{lim} = 5\chi_{0.2}$$
$$\alpha = 1.30 \quad \text{for} \quad \chi_{lim} = 10\chi_{0.2}$$

If the other variables are unchanged, the α/α_p ratio indicates the increase in strength of the conventional ultimate moment for a hardening material with respect to that of an elastic/perfectly plastic material. In the cases under consideration it varies from 5 to 22 percent.

Aluminum alloy structures

Fig. 6.44

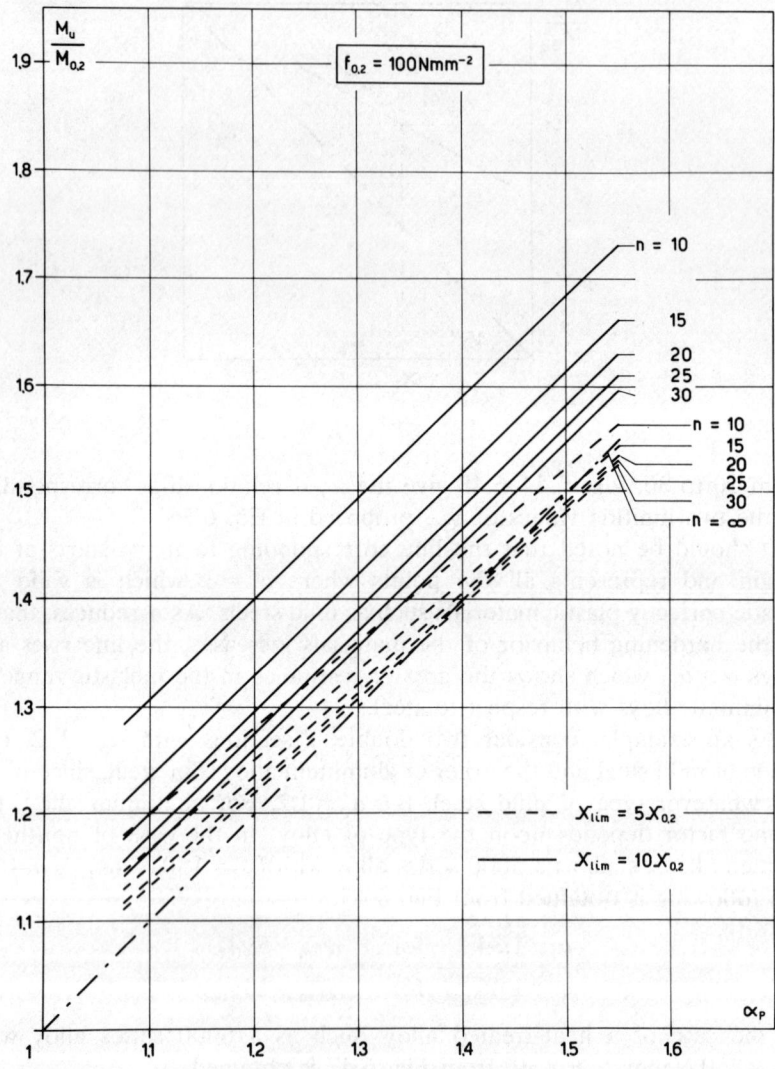

Figures 6.44 to 6.48 can also be used to indirectly obtain the moment–curvature relationship without using the general method based upon numerical simulation (see Section 6.3.4.2). In fact Eq. 6.54, which is exact in the case of two concentrated masses, can be extended to a real section whose geometrical properties can be identified through its shape factor α_p. For this purpose it has to be arranged in the Ramberg–Osgood

Fig. 6.45

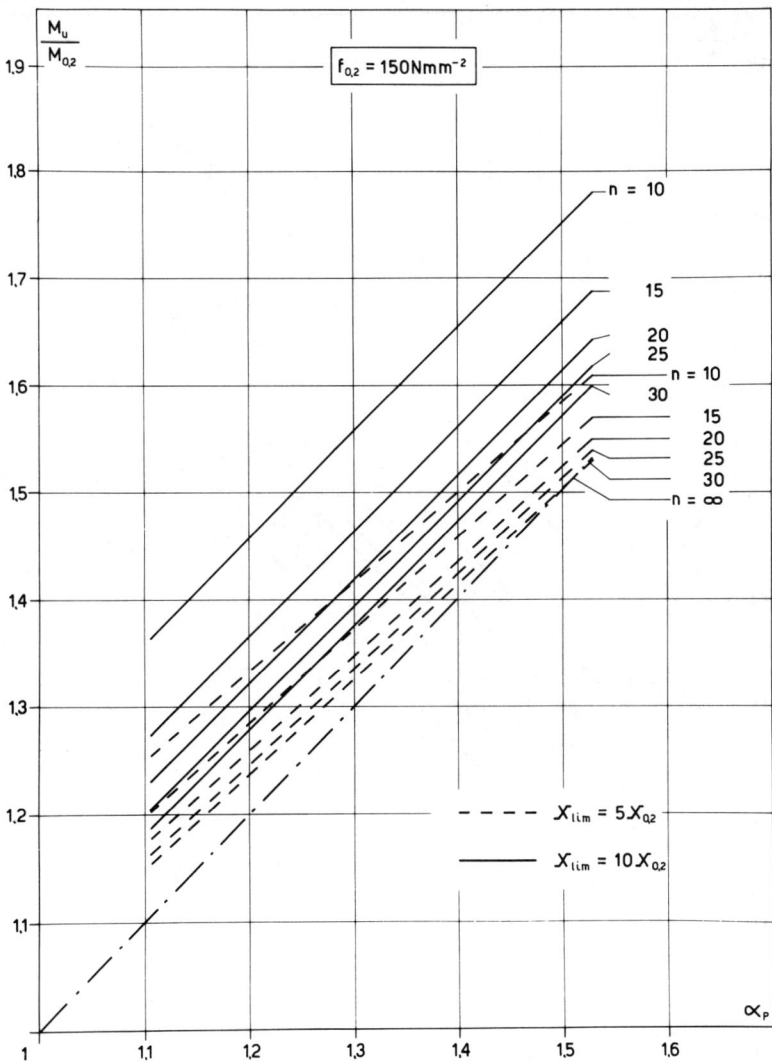

form:

$$\frac{\chi}{\chi_{0.2}} = \frac{M}{M_{0.2}} + k\left[\frac{M}{M_{0.2}}\right]^m \qquad (6.59)$$

where the k and m parameters which characterize the law can be obtained from Figs 6.44–6.48.

Aluminum alloy structures

Fig. 6.46

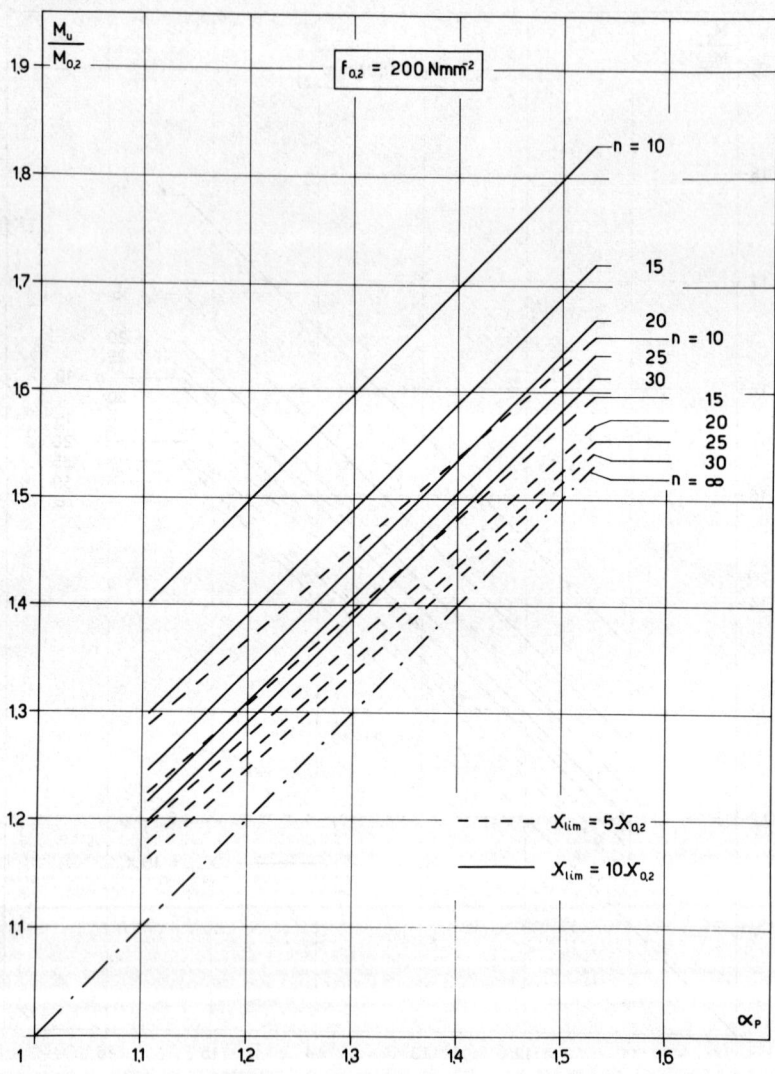

For a given cross-section of a given material, $f_{0.2}$, n and α_p are known. Thus the ratios $\alpha = M/M_{0.2}$ corresponding to the limit curvature ratios $\chi/\chi_{0.2} = 5$ and 10 are obtained from these curves. In fact the following result:

$$5 = \alpha_5 + k\alpha_5^m$$
$$10 = \alpha_{10} + k\alpha_{10}^m \tag{6.60}$$

Fig. 6.47

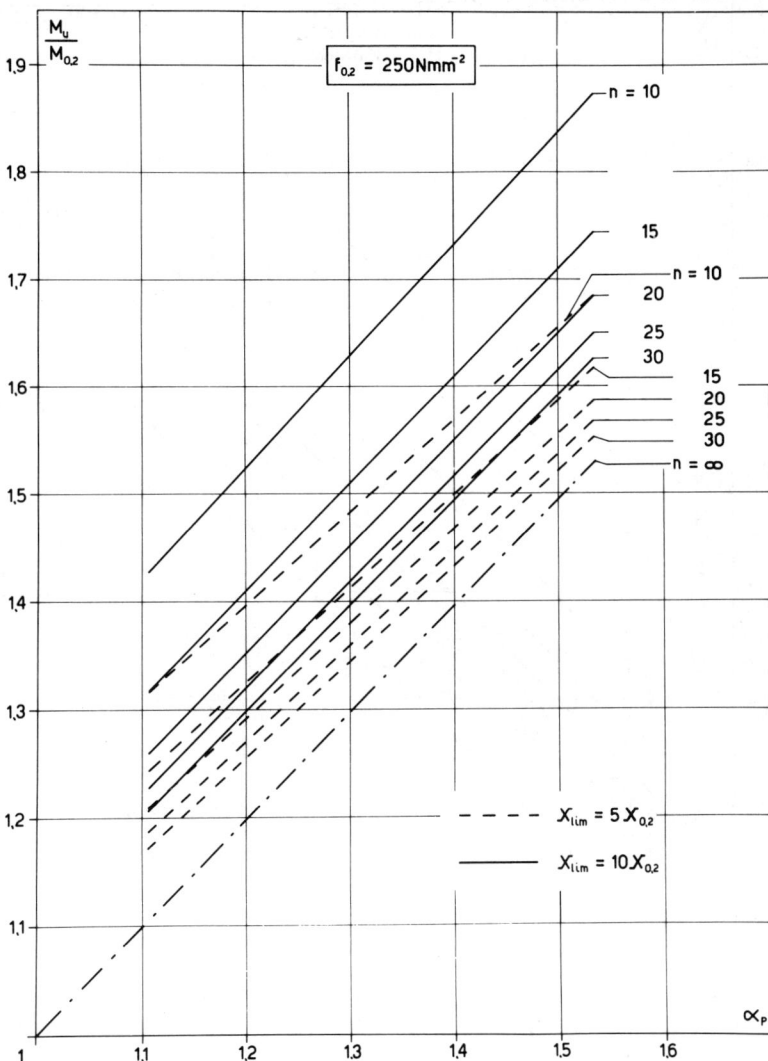

from which:

$$m = \frac{\log\left[(10-\alpha_{10})/(5-\alpha_5)\right]}{\log(\alpha_{10}/\alpha_5)}$$

$$k = \frac{5-\alpha_5}{\alpha_5 m} = \frac{10-\alpha_{10}}{\alpha_{10} m}$$

(6.61)

Aluminum alloy structures

Fig. 6.48

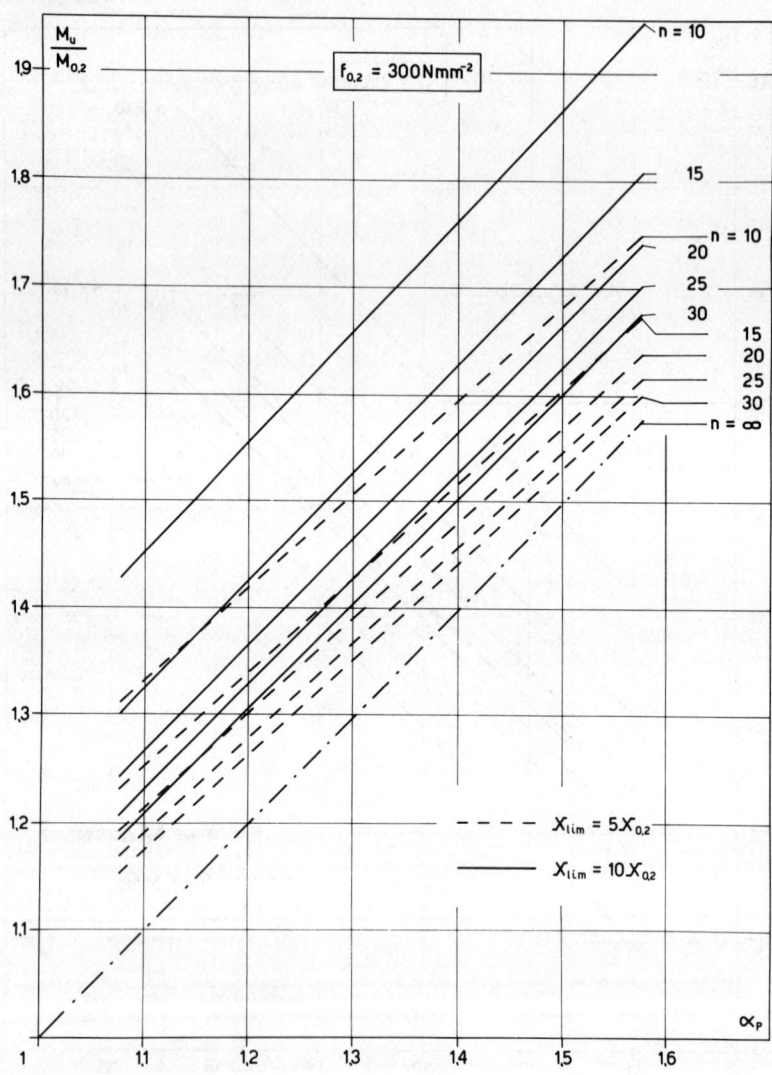

The agreement of the results obtained in this way is shown by the comparison of Fig. 6.49. The values of Fig. 6.49a have been used to compute from Eqs 6.58, 6.59 and 6.60 the points ($M/M_{0.2}$, $\chi/\chi_{0.2}$) given in Fig. 6.49b. These are compared with the curves from the numerical simulation, related to three different cross-sections with different values of the exponent n (8, 16 and 32). The agreement between the exact and the approximate method is excellent.

Strength of structural elements

Fig. 6.49a

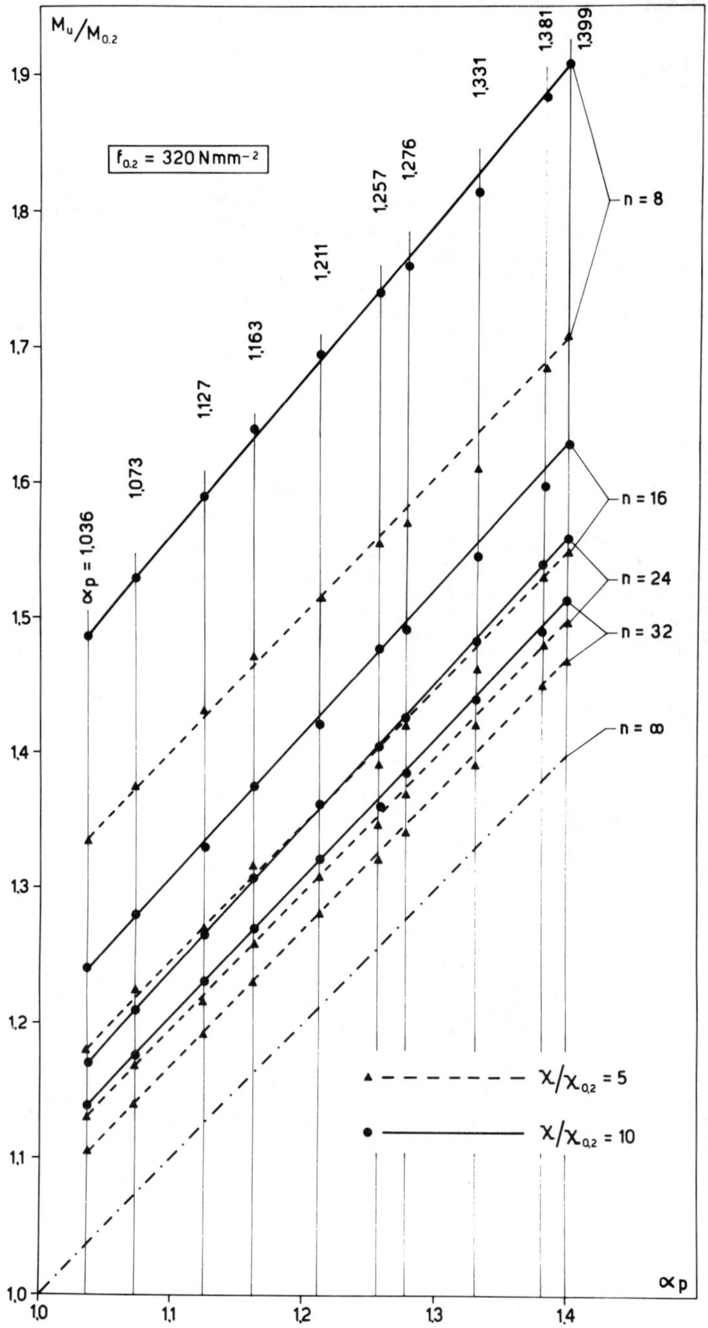

Aluminum alloy structures

Fig. 6.49b

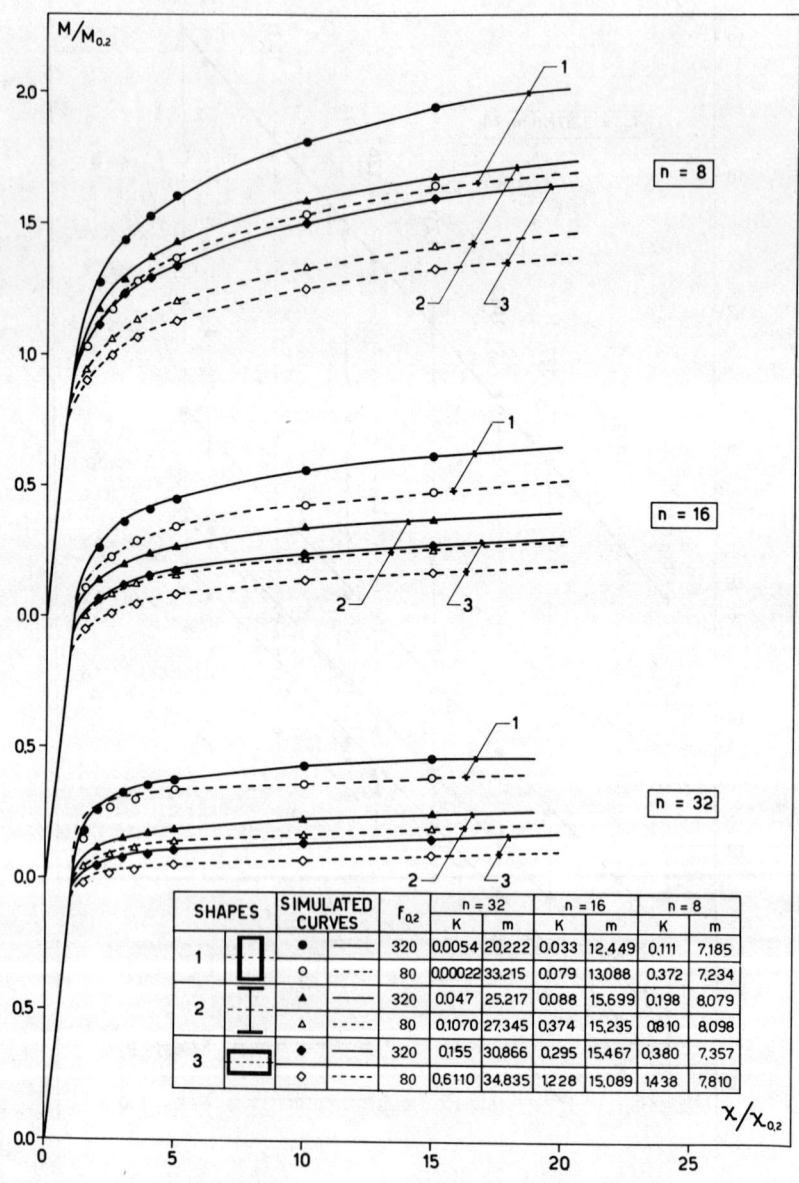

6.4 Behavior of the cross-section beyond the elastic limit

6.4.1 Definition of the Problem

The behavior of aluminum alloy beams in the inelastic range has been examined by some authors, from both the theoretical and experimental points of view [5–16]. The solution of the problem, in general terms, requires answers to the following questions:

(a) Which law of those in the literature best interprets the inelastic behavior of the structure to failure?
(b) What is the influence of the shape of the cross-section on the inelastic behavior of the structure?
(c) How reliable is a computer incremental procedure which simulates the behavior of the structure, step by step, to failure?
(d) What are the ductility requirements of the material to permit the redistribution of stresses in the post-elastic range?
(e) What is the relationship between the ultimate behavior of steel structures (elastic/perfectly plastic material) and that of aluminum alloys (elastic-hardening material)?

The answer to question (a) has already been given in Section 6.3.4.1, where the $\sigma-\varepsilon$ law proposed by Faella [14] based upon the extension of the Ramberg–Osgood law at collapse, represents the most logical basis for further analysis.

The influence of the cross-section (question b) has also been considered *a priori* through the definition of the generalized shape factor (see Section 6.3.4.6). In fact, in contrast to steel, in which it has only a geometrical meaning (see Section 6.3.4.4), the new shape factor allows the hardening behavior of the $\sigma-\varepsilon$ law of the material to be accounted for. In this way, for nonredundant schemes, the problem can be considered solved by applying the procedure given in Section 6.3.4 because the behavior of the structure is dependent upon the highly stressed section. In the more general case of redundant schemes, the possibility of carrying out an incremental solution, as required in (c), also allows the two previous aspects (material $\sigma-\varepsilon$ law and shape of the cross-section) to be considered, together with the presence of geometrical and mechanical imperfections. The features of this method due to Faella and De Martino [16] will be given in Section 6.4.2.

From the examination of the simulated load–deflection response of different bent structures with different loading and support conditions (see Sections 6.4.3 and 6.4.4), useful information on the ductility requirements of aluminum alloys, as required by question (d), can be obtained.

From the analysis of all the numerical results the information necessary

Aluminum alloy structures

to evaluate the real safety factor of aluminum alloy structures can be obtained. This provides an answer to question (e), which allows the usual limit design calculation methods based upon the formation of plastic hinges to be applied to aluminum alloy structures.

6.4.2 Calculation Method [16]

The method allows the solution of bent members such as continuous beams which are frequently used and are significant for the post-elastic behavior of metal structures.

The structure is discretized into fixed intervals in which the deformability is considered to be concentrated. The moment–rotation relationship is given by:

$$\varphi_i = \left[\left(\frac{M}{M_{0.2}}\right)_i + k\left(\frac{M}{M_{0.2}}\right)_i^m\right]\chi_{0.2} d \tag{6.62}$$

where d is the length of the single element.

The deformed shape of the continuous beam is defined by the vector of transverse displacements v_i of the ith element (Fig. 6.50). The absolute rotations of the fixed elements are related to the nodal displacements v_i through the relationships:

$$\beta_i = -\frac{(v_i - v_{i+1})}{d} \tag{6.63}$$

In matrix form we have the following relationship between the displacements and the relative rotations:

$$\varphi = \frac{1}{d}\boldsymbol{CBq} \tag{6.64}$$

where d is the length of the elements and $\boldsymbol{C}$ and $\boldsymbol{B}$ are transformation matrices. If we set

$$\boldsymbol{M} = \boldsymbol{D}\varphi \tag{6.65}$$

for the moment–rotation relationship, where $\boldsymbol{D}$ is a diagonal matrix with

Fig. 6.50

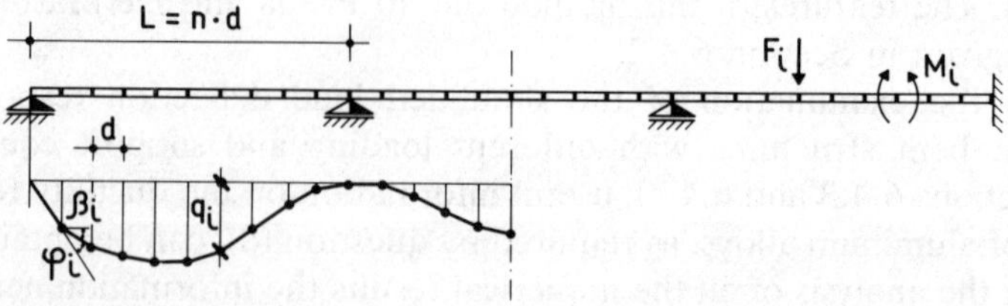

the generic term $D_{ii} = M_i/\varphi_i$ varying during the different phases of the loading process, through the dummy load method the following nonlinear system of equations results:

$$\frac{1}{d^2} \boldsymbol{B}^T \boldsymbol{C}^T \boldsymbol{DCBq} = \Lambda \boldsymbol{F} \qquad (6.66)$$

where Λ is the multiplier of the load system $\boldsymbol{F}$.

Nonlinearity of the system of Eq. 6.66 derives from the dependency of the diagonal matrix $\boldsymbol{D}$ upon the deformation of the structure.

The solution of the system is obtained through an incremental solution which increases the parameter Λ. For each increment the matrix $\boldsymbol{D}$ is computed by successive approximation starting with the value obtained from the previous increment.

The results of the method were shown to be reliable by comparing with experimental results (Fig. 6.51) [17].

6.4.3 Analysis of Results

The method described in Section 6.4.2 has been applied to several structural layouts of beams with three different cross-sections, covering the range of possible shape factors from the extreme case of two concentrated masses to that of the rectangular section.

Three materials with different hardening behaviors have been examined: strongly hardening ($n = 8$), medium hardening ($n = 16$), and weakly hardening ($n = 32$). The structural layouts considered were mainly symmetrical and asymmetrical continuous beams with two or three spans. Different values of span have been considered in order to provide different maximum-moment/minimum-moment ratios and hence different ductility requirements.

Figures 6.52, 6.53 and 6.54 show the behavior of the asymmetrical schemes. The cases under consideration are:

(a) Two equal spans
(b) Two unequal spans
(c) One span completely restrained at one end and simply supported at the other end

If we indicate with L_1 and L the lengths of the spans, cases (a), (b) and (c) correspond to the ratios $L/L_1 = 1, 0.5, 0.0$. The continuous lines represent the behavior of the simulated curve obtained by the method described in Section 6.4.2. The dashed lines represent the diagrams obtained for the same beams, with the assumption of elastic/perfectly plastic material, using the plastic hinge method, which assumes for the limiting moment

Aluminum alloy structures

Fig. 6.51

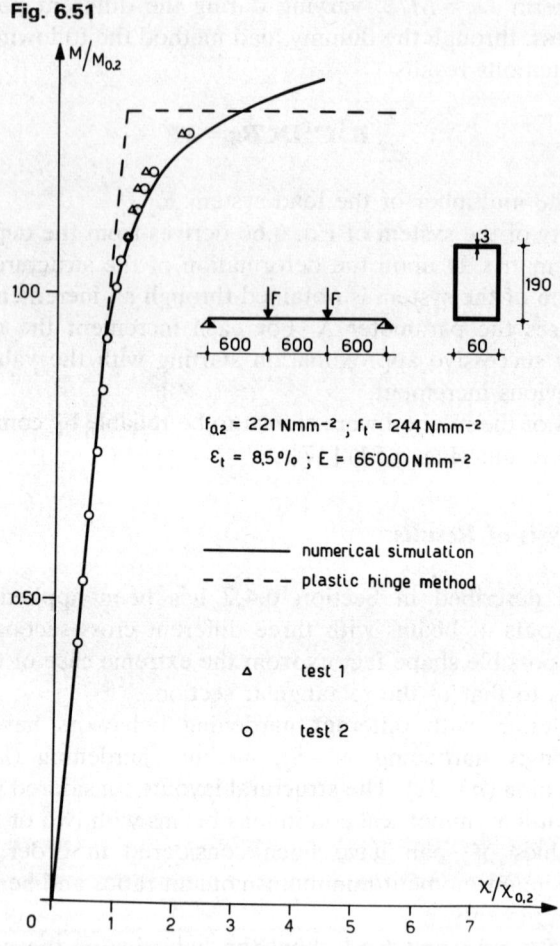

the fully plastic moment with $f_y = f_{0.2}$. The discontinuities of these diagrams correspond to the occurrence of plastic hinges.

The loads and displacements are normalized with respect to their corresponding elastic limit values, which are reached when the $f_{0.2}$ stress is attained in the highly stressed section. In order to make more significant comparisons, the curves which cross each F–v curve at the points in which the highly stressed sections reach a limiting dimensionless curvature equal to $\chi/\chi_{0.2} = 5$ and 10 have also been shown.

This second limit is particularly significant because many structural aluminum alloys exhibit a ductility which just allows this curvature. In

Fig. 6.52

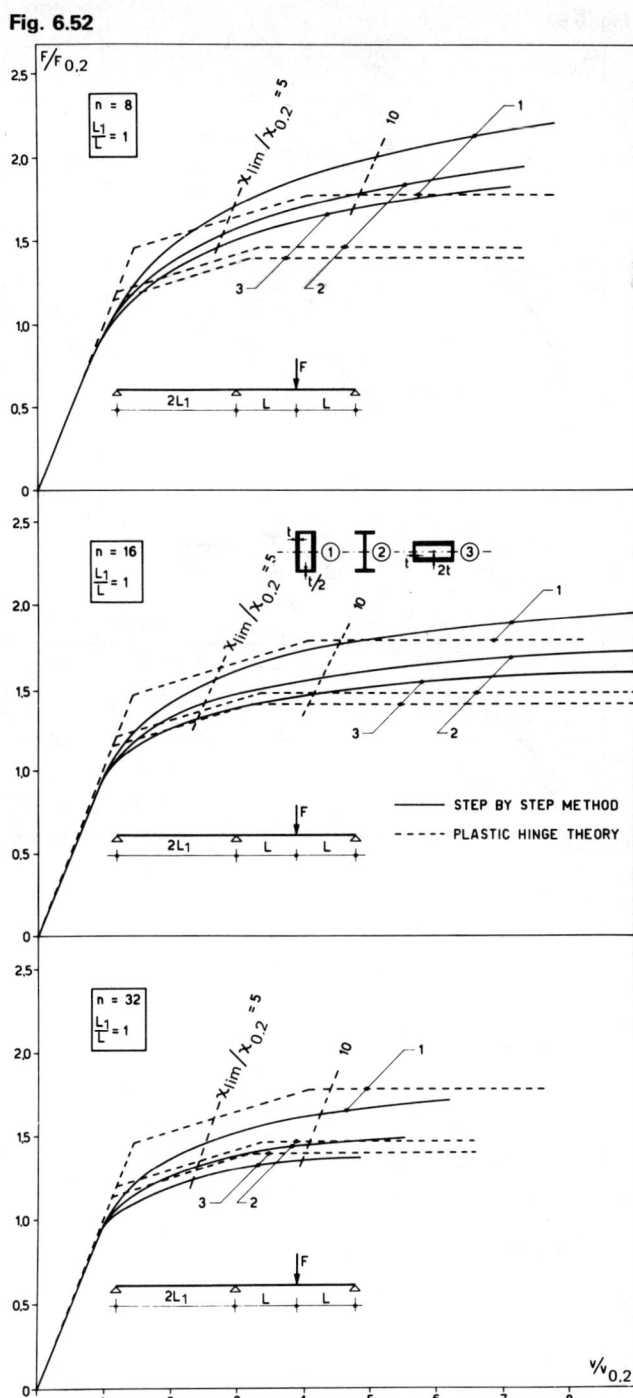

Fig. 6.53

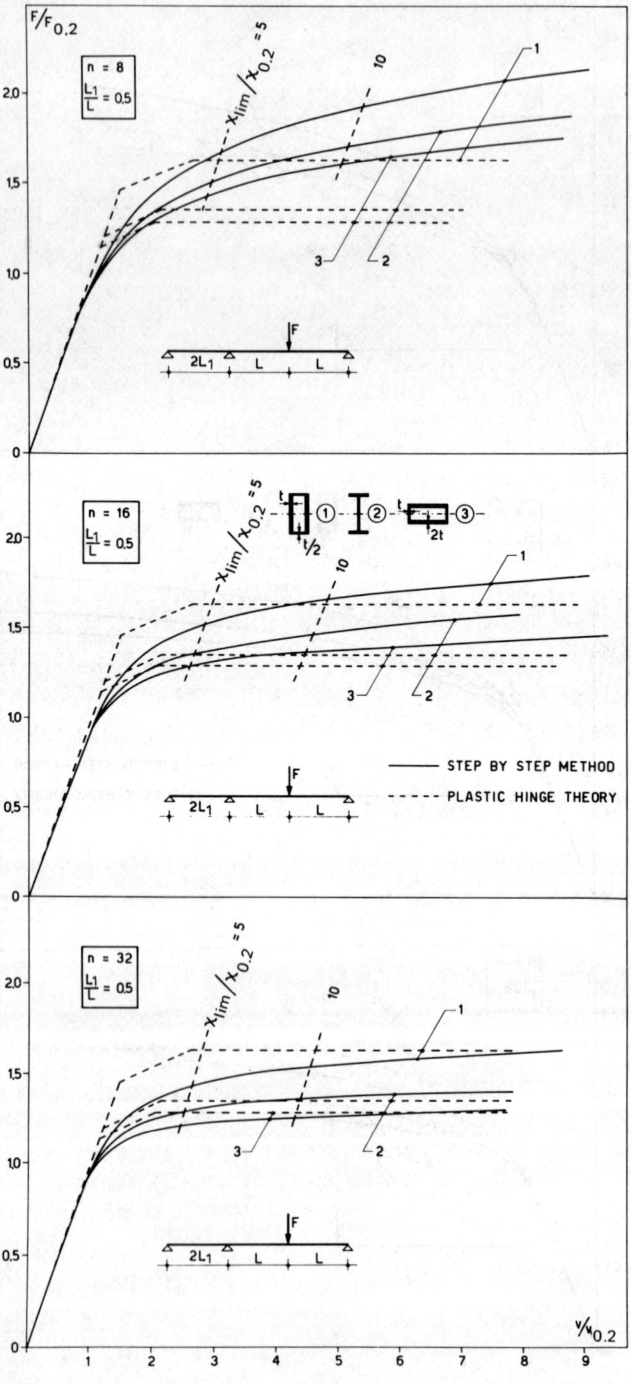

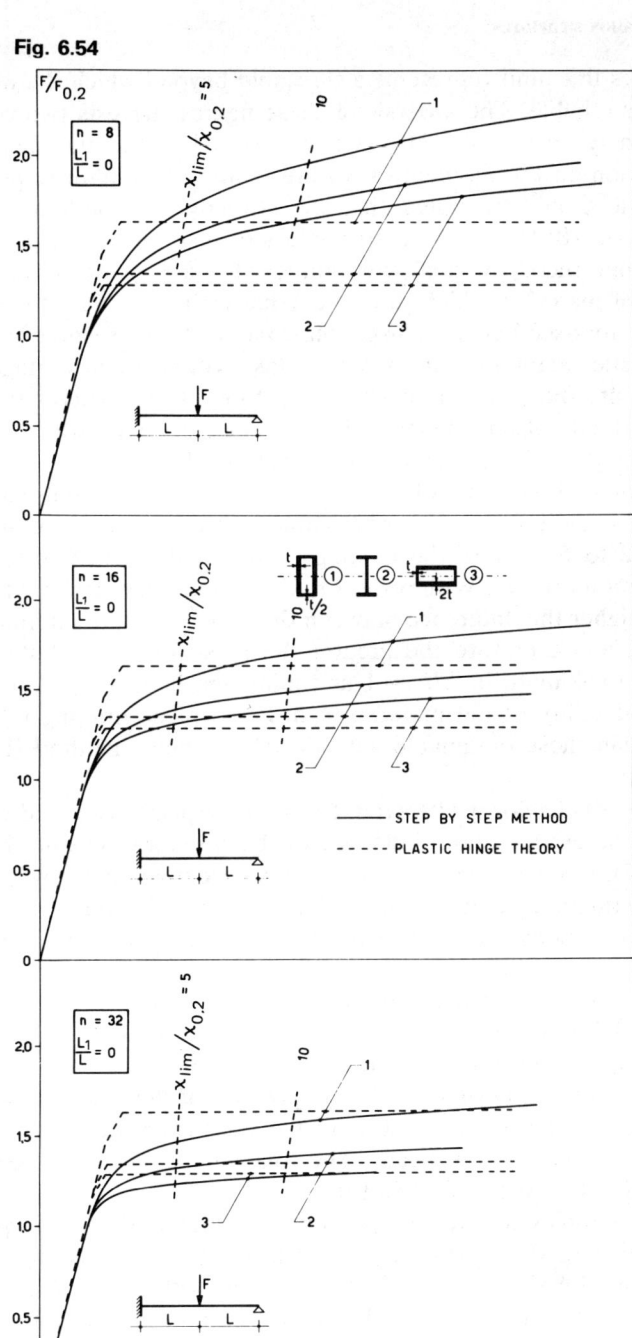

Fig. 6.54

these cases this limit represents a threshold beyond which the material is no longer reliable. The analysis of these figures suggests two considerations with respect to the effects of the $\sigma-\varepsilon$ law and the shape of the cross-section on the moment–curvature relationship. With respect to the material it should be noted that the strongly and medium hardening materials ($n = 8, 16$) reach greater values of resistance than predicted by plastic hinge theory at small curvatures. The same is not always true in the case of materials which show weak hardening ($n = 32$). On the other hand the cross-sections with large shape factors (cross-section 1) require a large plastic adaptation in order to take complete advantage of the strength of the section itself. Therefore, these curves cross the elastic/perfectly plastic curves at larger values of curvature.

The analysis of the ultimate behavior of aluminum alloy structures when compared with that of elastic/perfectly plastic materials can also be extended to the influence of other variables such as the structural layout (Figs 6.52 to 6.57) and the loading condition (Figs 6.58, 6.59) on the redistribution process. With respect to the first problem it should be noted that the higher the difference between the maximum and minimum elastic moments, and therefore the greater the need for redistribution of the moments (maximum in $L/L_1 = 1$ and minimum in $L = 0$), the higher the required ductility of the material in order to reach limiting load values greater than those obtainable with the plastic hinge method (Figs 6.52, 6.53, 6.54).

The same behavior is observed, for all the aspects examined ($\sigma-\varepsilon$ law, shape of the cross-section, influence of the structural scheme: Figs 6.55, 6.56, 6.57), in symmetrical layouts such as a completely restrained beam, a beam with three different spans and a beam with three equal spans. Two cross-sections are considered; the intermediate shape factor is excluded.

The next group of figures (Figs 6.58, 6.59) is related to the comparison between different loading conditions symmetrical with respect to the center of the beam, for the same structural layout of a beam completely restrained and simply supported. The position of the load is identified by the z/L ratio between the distance of the loads from the end of the beam and half the span of the beam. The following values have been considered: $z/L = 0.2, 0.4, 0.6, 0.8, 1.0$.

These examples confirm the previous considerations, and in particular the influence of the cross-section, which for the same loading condition requires a higher ductility in the sections with large shape factors, especially in the case of materials showing little hardening. The influence of the loading condition, in contrast, seems to be less even when varying the ratio between the maximum and the minimum moment from 1.19 to 1.57.

Strength of structural elements

Fig. 6.55

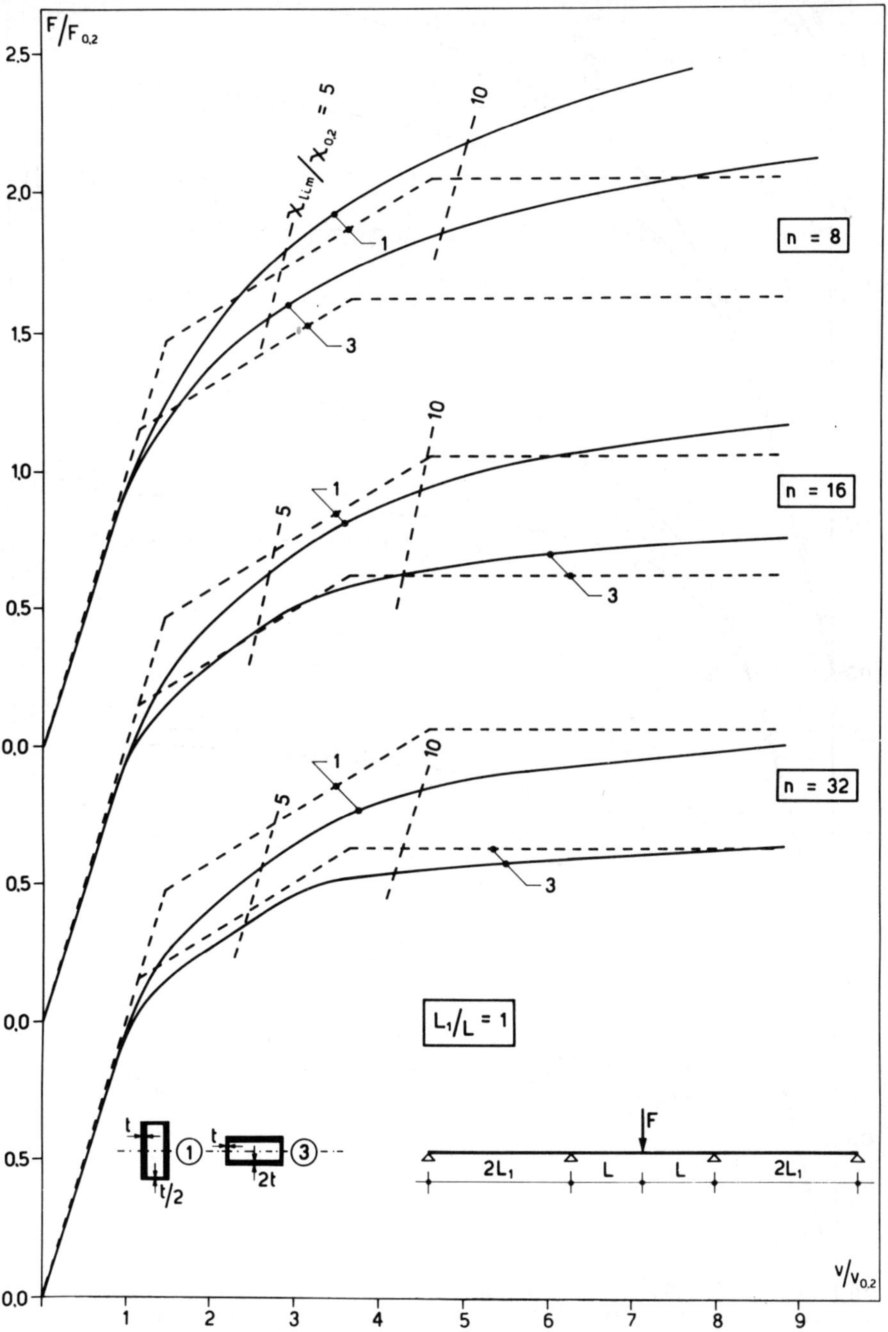

Aluminum alloy structures

Fig. 6.56

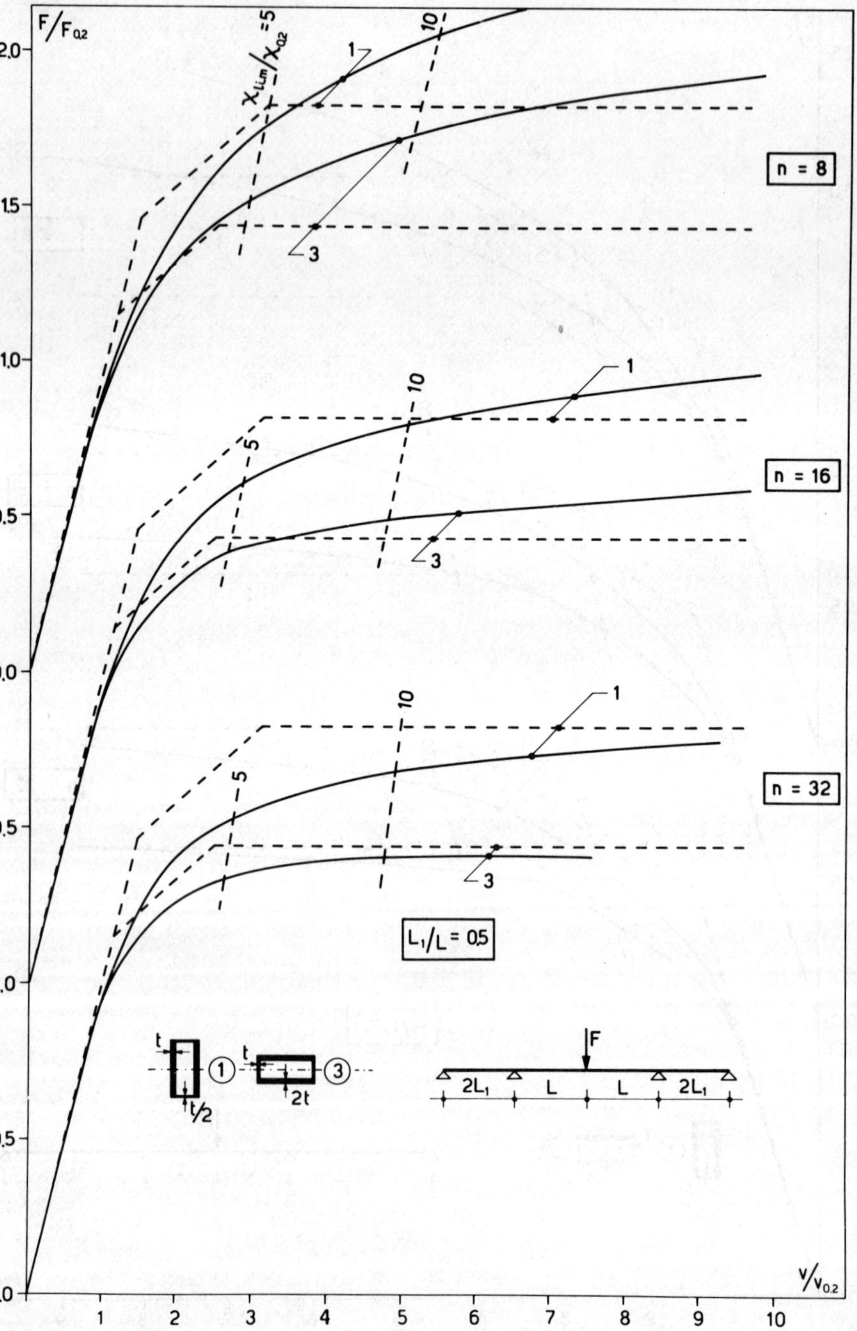

Fig. 6.57

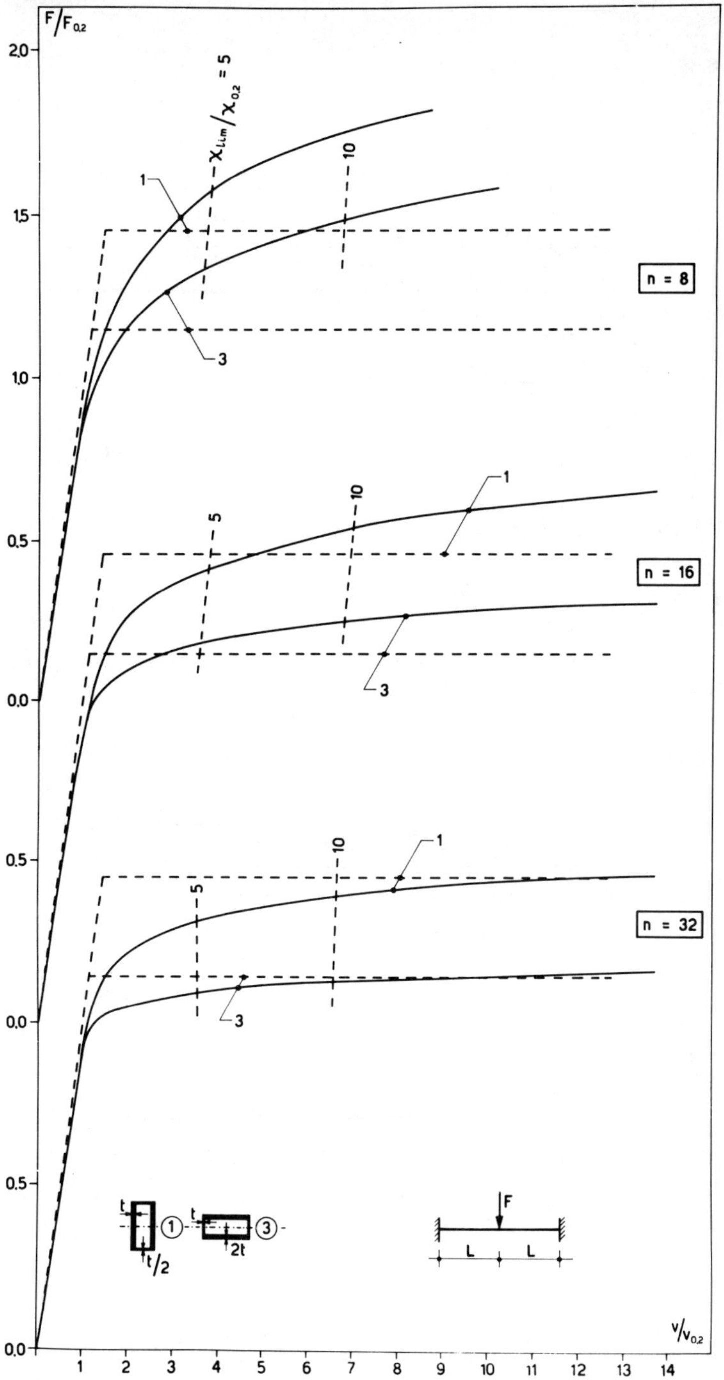

Aluminum alloy structures

Fig. 6.58

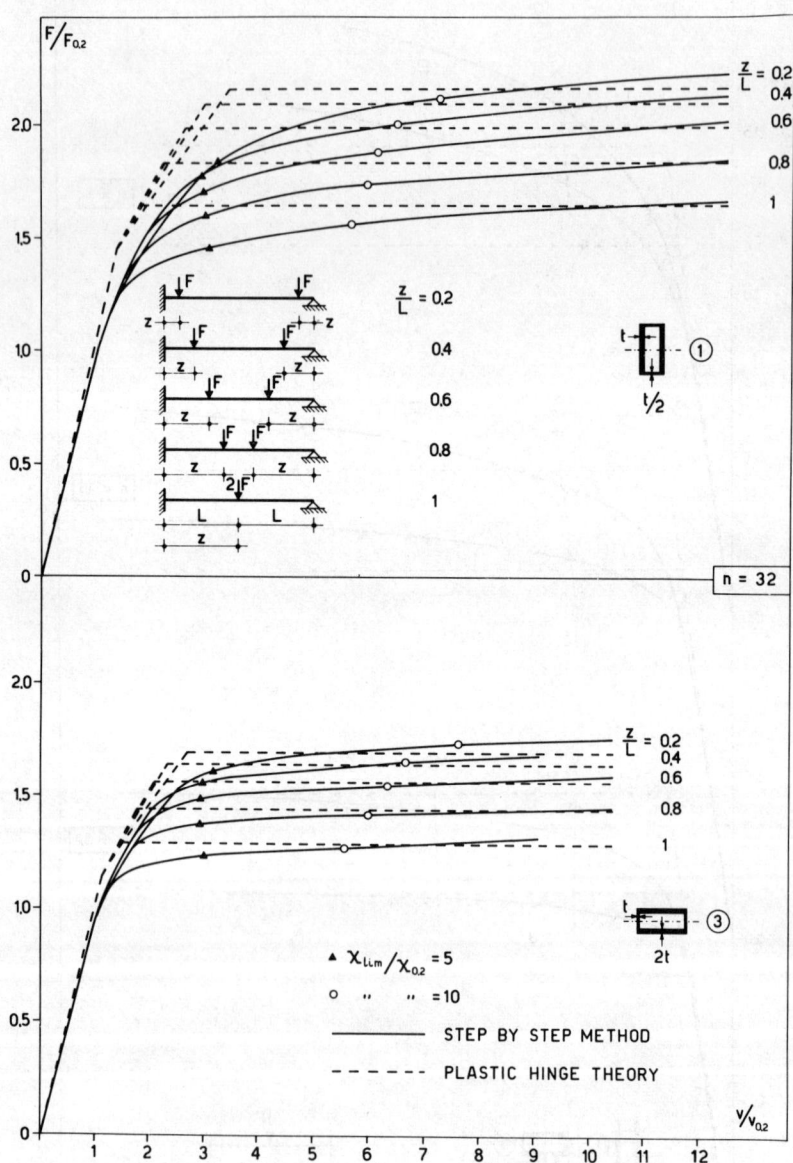

Fig. 6.59

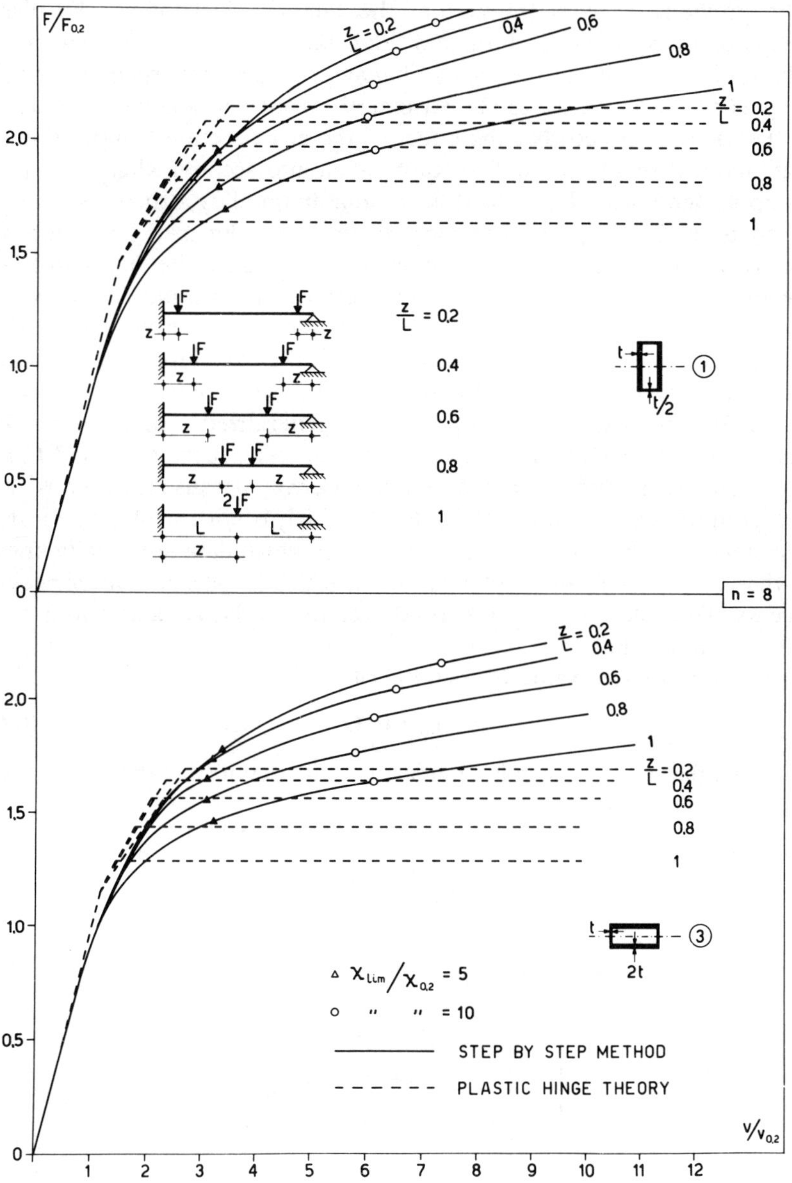

6.4.4 Ductility Requirements [17]

The previous analysis shows that the factors which affect the plastic behavior of continuous aluminum alloy beams can be classified in two different ways: with respect to the behavior of the cross-section and with respect to the influence of the structural layout and loading condition.

It should be noted that the effect of the cross-section is more significant, and it is therefore the first to be taken into account when attempting to apply the plastic hinge method to aluminum alloy structures.

As both the shape of the cross-section and the σ–ε law affect the behavior of the cross-section, it is useful to associate hardening and ductility parameters with all of the materials. This significant representation can be obtained by assigning to each alloy, through its characteristic parameters, a point on a diagram in which the axes represent hardening (n exponent of the σ–ε law) and brittleness, the inverse of ductility.

Whereas the exponent n is given by a generalized expression which is accepted and dependent upon known parameters, such as $f_{0.2}$ and f_t, it is more difficult to define ductility in a precise way because it depends upon the uniform deformation ε_u (Fig. 6.60), which is not usually known from the standard tensile test. The parameter for which data are usually known is the ultimate percentage elongation ε_t, which is measured after necking occurs. The value of ε_u is not strictly related to the ε_t value and is often 30 percent smaller.

If a conservative value is assumed of

$$\varepsilon_u \cong 0.5 \varepsilon_t \tag{6.67}$$

then the parameter which is significant to the ductility of the material can be assumed:

$$\Delta = \frac{\varepsilon_u}{\varepsilon_{0.2}} = \frac{\varepsilon_t E}{2 f_{0.2}} \tag{6.68}$$

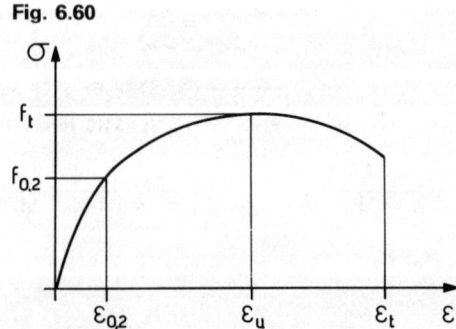

Fig. 6.60

Strength of structural elements

Fig. 6.61

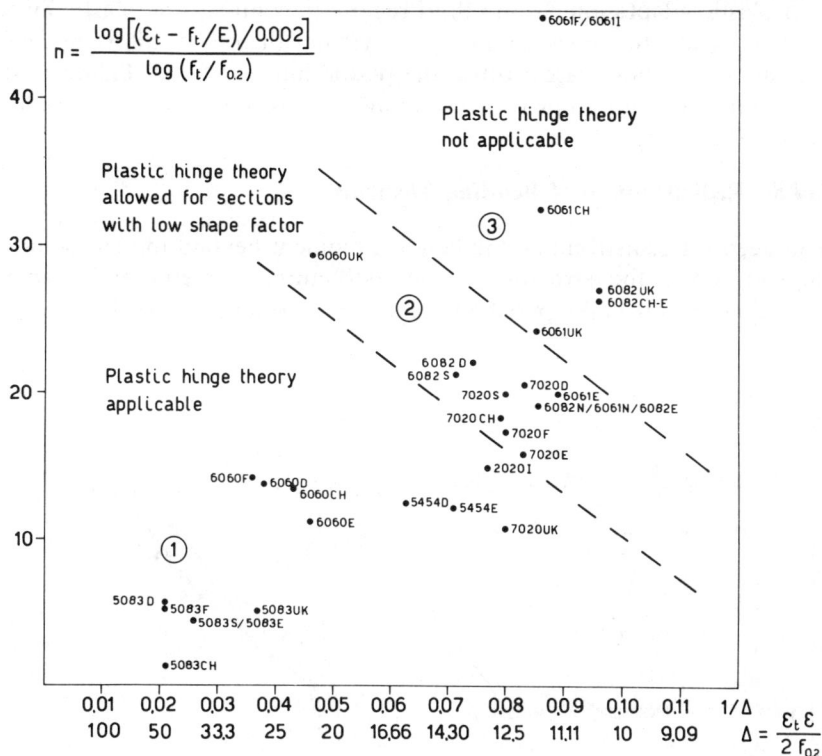

which numerically expresses the extension of the inelastic behavior of the material.

Under this hypothesis the alloys considered by the ECCS committee for Aluminum Alloy Structures have been drawn in Fig. 6.61. The data for these alloys are given in Fig. 3.4. Points of the alloys are also shown, with their numerical designation.

On Fig. 6.61 it is also possible to represent, for a given alloy, a number of points which correspond to the ductility requirements for different structural layouts, loading conditions and cross-section shapes in order to obtain an ultimate load at least equal to that given by the plastic hinge method.

This analysis, developed with respect to the cases examined in Section 6.4.3, leads to three zones being defined in the figure. In the first region the materials are characterized by sufficient ductility and hardening to apply the plastic hinge method in all cases. In the second region, characterized by less ductility and hardening, the applicability of the method is

275

Aluminum alloy structures

limited to those sections with a small shape factor which do not require high plastic adaptation. In the third region, in contrast, the plastic hinge method usually provides limiting values greater than those allowable, and it is unsafe in these cases to use the plastic hinge method. Figure 6.61, though qualitative, can provide useful indications of the ductility requirements of different alloys.

6.4.5 Redistribution of Bending Moments

The need for redistributing the bending moment beyond the elastic limit depends upon the structural layout (structure geometry and loading condition), whereas the possibility of so doing depends upon the ductility of the material.

Fig. 6.62

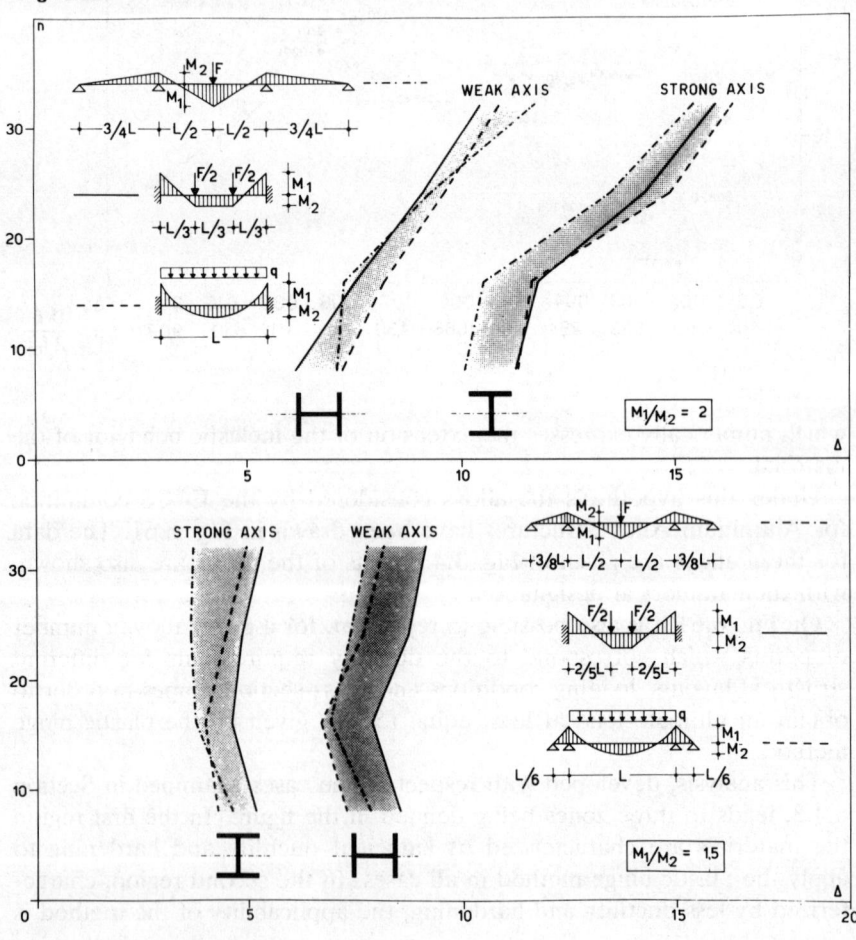

Fig. 6.63

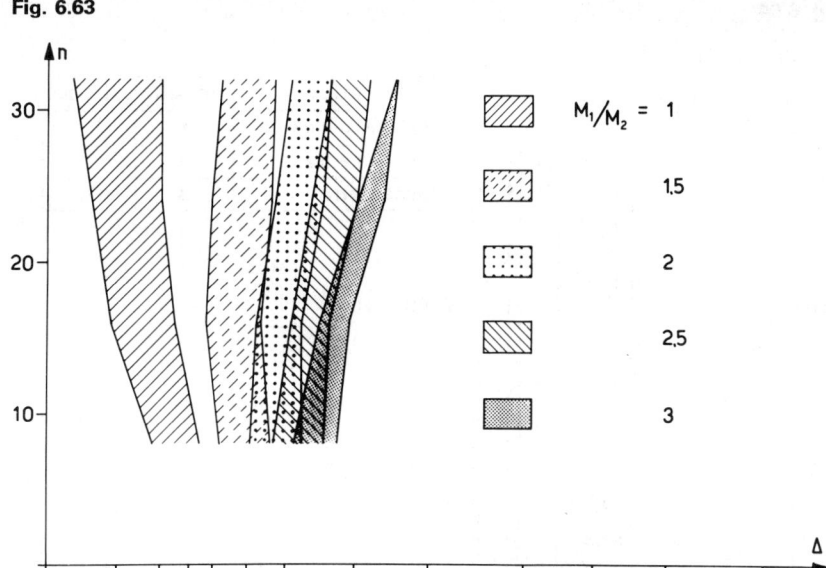

A parameter which can characterize at the same time the redistribution requirements and the ductility requirements can be identified as the ratio between the maximum and minimum moments, M_1 and M_2, representative of all the structural layouts (Fig. 6.62). From the analysis of different layouts, corresponding to M_1/M_2 values varying from 1 to 3 (Fig. 6.63) and for different values of the shape factor between the limits 1 and 1.5, it is evident that the representation in the $n-\Delta$ plane (which relates the hardening and ductility values of the materials) identifies regions which can be considered independent of n.

On the basis of this consideration, if we assume $10\ n = f_{0.2}$, it is possible to provide an alternative, even if complementary, representation to that of Section 6.4.4. In this representation (Fig. 6.64) the material ductility properties are superimposed on the redistribution requirements of the structural layout by means of the M_1/M_2 ratio. The symbols corresponding to several non-heat-treated (1 to 37) and heat-treated alloys (37 to 77) are plotted in the same figure, together with stars representing the different mild and high strength steels (78 to 88).

The mechanical properties corresponding to these materials are given in Fig. 6.65. Each point, representative of a structural material, lies on a given redistribution range (M_1/M_2). This indicates that any structural layout, characterized by the M_1/M_2 value, can be evaluated at collapse by means of the plastic hinge method without danger of early brittle fracture.

Aluminum alloy structures

Fig. 6.64

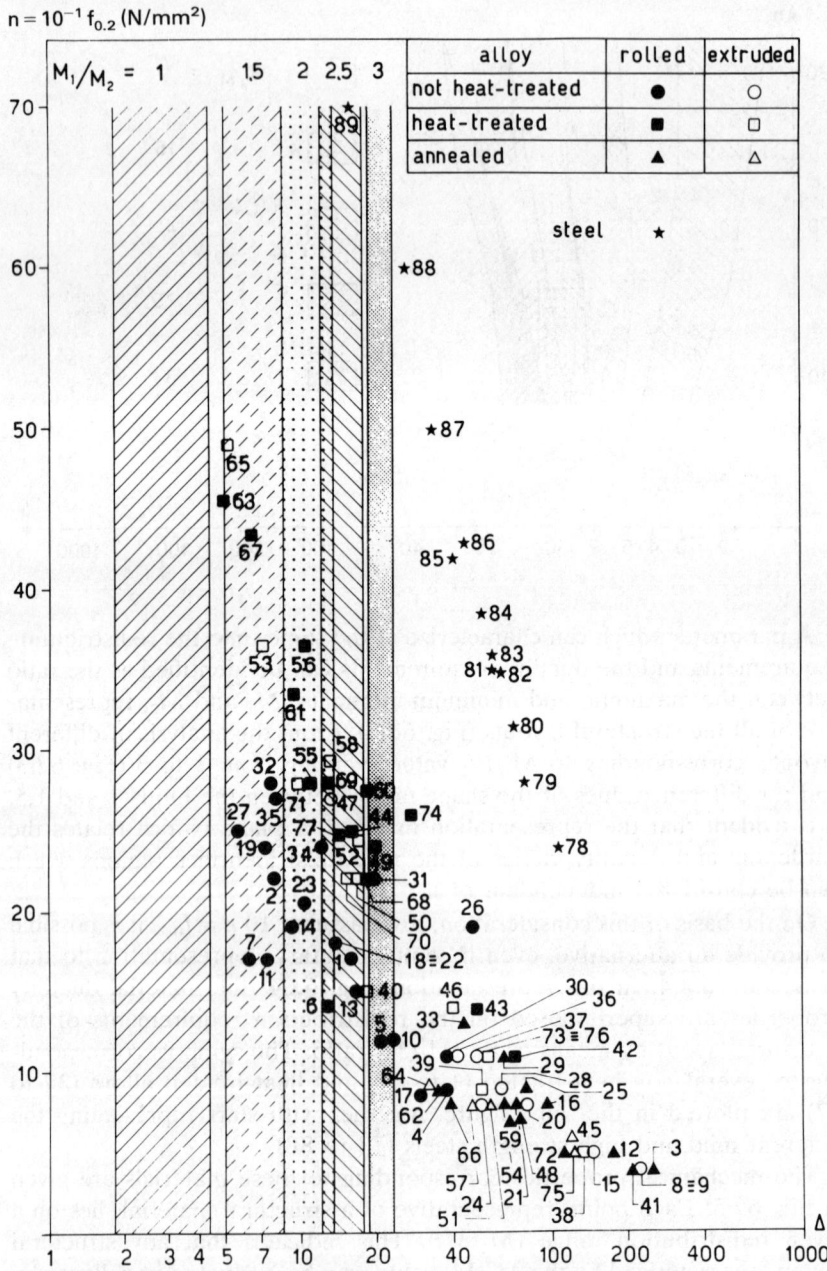

Fig. 6.65 Mechanical properties of metallic materials

	Designation						ε_t (%)	
No.	Formula	Numerical designation	Stage	Thickness (mm)	f_t (N mm^{-2})	$f_{0.2}$ (N mm^{-2})	As-rolled	Extruded
1	AlMn1.2Mg	3004	H25	0.6–3	220	170	6	
2			H60		260	215	4	
3	AlMn1.2Cu	3003	R	all	100	35	25	
4			H15	0.8–6	120	85	9	
5			H30	0.8–6	135	120	7	
6			H45	0.8–6	155	135	5	
7			H60	0.8–4	185	165	3	
8			Hp	—	100	40		25
9	AlMg0.8	5005	R	all	100	40	25	
10			H25	0.8–6	135	120	8	
11			H60	0.8–4	185	165	4	
12	AlMg1.5		R	all	135	50	25	
13			H25	0.8–6	165	145	7	
14			H60	0.8–4	195	185	5	
15			Hp	—	135	50		20
16	AlMg2.5	5052	R	all	155	80	22	
17			HL	8–75	165	85	7	
18			H20	0.8–6	215	175	7	
19			H40	0.8–4	255	235	5	
20			Hp	—	155	80		18
21	AlMg2.7Mn	5454	R	all	205	80	16	
22			H20	0.8–6	245	175	7	
23			H40	0.8–4	265	200	6	
24			Hp	—	80			12
25	AlMg3.5		R	all	205	90	20	
26			H20	0.8–6	245	185	12	
27			H35	0.8–4	285	245	4	
28			Hp	—	215	90		16
29	AlMg4.4	5086	R	all	245	110	20	
30			HL	8–75	255	110	12	
31			H15	0.8–6	285	215	12	
32			H30	0.8–4	325	275	6	
33	AlMg4.5	5083	R	all	275	130	16	
34			H10	0.8–6	305	235	8	
35			H20	0.8–4	345	265	6	
36			Hp	—	265	110		12
37	AlMg5	5056A	Hp	—	245	110		16
38	AlMgSi	6060	TaN	≤3	135	50		20
39			TaA14	≤3	155	90		14
40			TaA16	≤3	195	145		8
41	AlSi1MgMn	6082	R	all	90	40	25	
42			TN	0.8–4	205	110	22	
43			TA14	0.8–4	235	135	20	
44			TA16	0.8–4	295	245	11	
45			R	—	110	60		20
46			TA14	—	235	135		16

(contd.)

Aluminum alloy structures

Fig. 6.65—(continued)

	Designation						ε_t (%)	
No.	Formula	Numerical designation	Stage	Thickness (mm)	f_t (N mm^{-2})	$f_{0.2}$ (N mm^{-2})	As-rolled	Extruded
47			TA16	—	315	265		10
48	AlCu4MgMn	2017A	R	5–20	165	70	14	
49			THN	3.5–20	380	235	13	
50			TN	≤38	375	215		10
51	AlCu4.4SiMnMg	2014	R	all	185	80		12
52			TN	all	345	240		12
53			TA	9–20	410	370		7
54	AlCu4.5MgMn	2024	R	≤20	175	80	14	
55			THN	0.5–20	425	275	8	
56			TH06N	0.5–13	475	360	10	
57			R	all	195	80		11
58			TN	≤38	390	290		10
59	AlCu4.5MgMn		R	0.5–20	165	70	14	
60			THN	0.5–7	405	270	14	
61			TH06N	0.5–13	425	330	9	
62	AlZn4.8MgCu	7075	MR	≤20	185	90	9	
63			TA	0.5–20	520	445	6	
64			R	all	185	90		9
65			TA	≤38	540	480		7
66	AlZn5.8MgCu		R	0.5–20	185	80	9	
67			TA	1–13	495	425	8	
68	AlZn4.5Mg	7020	TaN	0.5–12	315	215	12	
69			TaA	0.5–12	355	275	10	
70			TaN	all	315	215		10
71			TaA	all	355	275		8
72	AlMg1SiCu	6061	R	all	100	50	16	
73			TN	0.8–6	205	110	15	
74			TA16	0.8–6	295	245	10	
75			R	—	110	60		16
76			TN	—	175	110		16
77			TA16	—	265	235		9
78	Fe360			≤16	360	235	26	28
79	Fe430			≤16	430	275	23	24
80	Ex-TEN 45				445	310	21	
81	Ex-TEN 50				480	345	20	
82	COR-TEN A				480	345	22	
83	Fe510			≤16	510	355	21	22
84	Ex-TEN 55				515	375	19	
85	COR-TEN C				550	410	20	
86	S420			<16	540	420	19	
87	S490			<16	570	490	16	
88	S590			<16	640	590	15	
89	S690			<16	780	690	14	

For example, alloy 67 can be used in all the layouts with $M_1/M_2 \leq 1.5$, but it cannot be used when $M_1/M_2 > 1.5$.

Since in the commonly used structural layouts the value M_1/M_2 is always less than 3, it can be concluded that all the materials which fall in the unshaded region on the right of the figure can be used in plastic analysis without any particular limitation, whereas for the other materials the ductility requirements given by Fig. 6.64, which depend upon the structural layout (M_1/M_2), have to be taken into account.

In the range of activity of ECCS Committee T2 (task group 2.2) further studies [20–23] in aluminum alloy plastic design are now progressing, giving a general settlement of the theoretical assumptions [20] and making numerical and experimental applications [21, 22] not only for extruded, but also for welded members [23].

References

1. Moore, R. L. and Holt, M., *Beam and torsion tests of aluminum alloy 61S-T tubing*, National Advisory Committee for Aeronautics, Tech. Note No. 867 ALCOA, 1942.
2. Cozzone, F. P., Bending strength in the plastic range, *J. Aeronautical Sciences*, **10**, No. 5, 1943, pp. 137–51.
3. Dwight, J. B., *An investigation into the plastic bending of aluminum alloy beams*, Aluminum Development Association Report No. 16, May 1953.
4. Baehre, R., Brochner, I. and Sjolund, J., *Untersuchungen zur Anwendung der plastischen Tragwerksbemessung bei Aluminium-konstruktionen*, Tekn. Dr. A. Johnson Ingenjorsbyra Tekn. Medd., No. 15, Stockholm, 1965.
5. Baehre, R., Das Tragverhalten von biegungsbeanspruchten statisch bestimmten und unbestimmten Balken aus elastoplastischem Material—theoretische und experimentelle Untersuchungen, *Acta Polytechnica Scandinavica*, 1968, No. 51.
6. Baehre, R., *Theoretische und experimentelle Untersuchungen uber die Bemessungsgrundlagen fur Tragwerke aus elastoplastischem Material*, Arne Johnson Tech. Paper, No. 21/22, Stockholm, 1968.
7. Marsh, C., Limit state design and collapse of aluminum structures, *Proceedings of the 7th Int. Light Metal Congress*, Vienna, June 26 1981.
8. De Martino, A., Faella, C. and Mazzolani, F. M., Sul calcolo plastico per strutture a comportamento incrudente (Plastic design for strain-hardening structures), *Giornale Italiane della Costruzione in Acciaio CTA*, Verona, 1977.
9. De Martino, A., Faella, C. and Mazzolani, F. M., *Plastic adaptation for alu-alloy members in bending*, ECCS Committee 16, Doc. 16-78-1, May 1978.
10. De Martino, A., Faella, C. and Mazzolani, F. M., The use of plastic adaptation coefficients in the limit state design of steel shapes, *Costruzioni Metalliche*, 1981, No. 2.

Aluminum alloy structures

11. De Martino, A., Faella, C. and Mazzolani, F. M., Considerazioni critiche sul metodo dell'adattamento plastico (Critical remarks on the plastic adaptation method), *Giornale Italiane della Costruzione in Acciaio CTA*, Palermo, October 1981.
12. De Martino, A. and Faella, C., *First results about the inelastic behavior of aluminum alloy structures*, ECCS Committee 16, Doc. 16-76-2, 1976.
13. Marincek, M., *Flexural behavior of aluminum alloy structures beyond the elastic limit*, ECCS Committee 16, Doc. 16-77-6, 1977.
14. Faella, C., Comportamento flessionale delle barre in alluminio al di là dei limiti elastici (Flexural behavior of aluminum members beyond elastic limits), *La Ricerca*, September–December 1976.
15. Mazzolani, F. M., Design bases and strength of alu-alloy structures, *Proceedings of Jornadas Tecnicas sobre Estructure en Aluminio*, Bilbao, November 16–17, 1978.
16. De Martino, A. and Faella, C., Il calcolo plastico delle strutture in leghe di alluminio (Plastic design of aluminum alloy structures), *Costruzioni Metalliche*, 1978, No. 2.
17. De Luca, A., *Inelastic behavior of aluminum alloy continuous beams*, Istituto di Tecnica delle Costruzioni, Quaderni di Teoria e Tecnica delle Strutture, Rep. No. 506, 1982.
18. DTU, *Règles de conception et de calcul des charpentes en alliages d'aluminium*, First edn, 1971; Second edn, 1976.
19. ECCS, *European recommendations for aluminum alloy structures*, First edn, 1978.
20. Mazzolani, F. M., *Plastic Design of Aluminum Alloy Structures*, 'Verba volant, scripta manent', pp. 295–313, Liegi, 1984.
21. Valtinat, G. and Dangelmaier, P., *Application of the plastic theory to non welded and welded aluminum members*, Doc. ECCS-T2, 1984.
22. Mazzolani, F. M., Capelli, M. and Spasiano, G., *Plastic analysis of aluminum alloy members in bending: comparison between theoretical and experimental results*, Doc. ECCS-T2, October 1984.
23. De Martino, A., Faella, C. and Mazzolani, F. M., *Inelastic behaviour of aluminum double-T welded beams: a parametric analysis*, Doc. ECCS-T2, October 1984.

7 Stability of structural elements

7.1 General principles

7.1.1 ECCS Activities [1–24]

Recent developments in the use of aluminum alloys for structural applications have led to an increasing research into the instability problems of aluminum alloy structures.

Since 1970 the ECCS committee on aluminum alloy structures has carried out extensive studies and research on this subject [1]. Several theoretical and experimental research programs have been undertaken in this field in order to investigate the mechanical properties of the materials, their imperfections and their influence on the instability of members. The numerous tests carried out have enabled a statistical evaluation of the results. These data have allowed, for the first time, the characterization of aluminum alloy members as industrial bars in accordance with the most recent trends of safety principles in metallic structures (see Section 2.3). These results were also extrapolated to other types of structure through a numerical analysis which makes use of computer simulation.

Among the research programs in this field, undertaken with the cooperation and support of several European countries, were the tests on extruded members carried out at Liège University (1970–1972) [2, 4, 5]. There were also tests on welded built-up members carried out at Liège University in cooperation with the University of Naples and the Experimental Institute for Light Alloys of Novara (1974–77) [10, 12, 15, 16].

The analysis of these experimental results demonstrated the major differences between the behavior of steel and aluminum. The computational methods of the European recommendations have been based on this analysis. In particular the critical curves (a), (b) and (c) (see later) valid for bars with different cross-sections and of different materials, have been defined.

Other important results have come about thanks to modern simulation methods [17, 18, 19, 25, 26], which allow all the geometrical and mechanical properties, together with their imperfections, to be taken into account. The widespread use of computers has allowed satisfactory results to be obtained in the study of the instability phenomena of columns and beam columns.

Research is still needed in other fields, such as the local buckling of thin plates and their interaction with the global behavior of the bar [23], the instability of two-dimensional elements (plates, stiffened plates, web panels), and the post-buckling behavior of deep plate girders. These matters are currently being examined by the recently established (1980) ECCS committee TG/T2-2 on the strength and stability of structural members (chairman Mazzolani). The results of these studies will be used in the next edition of the European recommendations [68].

7.1.2 Numerical Simulation

The classical methods of analysis of instability problems are based upon the bifurcation of equilibrium of 'ideal bars' made of homogeneous material and having perfect geometry. In reality, because of the fabrication process which introduces 'structural and geometrical imperfections' into the bar (see Sections 2.4 and 2.5), the 'ideal bar' is nonexistent.

The modern approach for studying instability problems is to take account of the actual imperfections present in the bar. This approach is not possible if the classical stability theory is adopted. Changing from the ideal bar to the industrial bar, as defined in Section 2.3, requires the following steps:

(1) Several tests are undertaken to identify the mechanical properties and imperfections of each structural member.
(2) An analytical method is determined which allows simulation of the actual behavior of the bar under given loads, incorporating the experimental data acquired in the previous step.

Having defined the industrial bar, the research has been primarily devoted to simulation methods because these seem to be the best method of accounting for all of the complex phenomena involved with instability – such as the inelastic strain-hardening behavior of aluminum alloys and the post-elastic behavior of the members.

Simulation methods are usually based upon numerical techniques, such as finite difference or finite element, which make use of mathematical models with fine discretizations. The most sophisticated of the numerous methods allow real structural behavior to be closely followed. In fact, it is possible to introduce the material $\sigma-\varepsilon$ law and to take account of the unavoidable geometrical and mechanical imperfections produced by the industrial fabrication process, which cause an actual structure to behave differently from the ideal one.

However, it is necessary to correlate the simulation methods with the experiments which provide the input data for the numerical analysis. When the input data represent the statistical information acquired from systematic experimentation, the output of the simulation method represents an equivalent test, and therefore the analysis can be considered as semi-experimental. Thus it is necessary to make a preliminary calibration of each simulation method so that the results conform with the experimental evidence.

If this check is satisfactory, the computer can be considered as a large universal testing machine which takes a structural element and applies a constant or variable load. The computer then pulls, bends, torques or compresses the element to failure and measures the stresses, the strains and the displacements. In a few seconds it analyzes all these results and gives us the answers requested. Such a test on the computer has great advantages over the analogous laboratory test: it is simpler, faster and more economical.

From what has been said it can be concluded that there are distinctive tasks to be performed by the laboratory and the computer. Laboratory tests characterize the real structural elements through the measurement of the σ-ε material law and the distribution of the imperfections. Computers will use these input data to test the structures and to extrapolate from the actual experiments to other types of material and geometrical shape, thus saving time and money.

The simulation method developed by Faella and Mazzolani (1974) is now given [17–19]. This method, although a general one, has been especially conceived for analyzing aluminum alloy bars and was used in order to confirm the calculation methods suggested in the ECCS European recommendations.

Consider a pin-ended bar subjected to an eccentric load N in a major bending plane (Fig. 7.1, part 1). The analysis of the stable and unstable behavior of the bar uses the following hypothesis:

(a) Cross-sections remain plane.
(b) Curvature can be approximated with the second derivative.
(c) The deformed shape of the bar remains in a given plane.

The first two hypotheses represent the classical beam theory hypothesis, and in particular the second neglects the effects of large displacements which can be considered negligible in this particular problem. The third assumption limits the deformation of the bar to a given plane defined by the z and y axes (Fig. 7.1, part 2), and does not take account of other effects such as flexural torsion and twist.

Aluminum alloy structures

Fig. 7.1

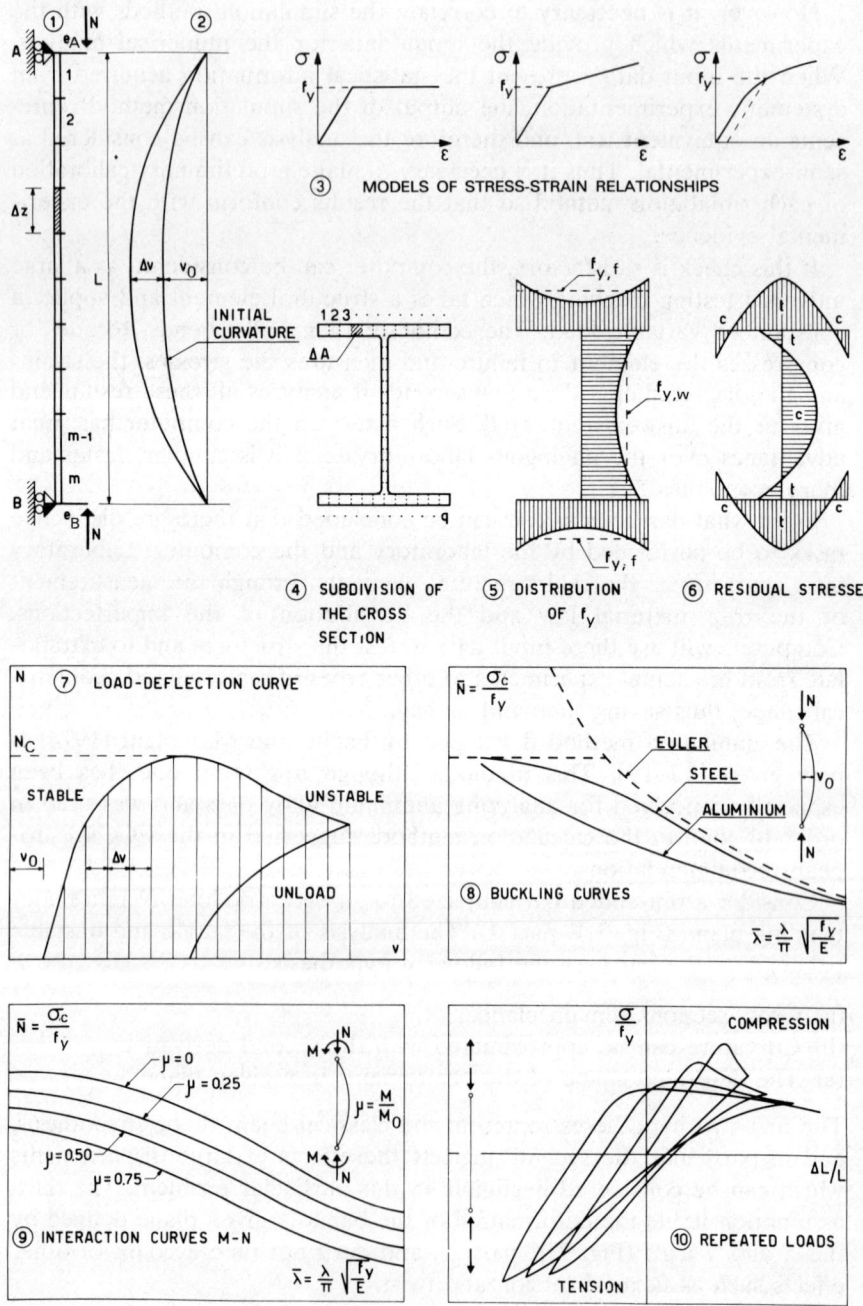

From the initial deformation $v(z)$, known for the unloaded bar, each variation of load causes the deformation to change. In each step of the incremental analysis, each part of the bar is in equilibrium under the axial load

$$N = N_{\text{applied}} \tag{7.1}$$

and the bending moment

$$M = N(v_0 + v) + Ne_A + N(e_B - e_A)\frac{z}{L} \tag{7.2}$$

For equilibrium:

$$N = \int_A \sigma \, dx \, dy \tag{7.3}$$

$$M = \int_A \sigma y \, dx \, dy \tag{7.4}$$

where A is the area of the cross-section of the bar referred to the principal axes x and y.

The deformation of the bar is characterized by a constant strain $\bar{\varepsilon}$ and a rotation χ about the center of gravity. The elementary strain is therefore

$$\varepsilon = \bar{\varepsilon} - \chi y$$

where, because of hypothesis (b),

$$\chi = \frac{d^2 v}{dz^2} \tag{7.5}$$

The material is characterized by a generic σ–ε law (Fig. 7.2), such as the Ramberg–Osgood general law (see Section 2.2.5.3):

$$\varepsilon = \frac{\sigma}{E} + \varepsilon_0 \left(\frac{\sigma}{f_{\varepsilon_0}}\right)^n \tag{7.6}$$

This law can interpret simpler models (Fig. 7.1, part 3) such as elastic/perfectly plastic materials (mild steel), elastic-hardening materials (high-strength steels) and inelastic materials (aluminum alloys). The law is used in those fibers in which the local deformation increases (region OAB of Fig. 7.2). Local unloadings are interpreted elastically up to the change of sign of stress (region BH of Fig. 7.2) and then follow the same law in the opposite region (region HDI of Fig. 7.2) appropriately translated in accordance with the Bauschinger effect (see Section 2.5.3).

The solution of the problem is given by solving the six Eqs 7.1–7.6 in the six unknowns M, N, $\bar{\varepsilon}$, χ, σ and ε. These equations are solved at each load increment by a procedure which discretizes the longitudinal and

Aluminum alloy structures

Fig. 7.2

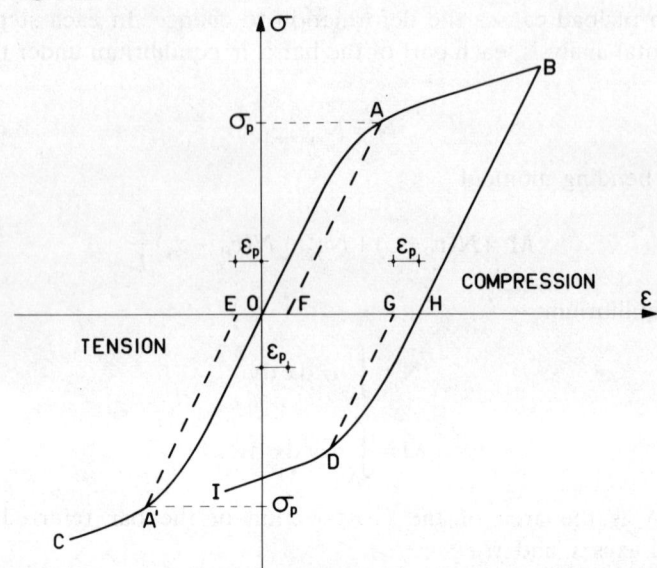

transverse cross-section of the bar and makes use of an iterative finite difference method. The bar is longitudinally cut into m parts of length $\Delta z = L/m$. Each of the $m-1$ internal sections obtained in this way is then divided into q elementary areas ΔA, in which stress and strain are assumed to be constant, with the load acting at the center of gravity of each area (see Fig. 7.1, part 4).

Each elementary area is associated with a value of the yield stress f_y (or $f_{0.2}$) in order to take into account the inhomogeneous strength distribution (see Fig. 7.1, part 5) and the local value of residual stress (see Fig. 7.1, part 6).

Whereas the longitudinal discretization only depends on the required accuracy of the solution, the cross-section cuts are governed by the necessity of considering the effects of structural imperfections.

The numerical simulation analysis consists of the following steps:

(a) An initial deformation of the bar $v_0(z)$ is assumed, which represents the deformed shape in equilibrium for $N=0$. This deformed shape can be of any configuration (parabolic, sinusoidal etc.) and is characterized by the value v_0 of midspan deflection.
(b) An appropriate parameter of deformation is increased (for example the midspan deflection in the case in which $e_A = e_B = e$ and $e_A = e_B = 0$, or the rotation of one end in the case in which $e_A \neq e_B$ and particularly when $e_A = -e_B$) by a quantity Δv. The deformation of the

bar is computed, starting from the values of the cross-section for the present values of N and M, i.e. those values determined from the previous increment. The incremental deformation is given by the following iterative formula:

$$v_{i-1} = 2v_i - v_{i+1} + \chi_i(M, N)\Delta z^2 \tag{7.7}$$

which allows the values of v_{i-1} to be identified when the values of v_i and v_{i+1} are known.

(c) The third step consists of applying Eqs 7.1 and 7.2 to guarantee translational and rotational equilibrium of the bar for a given value of the load N. Compatibility of the Δz portions is given by Eq. 7.7, although an external incongruency v^* at the supports can occur.

(d) In order to eliminate this incongruency v^*, the load N is varied by a quantity ΔN and the deformation is then checked until there is compliance with the external compatibility condition.

(e) When equilibrium and compatibility under the load $N+\Delta N$ have been obtained, the deformation is incremented in the manner previously explained. The incremental analysis carried out in this way gives the N–v relationship in the post-elastic stable and unstable range (see Fig. 7.1, part 7).

(f) In the cases in which only the collapse load (N_{max}) is required, the procedure can be stopped when the N–v curve begins to descend. In order to obtain this collapse load N_{max} more precisely, the last increment Δv must have a finer discretization $\Delta v'$. The value of N_{max} is usually used to obtain the N–λ buckling curves which are used in specifications to check the stability of columns (see Fig. 7.1, part 8).

(g) This procedure allows the N–λ curves to be obtained for a given value of the bending moment applied to the end of the bar (curves of bars under bending and compression), which are useful in specifications when checking bars subjected to compression and bending (see Fig. 7.1, part 9).

(h) The possibility of analyzing the behavior while unloading the bar, allows the post-elastic behavior of bars under alternating loads of opposite sign to be considered (see Fig. 7.1, part 10).

The main program makes use of some subroutines:

SHAPE describes the shape of the cross-section of the bar, which can be of any type
STIFFN relates the axial load N and the bending moment M to the deformation parameters of average strain $\bar{\varepsilon}$ and curvature χ
SIGMA takes into account the σ–ε law of the material and also the mechanical imperfections
DELTA establishes the values of the increment of the load ΔN with respect to the incongruency v^*, for use during step (d).

Aluminum alloy structures

The results of this method have been checked against other methods, such as that of Beer and Schulz [25], which has been used to obtain the European curves of instability for steel. It has also been checked against the method of Frey and Massonnet [26] in order to work out the European curves for aluminum alloy structures.

Beer and Schulz's method consists of two separate phases. First the moment–curvature relationship is established for given values of the axial load N. The shape factor of the cross-section is taken into account together with the residual stresses. The relationship between the midspan deflection and slenderness is then computed by a double iteration. In this way the maximum value of λ for each value of the assigned axial stress is found and therefore the N–λ curve can be drawn. This procedure, though very fast, is not particularly suitable for those more complex post-elastic applications in which unloading has to be taken into account. The method is therefore primarily suitable for technical purposes.

Frey and Massonnet's method obtains the N–v diagrams by the following steps. Starting from a generic equilibrium condition, an increment similar to the previous deflection is made, varying in this way the curvature of each section of the bar. The new curvature obtained in this way is used to determine the neutral axis position by iterating until equilibrium in each section is reached. This determines variable values of N along the axis. The curvature is then changed iteratively until the value of N is constant along the bar. This method allows all the imperfections and the local elastic unloadings to be taken into account. Even though it is different from the computational point of view, it gives the same results as Faella and Mazzolani's method.

Geometrical and mechanical properties of the cross-sections used for this comparison are given in Fig. 7.3, and are considered with no structural imperfections because the values of $f_{0.2}$, E and n have been

Fig. 7.3 Properties for cross-sections used in comparisons of methods for obtaining N–v curves

Sections	Dimensions (mm)		Materials		
		Designation	$f_{0.2}$ (N mm^{-2})	E (N mm^{-2})	n
A double T	flanges: web:	140×10 120×10	AlMg4.5Mn	157	70 276	7.997
B double T	flanges: web:	101.6×11.11 193.68×7.94	AlZnMg1	318	78 170	28.387
C tube	ext. diam.: thickness:	80 4	AlMg2.5Mn	162	69 210	10.242

experimentally evaluated by means of a stub column test. An initial out-of-straightness characterized by $v_0/L = 1/1000$ has been assumed.

The comparison is given in Fig. 7.4 in which the critical dimensionless stress $\bar{N} = \sigma_c/f_{0.2}$ is related to the slenderness ratio $\bar{\lambda} = \lambda/\lambda_0$, where $\lambda_0 = \pi\sqrt{(E/f_{0.2})}$. In order to have a significant comparison, the columns have been discretized in the same way in the longitudinal direction (m) and in the transverse direction (m_t). In particular the following values have been used: $m = 20$ for all the profiles and $m_t = 2 \times 10$ (flanges) + 10 (web) for double T profiles and $m_t = 20$ for the tube. The comparison has been satisfactory.

The influence of longitudinal discretization, varying from 2 to 20, has also been studied.

If it is assumed that the values obtained for $m = 20$ are exact then the following deviation can be defined:

$$s = \frac{N_{(20)} - N_{(m)}}{N_{(20)}}$$

as a function of the values of $N_{\max}$ obtained with the different discretizations (m).

Figure 7.5 shows that an approximation within ±1 percent can be obtained with $m = 8$. When the discretization is higher $(m > 8)$, the results are practically the same.

7.1.3 Experimental Results

7.1.3.1 Extruded profiles

Several tests on the unstable behavior of extruded bars have been undertaken by the European Convention for Constructional Steelwork (ECCS) at Liège University in 1970–72. Extruded double T profiles and box sections of different aluminum alloys from six countries were examined (see Fig. 7.6).

The following tests were carried out [2]:

(a) Measurement of geometrical imperfections
(b) Chemical analysis
(c) Tension tests
(d) Determination of the distribution of Young's modulus in the cross-section
(e) Determination of the residual stress distribution
(f) Stub column test
(g) Buckling test

The mechanical properties measured in these tests are given in Fig. 7.7.

Aluminum alloy structures

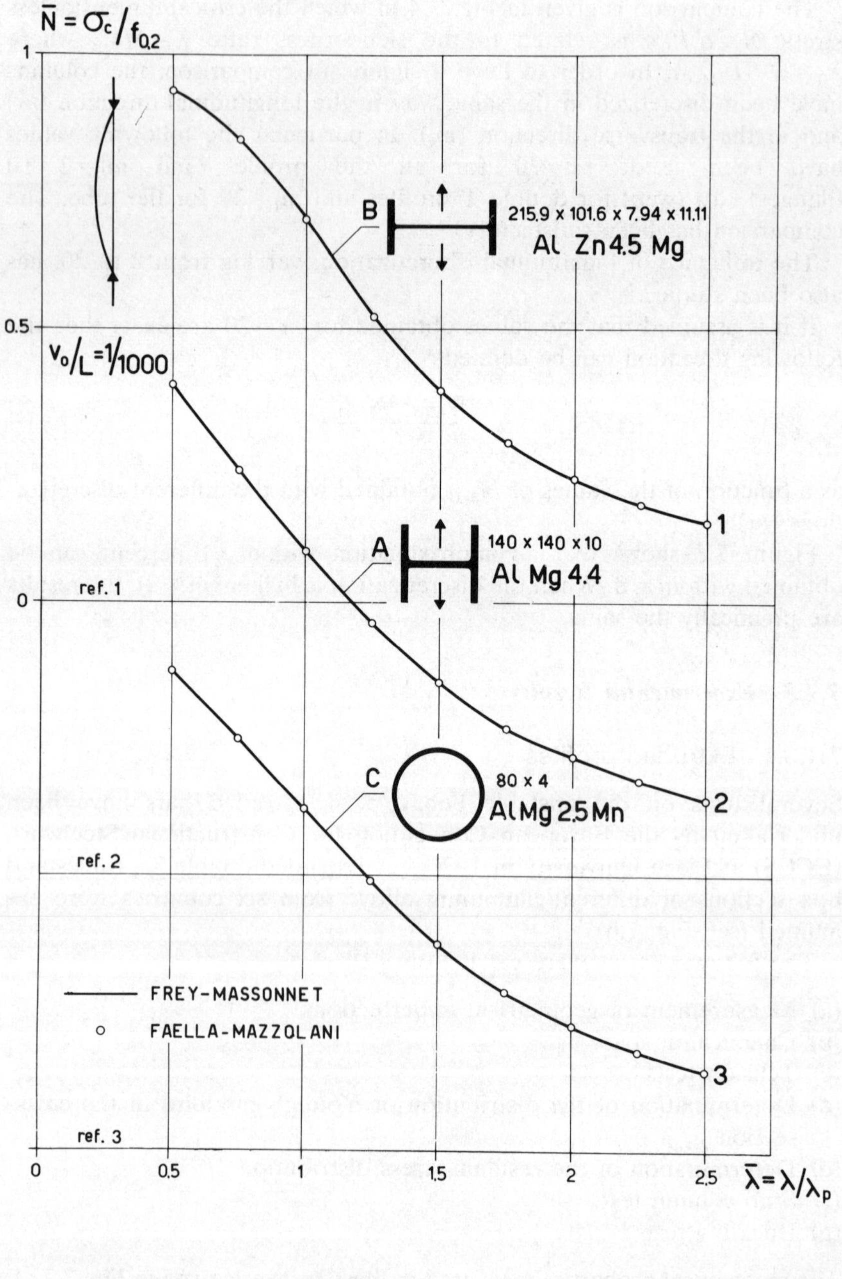

Fig. 7.4

Fig. 7.5

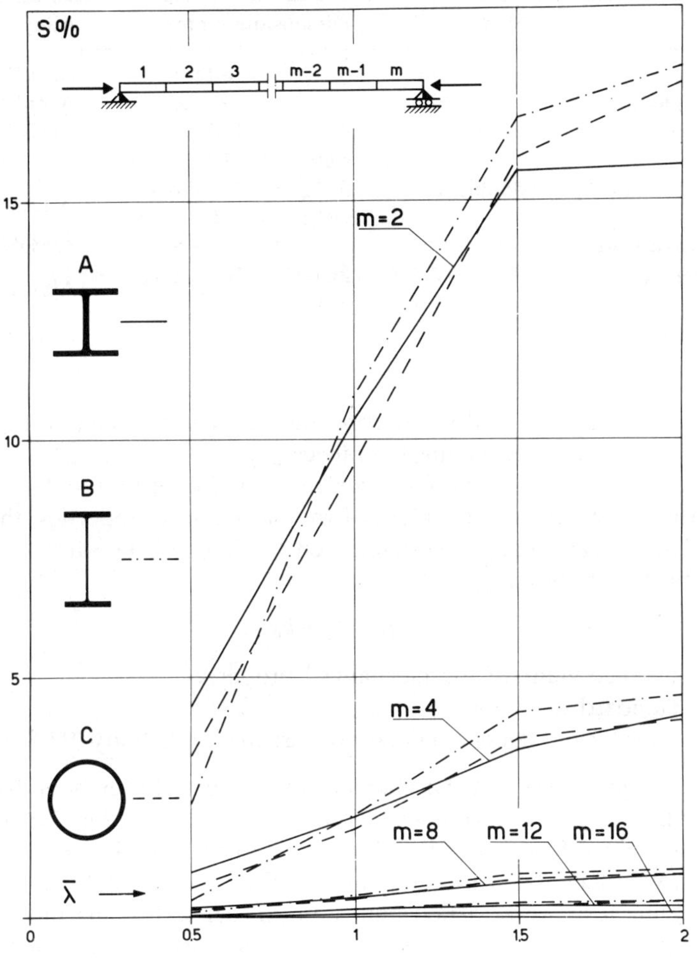

The comparison between the values of the elastic limit obtained in the tensile test and in the stub column test give deviations of opposite sign. The difference between the Young's modulus values in tension and compression were also significant. These differences can be explained by considering the Bauschinger effect caused by the tensile force applied in the industrial process to straighten the bars (see Section 2.5.3.2).

The load–displacement curve obtained from the stub column test describes the real behavior of the entire cross-section under compression and takes account of all mechanical imperfections. These experimental results also confirmed that the Ramberg–Osgood law (see Section 2.2.5)

Aluminum alloy structures

Fig. 7.6 Extruded profiles tested by ECCS

Country	Shape	Dimensions (mm)		Alloy
Italy	round tube		80×4	AlMg2.5Mn
Belgium	round tube		95×4	AlZnMg1
France	double T	flanges:	140×140	AlMg4.5Mn
		web:	10	
Switzerland	double T	flanges:	216×102	AlZnMg1
		web:	8	
Switzerland	round tube		90×5	AlMgSi1
Sweden	double T	flanges:	100×100	AlSiMg
		web:	7	
Norway	round tube		90×4.5	AlMgSi1

can conveniently be used to interpret the stress–strain relationship within the accuracy of experimental tolerances [3].

The statistical evaluation of twelve tests for each profile led to a definition of characteristic values of the mechanical properties, the limits of the statistical range being defined on the basis of the safety principles used by ECCS, which give

$$f_k = f_m \pm ks$$

f_m average value of the mechanical property
s standard deviation
k a coefficient depending upon the assumed reliability (97.5 percent)

The average values and the characteristic values (lower bounds) of the elastic limit $f_{0.2}$ and of Young's modulus E are given in Fig. 7.8, together with the values B and n of the Ramberg–Osgood law (in the form of Eq. 2.21) which have been used to simulate the experimental curves.

Buckling tests were carried out in an Amsler testing machine of 500 t capacity, with special hinges at the ends to produce an eccentricity of less than 0.2 mm.

The corresponding test results have been analyzed to give the average stresses ($f_{c,m}$), the standard deviations (s) and the characteristic stresses ($f_{c,k}$). Their values are given in Fig. 7.9.

7.1.3.2 Welded profiles

Various research programs have been devoted to the instability of welded columns, and preliminary experimental research has been carried out on welded joints (ref. [15] of Chapter 2). The subsequent program on welded profiles was carried out in Italy at the ISML of Novara (refs [16, 17, 18] of Chapter 2), where residual stresses and mechanical properties were

Fig. 7.7 Measured properties in extruded profile tests

Shape	Country	Tension test (N mm^{-2})						Stub column test (N mm^{-2}) $f_{0.2,c}$	Scatter (N mm^{-2}) $f_{0.2,t} - f_{0.2,c}$
		Standardized			Nonstandardized		$f_{0.2,t}$ (average)		
		Flanges	Web		Flanges	Web			
Double T	France	181.2	170.4		166.6	157.8	169.0	157.2	+11.8
	Switzerland	321.2	305.0		300.9	292.0	304.8	335.2	−31.4
	Sweden	291.7	269.6		262.6	263.6	271.9	282.0	−10.2
Round tubes	Italy		192.1			166.6	179.3	162.7	+16.7
	Belgium		305.6			303.8	304.7	340.1	−35.4
	Switzerland		289.4			268.5	278.9	298.8	−19.9
	Norway		236.4			243.0	239.7	244.9	−5.2

Fig. 7.8 Statistical values of the parameters of the Ramberg–Osgood law deriving from tests

Shape	Country	$f_{0.2}$ (N mm^{-2})		E (N mm^{-2})		B (N mm^{-2})		n	
		Mean	Characteristic	Mean	Characteristic	Mean	Characteristic	Mean	Characteristic
Double T	France	157.2	155.9	7028	6631	341.9	315.4	79.97	88.21
	Switzerland	335.2	326.4	7580	7054	433.6	394.9	241.5	325.9
	Sweden	282.0	255.3	7364	7140	347.6	284.7	297.0	570.3
Round tube	Italy	162.7	158.6	6921	6679	298.4	280.0	102.4	109.3
	Belgium	340.1	325.8	7210	6834	404.5	383.7	357.8	379.8
	Switzerland	298.8	292.5	6724	6513	369.0	358.7	294.5	304.7
	Norway	244.9	213.8	7458	7146	334.4	277.6	199.4	237.9

Aluminum alloy structures

Fig. 7.9 Results for buckling tests

Shape	Country	Slenderness	Number of tests	Coefficient k	$f_{c,m}$ (N mm^{-2})	s (N mm^{-2})	$f_{c,k}$ (N mm^{-2})
Double T	France	69	9	2.306	104.2	0.868	84.6
	Switzerland	64.5	8	2.365	185.2	2.52	126.9
		48.5	8	2.365	263.2	1.51	228.1
		32.4	8	2.365	293.5	0.953	271.5
		48.3	4	3.182	255.6	0.223	194.7
Round tubes	Italy	34.4	3	4.303	137.4	0.50	116.2
	Belgium	41.15	5	2.776	270.0	0.54	255.3
	Switzerland	51.2	8	2.365	203.5	0.77	185.6
	Norway	28.9	4	3.182	199.6	2.23	194.7
		51.9	8	2.365	203.5	0.77	185.6

measured (see Chapter 2), and at Liège University [10], where stub column tests and buckling tests were carried out.

Longitudinally welded members with three different cross-sections of AlSiMg alloy (6082-T6) were used (Fig. 2.3):

Type P   double T profile built up by two extruded flanges butt welded to rolled plates (webs)
Type T   double T profile made of rolled plates fillet welded together in a similar way to steel cross-sections
Type C   box section made of C extruded profiles butt welded to rolled plates

The experimental program was divided into the following parts:

(a) Measurement of geometrical properties
(b) Measurement of longitudinal residual stresses
(c) Measurement of the distribution of σ–ε curves in tension and in compression in the cross-section
(d) Measurement of the specific weight
(e) Buckling tests on columns with slenderness ratios equal to 1
(f) Stub column tests

The extruded parts came from Switzerland (Alusuisse) and the plates from Norway (Nordisk Aluminum), with the welding being undertaken in France (Centre Téchnique de l'Aluminium, Paris). Tests (a) and (f) were carried out at Liège University (Belgium), and tests (b), (c) and (d) were carried out at the Experimental Institute for Light Metals (ISML) in Novara.

The nominal values of the properties of the cross-sections are given in Fig. 7.10. The nominal length of the specimens is equal to 1100 mm for the P and T columns and is equal to 2700 mm for the C columns.

The principal dimensions of each cross-section were measured at the ends and center of each specimen. The average values of the geometrical properties (A, I, i) corresponding to the measured dimensions are given in Fig. 7.11; the differences between the nominal values and the average values were insignificant.

Fig. 7.10 Properties for welded profile tests

Type	Area A_{nom} (cm^2)	Inertia moment around weak axis I_{nom} (cm^4)	Minor radius of gyration $i_{nom} = \sqrt{(I_{nom}/A_{nom})}$ (cm)	Weight q_{nom} (N m^{-1})
P	40.0	202	2.25	105.84
T	38.7	201	2.28	102.41
C	45.5	1347	5.44	120.54

Fig. 7.11 Average values of geometrical properties

Type	A_m (cm^2)	I_m (cm^4)	i_m (cm)
P	40.1	202	2.25
T	39.35	202	2.27
C	45.3	1379.5	5.25

The V deformation of the flanges of the T profile (due to the weld) is within acceptable limits (see Fig. 2.25). The eccentricity of the web in the P and T sections was measured at the ends of each specimen and the average values of this eccentricity and of the corresponding load eccentricities have already been given in Fig. 2.26.

The initial out-of-straightness of each specimen was measured in the two principal planes of each cross-section. Measurements for the double T sections for minor axis bending were made at the extremities of the flanges.

The corresponding results for the maximum initial midspan deflection have already been given in Fig. 2.23 for box sections (type C). The ratio between the maximum displacement and the span is always less than 1/1000, even if account is taken of the load eccentricity due to the eccentricity of the web in double T sections.

The length L and weight Q of each profile are given in Fig. 7.12.

The measured specific weight γ is equal to 26.362 and 26.754 kN m^{-3} for rolled plates and extruded profiles respectively. The average specific weight measured on the entire cross-section is equal to 26.362 kN m^{-3} for double T profiles and to 26.558 kN m^{-3} for P and C profiles. These two values have been used to calculate the 'experimental' area A_{exp}, given by

$$A_{exp} = Q/\gamma L$$

The measurement of the residual stresses and the analysis of the results led to the definition of the models explained in Section 2.5.1.4 and shown in Figs 2.41–2.43.

Measurements of the distribution of tensile properties in the cross-sections of the tested profiles confirmed that there is insignificant variation far from the weld but a significant decrease of mechanical properties close to the weld. However the region affected by the heat input, in which there is a decrease in strength (reduced-strength zone), extends approximately 20 mm from each side of the weld. The minimum values of decreased strength correspond to stage T4 (naturally aged stage) for the base metal (AlSiMg) and to stage O (annealed) for the weld metal (5183). The values corresponding to the three zones with different strengths are given in Fig. 2.49.

Fig. 7.12 Profile characteristics

Specimen	Length L (mm)	Weight Q (N)	Density (kN m^{-3})	'Experimental' area A_{exp} (cm^2)
P11	1096	116.33	↑	39.95
12	1107	117.43		39.95
14	1107	117.56		40.00
21	1096	116.15		39.90
22	1107	117.30		39.90
24	1107	117.09	26.558	39.80
31	1107	117.59		40.00
32	1107	117.69		40.00
33	1107	117.79		40.05
41	1107	117.53		40.00
42	1097	116.27	↓	39.90
T11	1105	115.64	↑	39.70
12	1104	115.55		39.70
14	1104	115.59		39.70
21	1106	116.56		40.00
22	1106	115.72		39.70
24	1106	116.04	26.362	39.80
31	1106	115.66		39.65
32	1106	114.99		39.45
33	1106	115.29		39.55
41	1096	114.65		39.70
42	1106	115.42	↓	39.60
C11	2687	320.46	↑	44.90
12	2687	322.42		45.20
21	2687	321.93		45.10
22	2687	319.97		44.85
31	2687	320.46	26.558	44.90
32	2688	321.93		45.10
41	2687	322.91		45.25
43	2687	321.93		45.10
51	2687	321.44		45.05
53	2685	321.44	↓	45.10

Buckling tests have been carried out in the Amsler testing machine at the Laboratoire du Genie Civil in Liège, under perfectly hinged conditions. The results are given in Fig. 7.13: the average slenderness is computed using the lengths of Fig. 7.12 and the measured radii of gyration (see Fig. 7.11). The failure stress σ_c is computed by the collapse load N_c and the 'experimental' area A_{exp} of Fig. 7.12.

Aluminum alloy structures

Fig. 7.13 Buckling test results

Specimens	Slenderness (average)	Maximum load N_c (kN)	Ultimate strength $\sigma_c = N_c/A_{exp}$ (N mm^{-2})	Statistical values			
				$\sigma_{c,m}$ (N mm^{-2})	s (N mm^{-2})	k_t	$\sigma_{c,k}$ (N mm^{-2})
P11		92 000	230				
P12		95 000	238				
P14		95 000	237				
P21		88 000	221				
P22	49.06	93 000	233	226.38	10.24	2.262	202.8
P24		92 000	231				
P31		97 000	242				
P33		98 000	245				
P41		84 000	210				
P42		90 000	225				
T11		76 000	191				
T12		77 000	194				
T14		72 000	181				
T22		71 000	179				
T24	48.66	85 000	214	186.20	10.11	2.262	162.6
T31		71 000	179				
T32		73 000	185				
T33		75 000	189				
T41		76 000	191				
T42		77 000	194				
C11		93 000	207				
C21		91 000	202				
C22		86 000	192				
C31		96 000	214				
C32	48.68	95 000	211	203.84	8.06	2.306	185.2
C41		91 000	201				
C43		96 000	213				
C51		95 000	211				
C53		99 000	219				

The statistical analysis was undertaken using Student's t test, which gave a deviation k_t always greater than 2. The characteristic values of the failure stress are given by:

$$\sigma_{c,k} = \sigma_{c,m} - k_t s$$

where $\sigma_{c,m}$ is the average stress and s is the standard deviation.

The non-dimensionalized failure stresses have been calculated using the average elastic limit $f_{0.2}$ under compression and it has been assumed that Young's modulus is equal to

$$E = 70\,000 \text{ N mm}^{-2} \quad \text{for rolled plates}$$
$$E = 72\,500 \text{ N mm}^{-2} \quad \text{for extruded profiles}$$

The nondimensionalized values of the parameters

$$\bar{N}_{c,m} = \frac{\sigma_{c,m}}{f_{0.2}} \qquad \bar{N}_{c,k} = \frac{\sigma_{c,k}}{f_{0.2}} \qquad \bar{\lambda} = \frac{\lambda}{\pi}\sqrt{\frac{f_{0.2}}{E}} \qquad (7.8)$$

are as follows:

Type	T	P	C
$\bar{N}_{c,m}$	0.62	0.69	0.625
$\bar{N}_{c,k}$	0.54	0.62	0.57
$\bar{\lambda}$	1.02	1.06	1.05

The experimental load–displacement diagrams showed that all 2.70 m box section columns (C profiles) experienced progressive instability. All columns experienced instability in the same plane of the initial out-of-straightness.

Double T columns (P and T types) showed sudden buckling more frequently than progressive buckling owing to the small initial out-of-straightnesses, which lead to bifurcation instability and thus high values of the collapse load. Even though there is no relationship between the plane of buckling and the plane of initial out-of-straightness, the web eccentricity appears to compensate for the initial out-of-straightness.

7.2 Members under compression

7.2.1 Physical Behavior

Depending upon the shape of the cross-section, axially loaded bars can buckle in three different ways:

Plane buckling
Twist buckling
Flexural–torsional buckling

Plane buckling occurs when the deflection of the bar remains in a given plane. It is typical of cross-sections with two planes of symmetry in which buckling occurs in a principal plane which is:

The minor axis bending plane if the boundary conditions are the same in both directions
The plane characterized by the highest slenderness ratio if there are different boundary conditions

If the bar does not have two planes of symmetry, plane buckling can still occur if the bar is prevented from twisting in order to avoid flexural–torsional buckling.

Aluminum alloy structures

Pure twist buckling is of major concern in those sections which have negligible warping torsional rigidity (e.g. the + section), because of the convergence of the elements of the cross-section at the same point.

In this chapter plane buckling of extruded columns (Section 7.2.3), welded columns (Section 7.2.4) and built-up columns (Section 7.2.5), is considered. Flexural–torsional buckling is also mentioned (Section 7.2.6).

First, the analysis of the main parameters affecting buckling is given in Section 7.2.2 by means of simulation calculations. There follows the definition of industrial bar, which emphasizes the differences between the actual behavior and the behavior of the ideal bar.

In fact geometrical imperfections (initial out-of-straightness, eccentricities) as well as mechanical imperfections (residual stresses, variations of the elastic limit), previously defined in Sections 2.4 and 2.5, play an important role in the buckling behavior of columns. There are also other parameters affecting this type of buckling, such as the average value of the elastic limit $f_{0.2}$ and the shape of the cross-section.

In the same way as for steel structures [27, 28], the simulated buckling curves for simply compressed members are given in the dimensionless plane $\bar{N}$–$\bar{\lambda}$, with the definitions

$$\bar{N} = \frac{\sigma_c}{f_{0.2}} \qquad \bar{\lambda} = \frac{\lambda}{\pi\sqrt{(E/f_{0.2})}} = \frac{\lambda}{\lambda_0} \qquad (7.9)$$

$f_{0.2}$ average conventional elastic limit on the entire cross-section, or the design strength of the material (f_d)
E Young's modulus of the material
σ_c failure stress, which corresponds to the load N_c that causes buckling of the bar in the considered plane
λ slenderness of the bar, given by $\lambda = L_c/i$
L_c effective length in the buckling plane, depending upon the support conditions at the ends of the bar of length L ($L_c = \beta L$)
i radius of gyration of the cross-section in the same plane in which L_c is computed

In order to compute the coefficient β which transforms the geometrical length into the effective one L_c, technical texts on steel structures can be referred to because the problem is mainly related to the structural systems, which are independent of the material.

7.2.2 Analysis of the Main Variables

7.2.2.1 Material

The analysis of the influence of the material (through the parameters of the σ–ε law) without geometrical and mechanical imperfections can be carried out by means of Shanley's method, which gives the collapse stress [3]:

$$\sigma_c = \frac{\pi^2 E_{t,m}}{\lambda^2} \tag{7.10}$$

where $E_{t,m}$ is the average tangent modulus of the entire cross-section obtained by the stub column test, and λ is the slenderness of the bar.

This stress can be expressed in the following nondimensionalized form if the definition Eq. 7.9 is made:

$$\bar{N} = \frac{E_{t,m}}{E} \frac{1}{\bar{\lambda}^2} \tag{7.11}$$

If it is assumed that for the stub column test the Ramberg–Osgood law is valid, the expression for $E_{t,m}$ in Eq. 7.11 can be directly derived from Eq. 2.21, which gives

$$\frac{d\varepsilon}{d\sigma} = \frac{1}{E} + n\frac{\sigma^{n-1}}{B^n} \tag{7.12}$$

from which

$$E_{t,m} = \frac{d\sigma}{d\varepsilon} = \frac{EB^n}{B^n + n\sigma^{n-1}E} \tag{7.13}$$

If Eq. 2.31 is considered, then

$$B^n = \frac{f_{0.2}^n}{0.002} \tag{7.14}$$

Therefore

$$\frac{E_{t,m}}{E} = 1 \bigg/ \left[1 + \frac{0.002nE}{f_{0.2}}\left(\frac{\sigma}{f_{0.2}}\right)^{n-1}\right] \tag{7.15}$$

which with Eq. 7.9 becomes

$$\frac{E_{t,m}}{E} = 1 \bigg/ \left[1 + \frac{0.002}{\pi^2} n\lambda_0^2 \bar{N}^{n-1}\right] \tag{7.16}$$

Substituting Eq. 7.16 in Eq. 7.11, we obtain

$$\bar{\lambda}^2 = 1 \bigg/ \left[\bar{N} + \frac{0.002}{\pi^2} n\lambda_0^2 \bar{N}^n\right] \tag{7.17}$$

or more simply

$$\bar{\lambda}^2 = \frac{1}{\bar{N} + nk\bar{N}^n} \tag{7.18}$$

Aluminum alloy structures

having assumed that

$$k = \frac{0.002}{\pi^2} \lambda_0^2 = \frac{1}{500} \frac{E}{f_{0.2}} \qquad (7.19)$$

This nondimensionalized form for buckling of aluminum alloy bars without imperfections relates the collapse load N to the slenderness $\bar{\lambda}$ through the parameters n and k, thus yielding two families of curves as a function of these parameters. The analogous analysis for steel structures leads to one curve which is used after dimensionalizing with respect to f_d and E. In the case of aluminum alloys the curves are also a function of other parameters related to n and k. In fact n is a function of the strain hardening parameter $f_{0.2}/f_{0.1}$, and k is a function of $E/f_{0.2}$. The last parameter makes the classification of the material with respect to their buckling behavior more complex.

Figure 7.14 shows the influence of n, for $k = 1$, on the behavior of the $\bar{N}$-$\bar{\lambda}$ curves. The behavior of the curves varies gradually with the increase of n (from 2 to 40), the rate of increment being very high for small values of n ($n < 10$) and almost negligible for large values of n ($n > 20$). This further relationship makes the exponent n an important parameter for classifying the unstable behavior of aluminum alloys.

Fig. 7.14

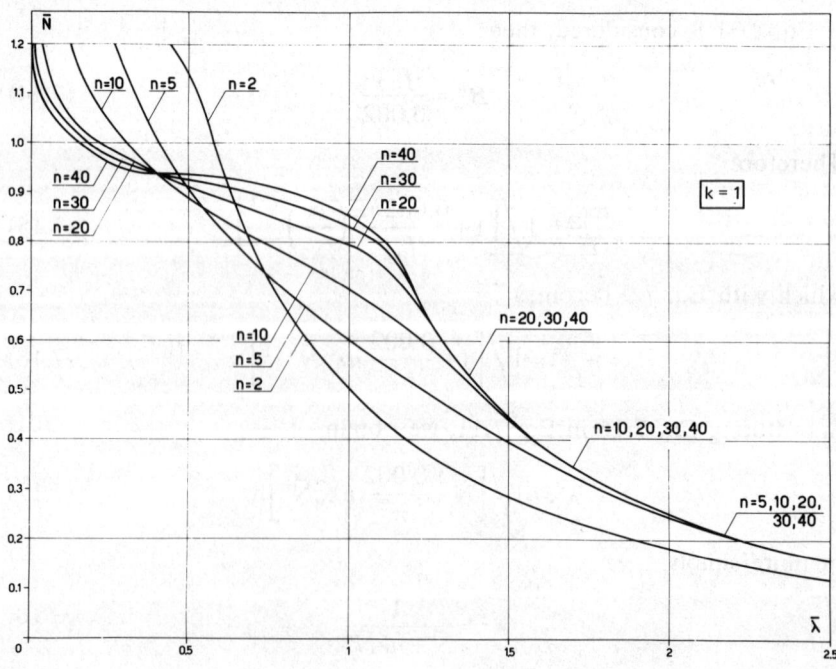

Stability of structural elements

From the curves two main families can be identified according to the variation of n: the non-heat-treated ($n < 10$) and the heat-treated ($n > 20$) families.

The coefficient k for the commonly used alloys can be placed in the following ranges:

$$k = 0.7 \text{ to } 1.6 \text{ for non-heat-treated alloys}$$
$$k = 0.4 \text{ to } 1.0 \text{ for heat-treated alloys}$$

The influence of k on the $\bar{N}-\bar{\lambda}$ curves is shown in Fig. 7.15 with $n = 8$ and 10 (k being equal to 0.75–1.6) and with $n = 20$ and 40 (k being equal to

Fig. 7.15

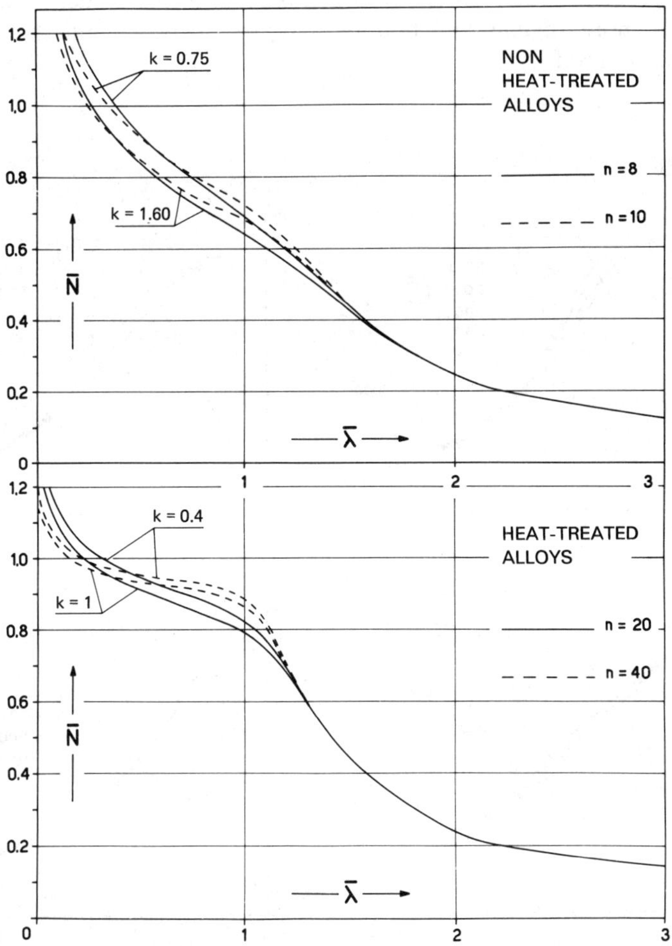

Aluminum alloy structures

0.4–1.0). The separation of the curves with respect to k is evident for values of $\bar{\lambda} \leq 1.5$. With the increase of n the influence of k is smaller. For $\bar{\lambda} > 1.5$ parameters k and n do not affect the behavior of the curves.

The influence of the ratio f_0/E (f_0 as defined in Fig. 2.21) on the behavior of the nondimensionalized curves has been shown by Marincek (ref. [7] in Chapter 2) in the $\bar{\sigma}$–$\bar{\lambda}$ plane, with

$$\bar{\sigma} = \frac{\sigma}{f_0} \qquad \bar{\lambda} = \frac{\lambda}{\pi \sqrt{(E/f_0)}} \qquad (7.20)$$

Figure 7.16 shows the corresponding curves for the values $n = 5, 11, 31$ and ∞, and $f_0/E = 0.001, 0.002, 0.005$ and 0.010 in the case of a rectangular cross-section with initial out-of-straightness equal to $L/1000$.

When this geometrical imperfection is present, the influence of the f_0/E ratio is more regular and bounds a region in which fall the materials commonly used.

Fig. 7.16

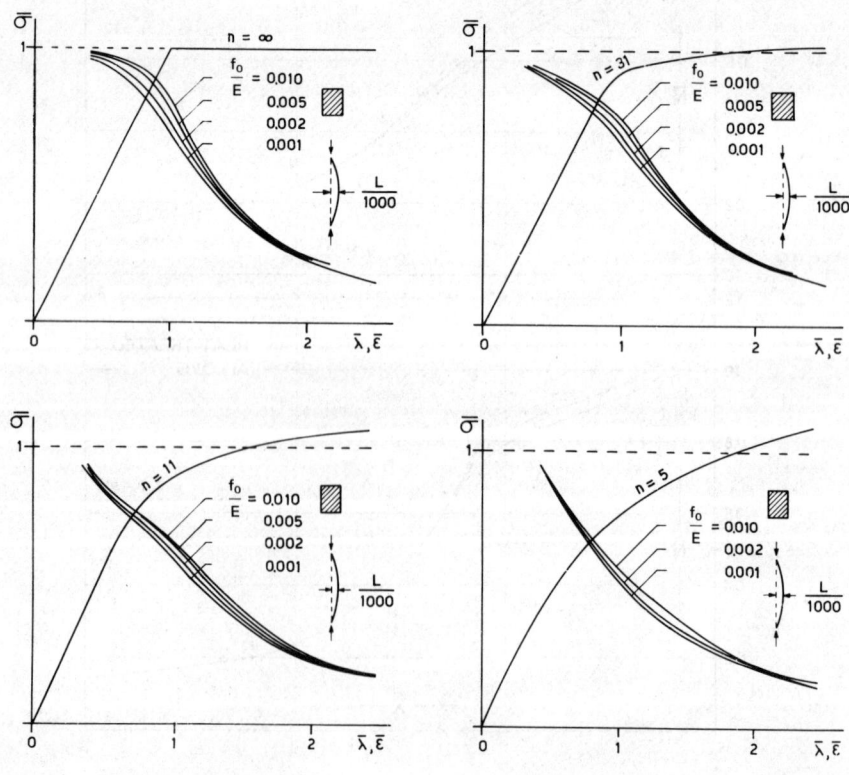

7.2.2.2 Initial out-of-straightness [20]

As already anticipated in Section 2.4.1, the initial geometrical imperfection of the axis of the bar is characterized by a sine curve with a midspan value equal to $L/1000$. This value is used in the simulation analysis given in the following sections.

The influence of two other values of v_0 is shown in Fig. 7.17 with respect to the two buckling planes (lateral and diagonal) for the same square box section, both for a non-heat-treated alloy (Fig. 7.17a) and a heat-treated alloy (Fig. 7.17b).

It is seen from these curves that the behavior is affected not by the buckling plane but by the type of material. Non-heat-treated alloys are more sensitive to these imperfections, especially to high values of slenderness, and this can be explained by looking at the σ–ε diagram which strain hardens starting from lower values of stresses.

7.2.2.3 Shape [6, 20]

The influence of the shape of the cross-section on the unstable behavior of columns has been thoroughly analyzed by Faella [20] by means of numerical simulation. First he analyzed the case $\bar{\lambda} = 1$ which identifies, for steel structures, the region where the shape of the cross-section has the largest influence on the load-bearing capacity.

The geometrical properties of the cross-sections considered are given in Figs 7.18 and 7.19 and are referred to with numbers varying from 1 to 13. They adequately represent all of the available shapes with double symmetry.

The materials considered are characterized by $10n = f_{0.2} = 80 \text{ N mm}^{-2}$ (non-heat-treated alloy), $10n = f_{0.2} = 300 \text{ N mm}^{-2}$ (heat-treated alloy) and $E = 70\,000 \text{ N mm}^{-2}$. The results are given in Figs 7.20 and 7.21, for the double T and box profiles respectiely. The diagrams are non-dimensionalized with respect to the failure stress $\sigma_{c,1}$ of section 1 (I profile with $t_f/t_w = 1$ and $h/t_w = 15$; minor axis bending). The progressive increase of load-bearing capacity is observed in double T sections with decrease of the dimensions of the web ($b/h = 2$; $t_f/t_w = 2$) and with the increment of lips ($d/b = 0.30$).

The following comparisons are given in those cases which were identified as limit cases in the previous analysis. They are sections 1 and 9 for major axis buckling (Fig. 7.22). Figure 7.23 shows the influence of the h/t_f ratio for these sections. However, the variations are within the bounds already found for $h/t_f = 15$.

At first glance it seems that for $\bar{\lambda} = 1$ the materials with high strain hardening capacity ($n = 8$) are not affected by the shape of the cross-section since the highest deviation of the collapse load was less than 1

Aluminum alloy structures

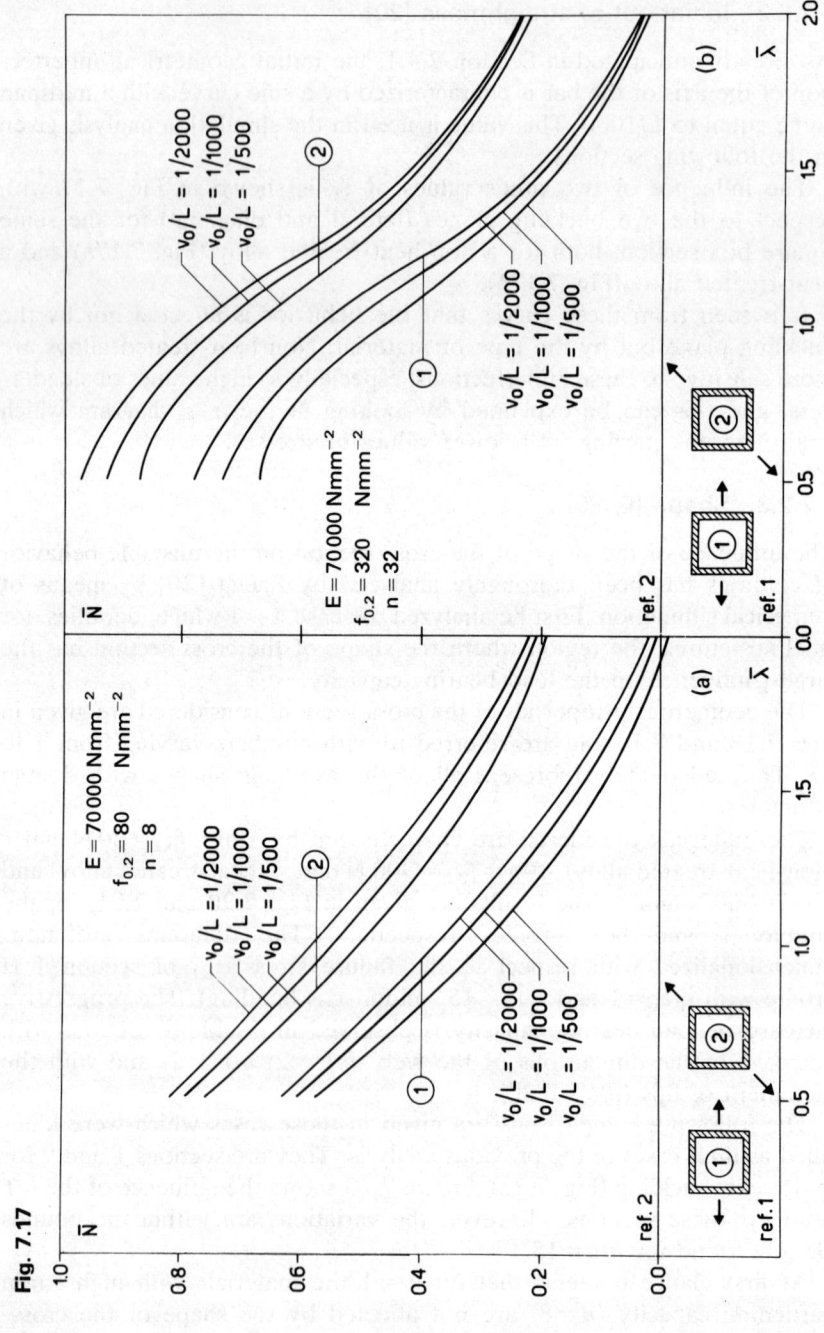

Fig. 7.17

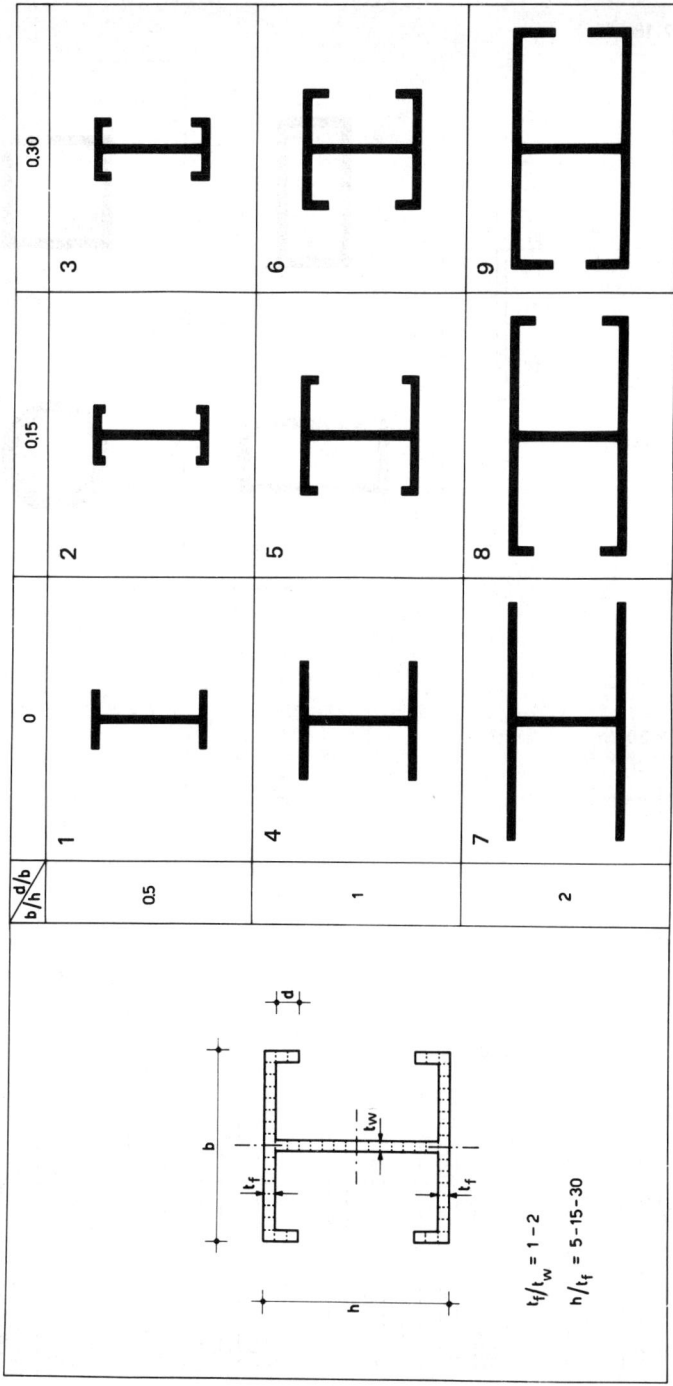

Fig. 7.18

Aluminum alloy structures

Fig. 7.19

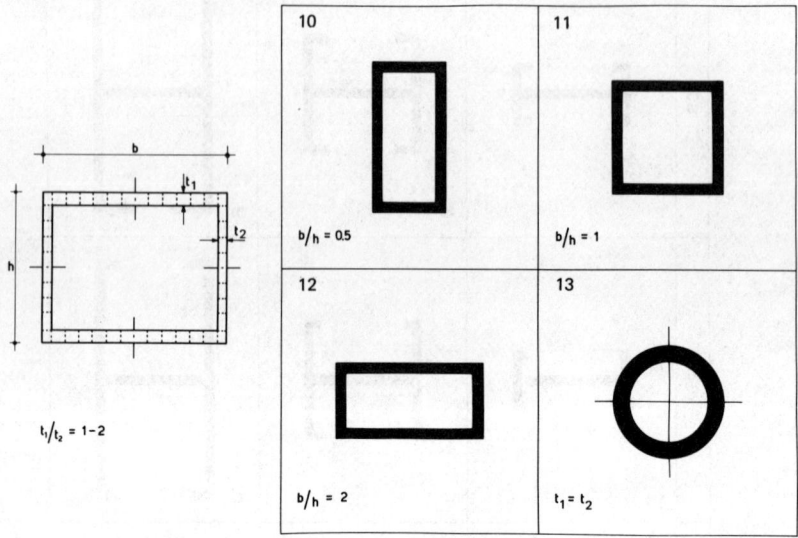

Fig. 7.20

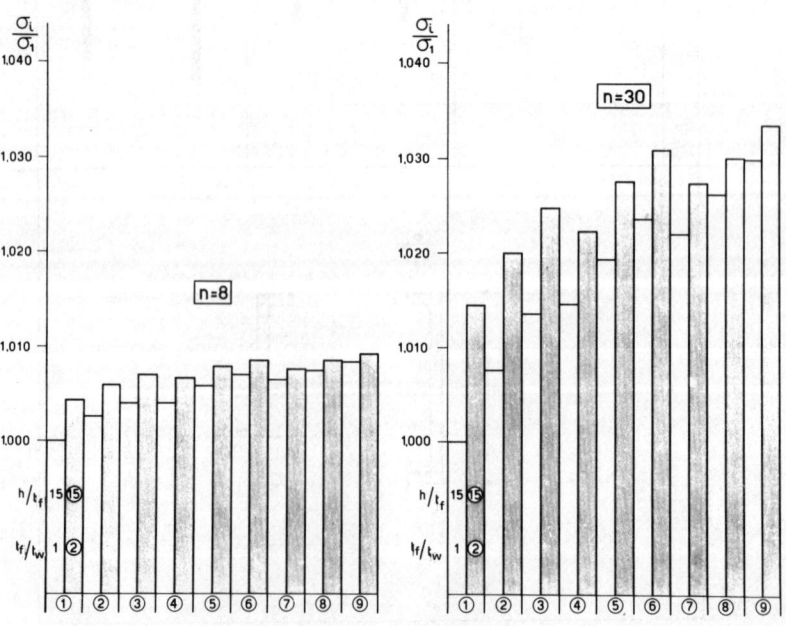

Stability of structural elements

Fig. 7.21

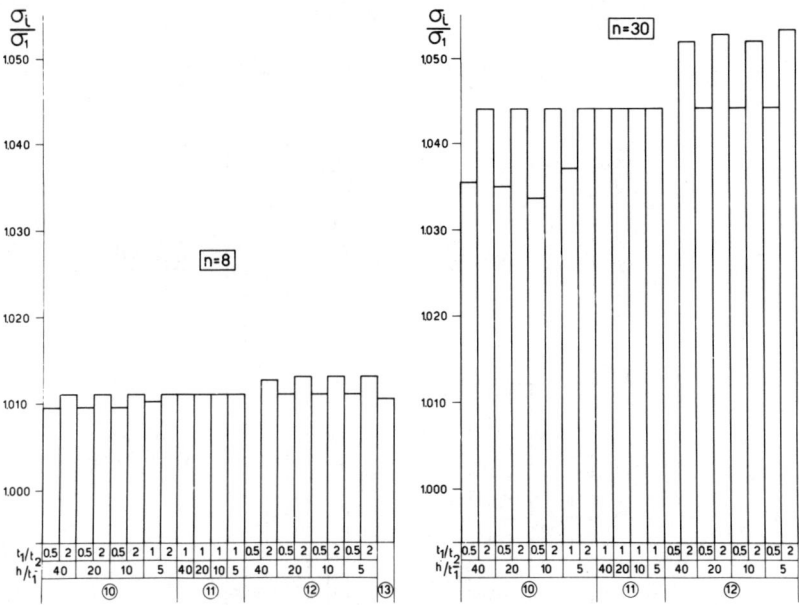

Fig. 7.22

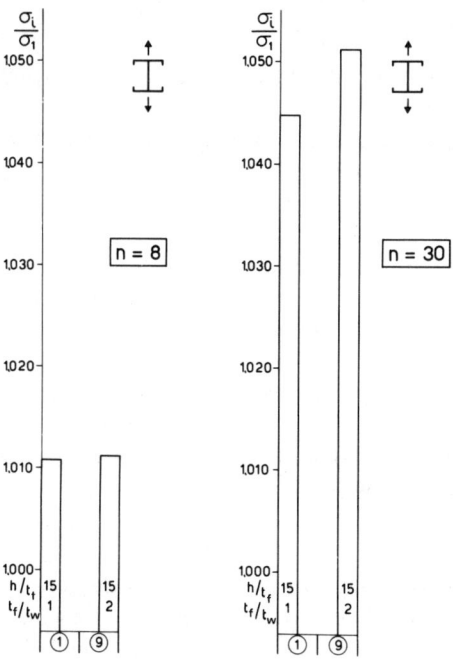

Aluminum alloy structures

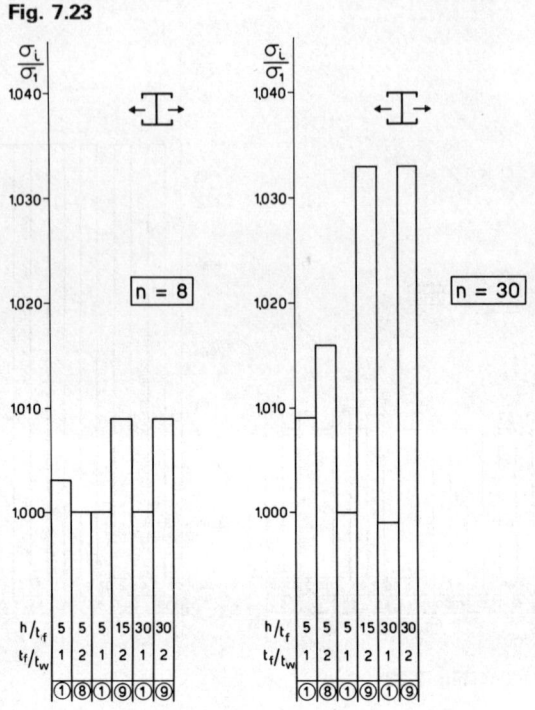

Fig. 7.23

percent. The heat-treated alloys show larger variations, up to 5–6 percent for $n = 30$.

Figure 7.24 shows the curves for profiles 1 and 9 for buckling due to major and minor axis bending. The tubes (shapes 10–12) are analyzed together with the cross-section which showed the most peculiar behavior (shapes 1–9). The analysis of the two curves shows that the difference between the two families of materials does not depend upon the shape, it depends instead upon the slenderness. In fact the largest deviations due to the shape of the profile are observed for non-heat-treated alloys in the bars which are not very slender ($\bar{\lambda} = 0.5$), whereas in the case of heat-treated alloys the largest deviations are observed in slender columns ($\bar{\lambda} = 1$).

At the same time the value of slenderness at which there is no influence of the shape of the cross-section, at the intersection of the $\bar{N}$–$\bar{\lambda}$ curves which identify the extreme behavior (Fig. 7.24), varies from $\bar{\lambda} = 0.95$ for non-heat-treated alloys to $\bar{\lambda} = 0.72$ for heat-treated alloys.

In conclusion we can say that in the case of definite hardening materials there is an inversion of behavior in the range of slendernesses of major interest ($\lambda \leq 100$) both for cross-sections with small shape factor (tubular

Stability of structural elements

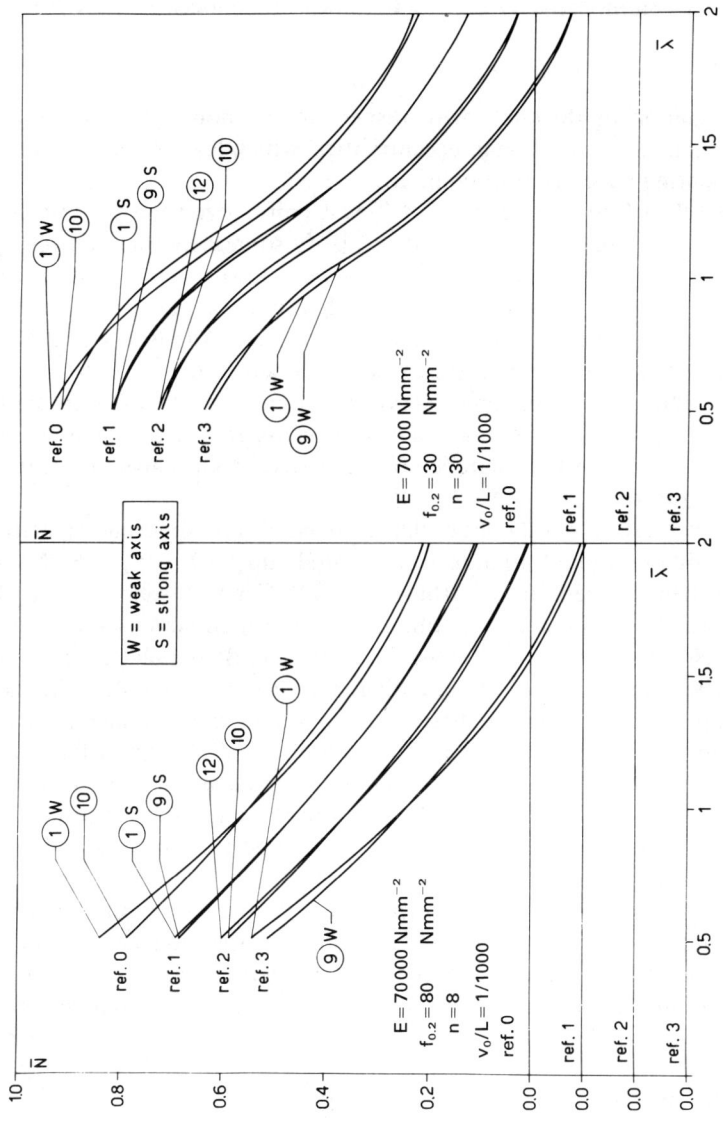

Fig. 7.24

sections with high B/t and s/t ratios) and for cross-sections with large shape factor (double T profiles in minor axis bending). This change of behavior is also observed for those materials not possessing a very high strain hardening capacity for small values of slenderness ($\lambda < 40$).

7.2.2.4 Thickness deviations

The variation of thickness with respect to the nominal dimensions in a profile can cause accidental 'eccentricities' which lead to a decrease in the load-bearing capacity of a column.

This effect (see also Section 2.4.2) was emphasized for the first time by Bernard and Frey [2, 4] who had to justify some experimental results on cylindrical tubes. Experimental values were lower than the values computed by numerical simulation on a bar with a perfect cross-section. When an appropriate value of the 'eccentricity' $\Delta t/t$ (see Eq. 2.42) was introduced, lower bound curves were obtained (see Fig. 7.25).

The influence of this imperfection on square tubes has been analyzed by Faella [20], who assumed a variation of thickness equal to $\pm t/10$, which is the permitted tolerance for extruded profiles given by European manufacturers.

Figures 7.26 and 7.27 show the decrease in load-bearing capacity due to this imperfection, for a non-heat-treated alloy ($10n = f_{0.2} = 80 \text{ N mm}^{-2}$) and a heat-treated alloy ($10n = f_{0.2} = 320 \text{ N mm}^{-2}$) respectively. Both cases were examined for an initial out-of-straightness $v_0 = L/1000$ and with respect to the two different buckling planes: parallel to the sides of the tube (cases 1 and 2) and diagonal (cases 3 and 4). The largest decrease is observed for heat-treated alloys buckling in a diagonal plane.

From these results it appears that the effect of change in thickness on the load-bearing capacity of a column is significant and should be taken into account. The variation of thickness in the cross-section, especially in the case of box sections, seems to be more important than the influence of the shape itself (Section 7.2.2.3).

However, in both cases the negative effects increase with higher values of n, that is when the material behaves more closely to steel. On the other hand, the effects of the shape of the cross-section and its deviations tend to decrease for strain-hardening materials which are characterized by smaller values of n.

7.2.2.5 Mechanical imperfections [21]

It has already been observed that the magnitude of mechanical imperfections (Section 2.5), residual stresses (Section 2.5) and inhomogeneous distribution of the elastic limit in the cross-section of the bar (Section 2.5.2) essentially depend upon the manufacturing process.

Stability of structural elements

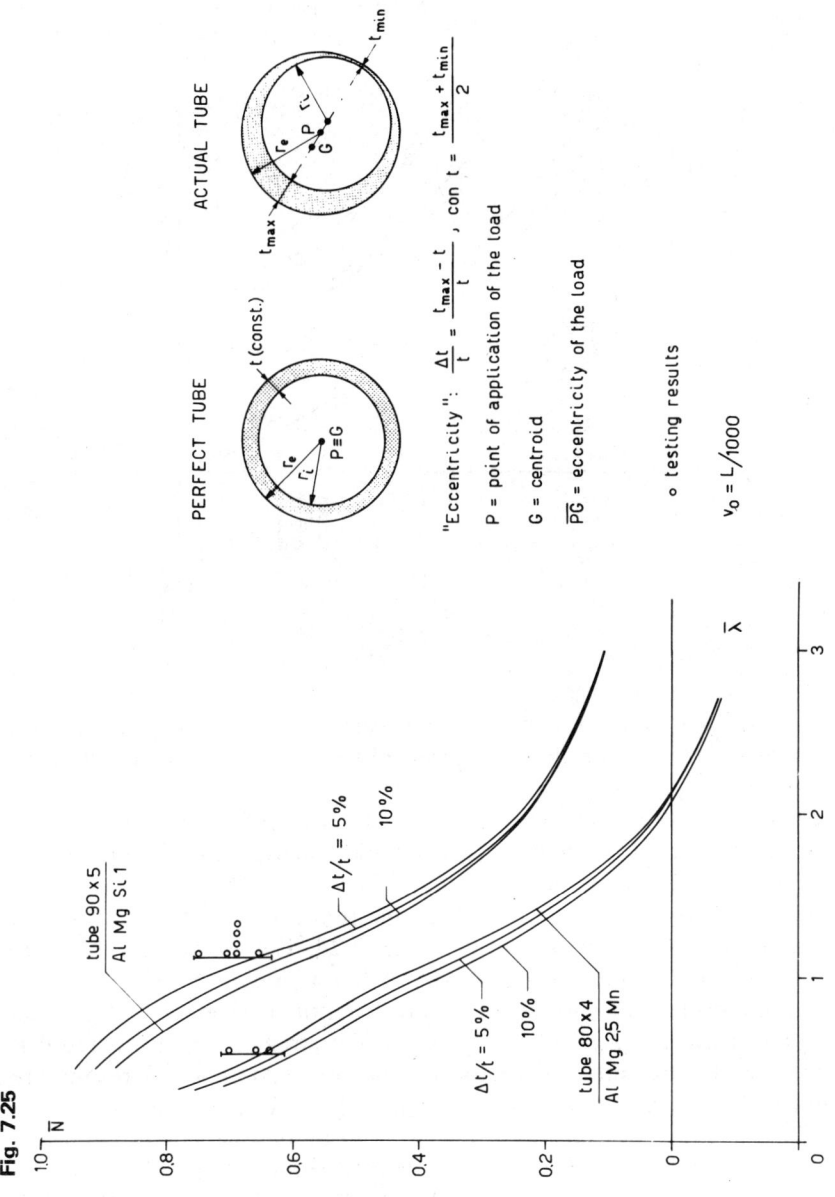

Fig. 7.25

Fig. 7.26

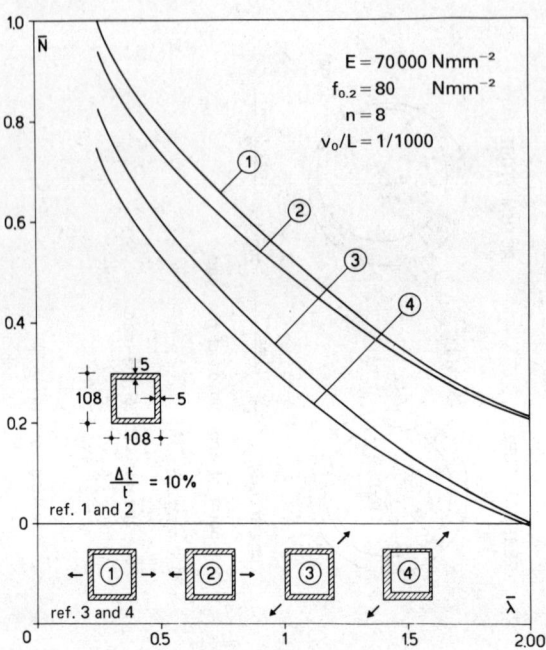

It is known that both imperfections are present in welded profiles, whereas in extruded profiles only residual stresses are observed and these are usually negligible.

In order to emphasize the effects of these imperfections on the load-bearing capacity of columns, some simulation analysis based upon experimental values has been carried out.

Residual stresses on a double T profile produced in France led to the definition of the models B and C of Fig. 7.28. The comparison between the corresponding buckling curves shows that both models practically give the same results and that the deviations from the curves without residual stresses (solid line A) are smaller when the strength of the material is increased. In the same figure the points from experimental results are shown; these correspond to the $f_{0.2}$ values given in parentheses.

The influence of mechanical imperfections on double T welded profiles has been emphasized for non-heat-treated alloys (Fig. 7.29) and for heat-treated alloys (Fig. 7.30). It is known that in the first case the effect of the decrease of $f_{0.2}$ in the so-called reduced-strength zone close to the welds is practically negligible (<10 percent), although the same is not true

Fig. 7.27

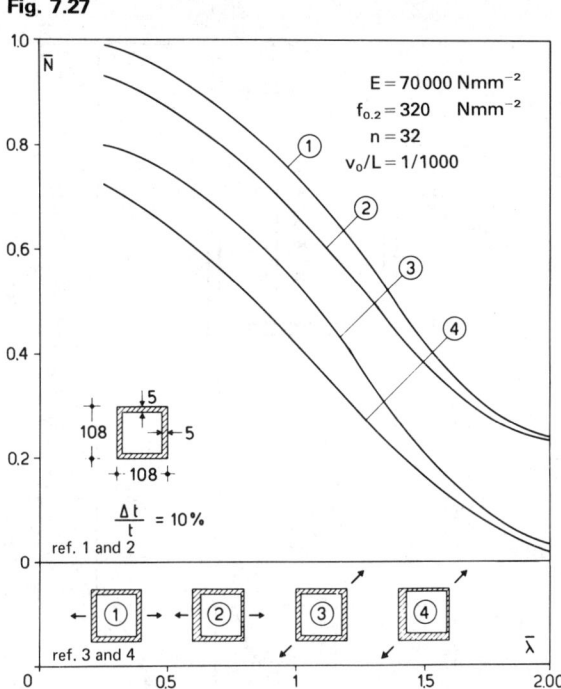

for heat-treated alloys (40 percent). In both cases the distributions of residual stresses are of the same order of magnitude. The models used to interpret these mechanical imperfections are all based upon experimental results.

The comparison between the corresponding buckling curves allows an assessment to be made of the influence of the imperfections, separated or combined, on the buckling behavior of the column.

Figure 7.30 shows that in the case of buckling about the major axis, residual stress is a beneficial effect since it provides further compression strength in the reduced-strength zone where the elastic limit is equal to $0.6f_{0.2}$, because of the tensile stresses. This combined effect produces a virtual increase of the elastic limit in the flange of the profile which improves the position of the corresponding curve (solid line) with respect to that corresponding to the variation of $f_{0.2}$ only (dashed line).

From the analysis of these results it can be concluded that the effects of mechanical imperfections on the load-bearing capacity of columns can be considered negligible in extruded profiles, whereas they must be accounted for in welded profiles.

Aluminum alloy structures

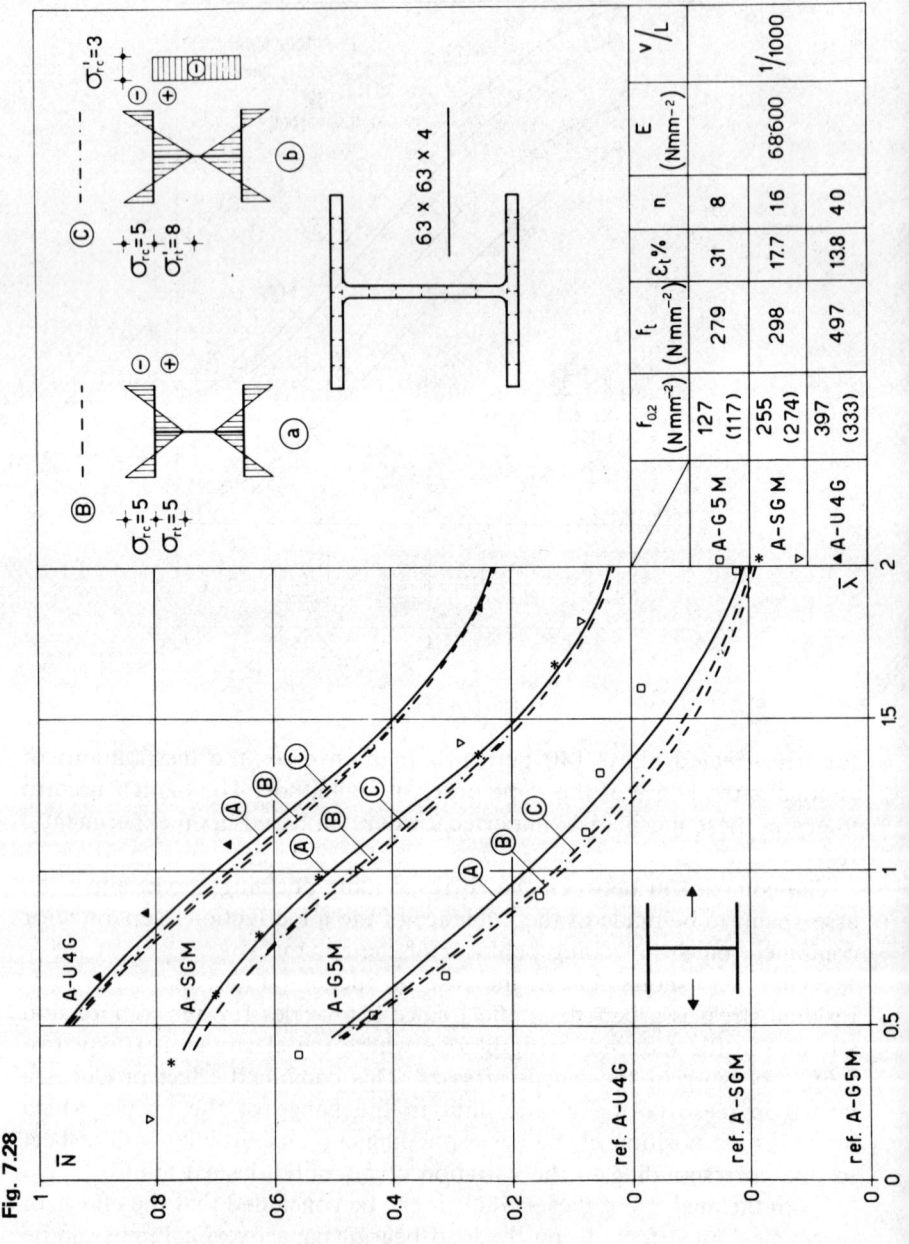

Fig. 7.28

Stability of structural elements

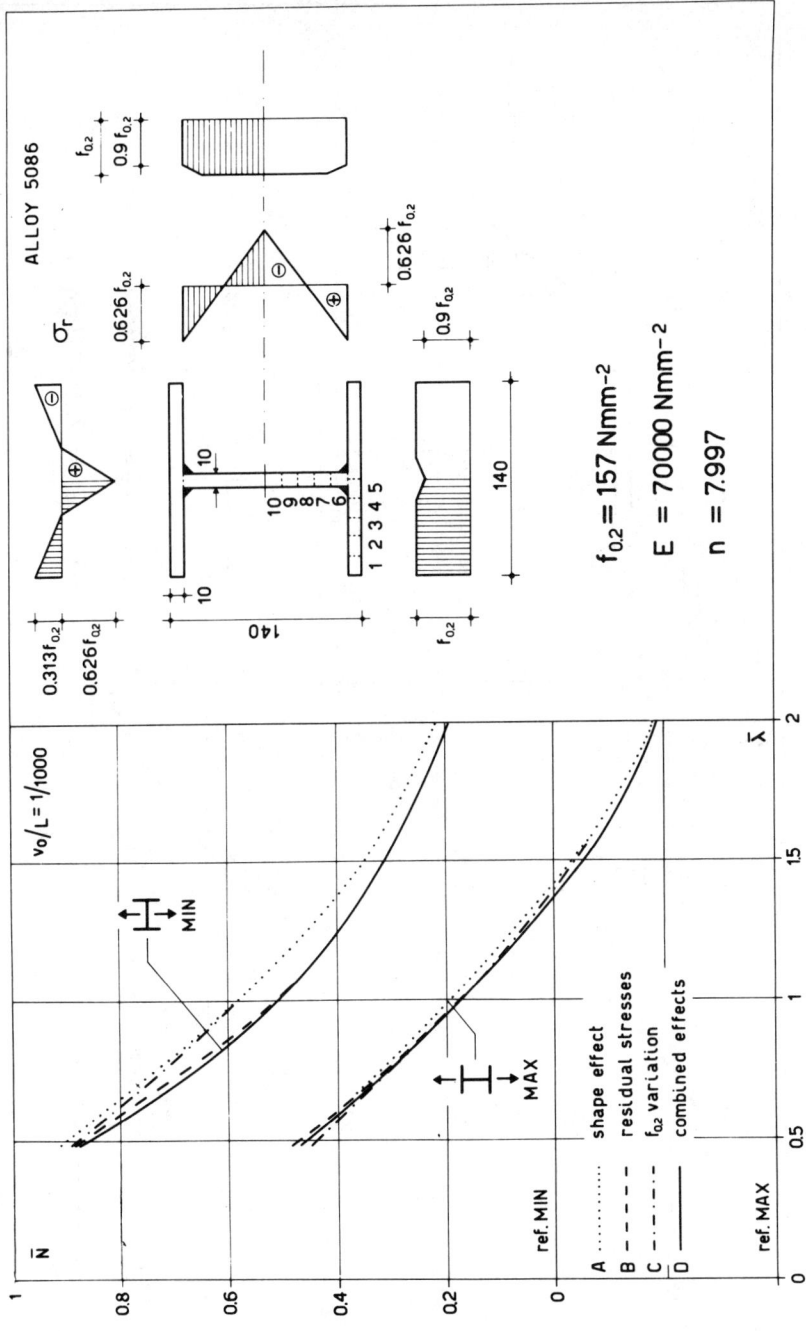

Fig. 7.29

Aluminum alloy structures

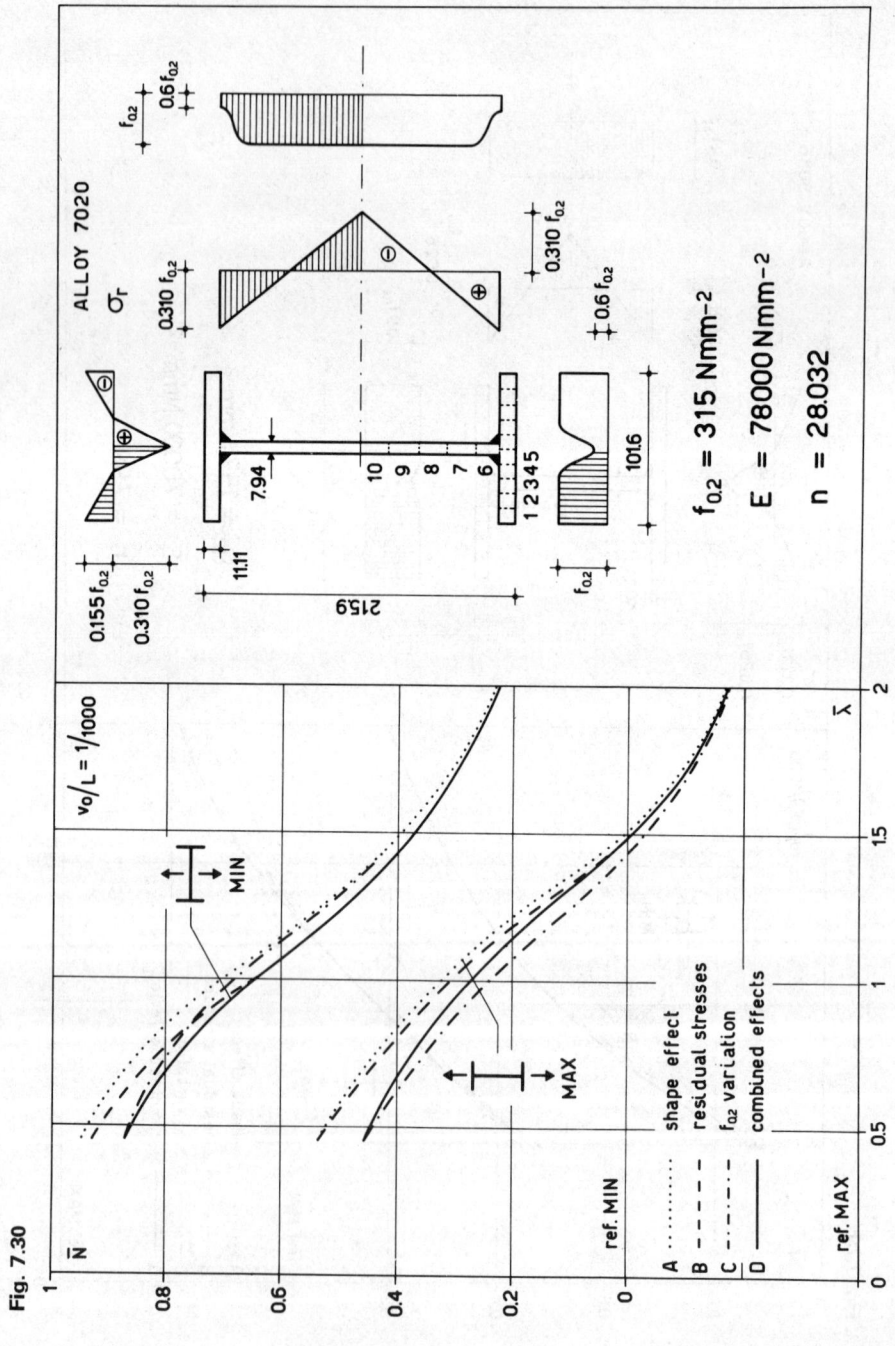

Fig. 7.30

7.2.3 Extruded Bars

7.2.3.1 Basic data

The method which is referred to here is that detailed by ECCS committee 16 (1977) and used in its recommendations (1978) [64].

The buckling curves used by ECCS have been based upon the available experimental and theoretical results and the following assumptions [9, 14]:

(a) A geometrical imperfection is usually assumed to be represented by an initial out-of-straightness with a midspan deflection $v_0/L = 1/1000$ (see Section 2.4.1).

(b) The influence of the shape of the cross-section and of its variations of thickness (see Sections 7.2.2.3 and 7.2.2.4):
   (i) is smaller than 5 percent and can therefore be neglected if the profile is an open symmetrical section (see Section 2.4.2)
   (ii) cannot be neglected if the profile is an open asymmetrical section (T, U etc.) or if it is a box section with an eccentricity equal to that defined in Section 2.4.2. It has been calculated that the decrease of strength is equal to 12 percent in T profiles, 13 percent in tubes and 16 percent in square box sections with an 'eccentricity' equal to 10 percent. The worst case is represented by the additive action of eccentricity and asymmetry.

(c) The $\sigma-\varepsilon$ law plays an important role in the shape of buckling curves, in contrast to steel structures (see Section 7.2.2.1). The Ramberg–Osgood law can be conveniently used (Eq. 2.32) with the values $E = 70\,000$ N mm^{-2} and $10n = f_{0.2}$ ($f_{0.2}$ in N mm^{-2}). This assumption is approximate but has the advantage of reducing the parameters of the law to just the conventional elastic limit.

(d) Since the elastic limit of commercial aluminum alloys (for structural use) can vary from 100 to 300 N mm^{-2}, it seems logical to use a nondimensionalized form for buckling curves. However, even in this case buckling curves are strongly affected by the exponent n which is related to $f_{0.2}$ as a consequence of the previous assumption (Fig. 7.31). Nevertheless, experience and computation (Section 7.2.2.1) showed that the behavior of the alloys commonly used can be mainly divided into two classes. The first comprises those heat-treated alloys with large values of $f_{0.2}$ (from 200 to 300 N mm^{-2}), and the second non-heat-treated alloys having small values of $f_{0.2}$ (about 100 N mm^{-2}). Hence we can distinguish two different types of behavior from the buckling curves.

(e) Residual stresses and variations of the elastic limit along the cross-section are negligible imperfections in extruded members (see Section 7.2.2.5).

Fig. 7.31

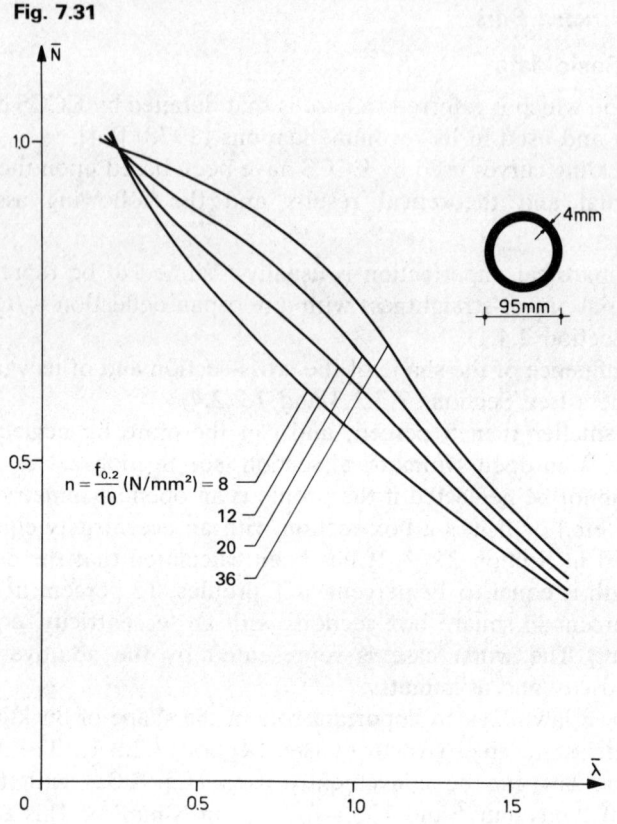

(f) Since experimental results were in very close agreement with the numerical analysis, simulation computations can be systematically used to elaborate the basic data and then extrapolate them in order to use these results for all the common cases.

7.2.3.2 Choice of the dimensionless curves

Starting from these statements, the main problem was how to select buckling curves. At first glance, it seemed that the principal aspects of the problem meriting consideration were:

(a) The cross-section: box or open section
(b) The cross-section: symmetrical or asymmetrical
(c) The material: heat-treated or non-heat-treated alloy

If all these variables were taken into account, too many curves would have resulted. A reduction can be obtained if the following are considered:

The necessity to cover the commonly used alloys for structural applications

The advantage of having as few buckling curves as possible

The scatter of the buckling curves in the $\bar{N}-\bar{\lambda}$ plane from the worst to the most favorable case

The practical deviations of the main properties such as $\sigma-\varepsilon$ law, yield stress, shape of the profile etc.

As a result of this study, the ECCS committee decided to adopt three nondimensional buckling curves (a), (b) and (c) (see Fig. 7.32), which cover all of the extruded aluminum bars with a guaranteed elastic limit greater than or equal to $100\,\mathrm{N\,mm^{-2}}$. These three curves have been computed with the following data (Fig. 7.33):

(a) Double T cross-section on the minor axis bending with $n = 20$
(b) Tubular section with an eccentricity of 10 percent and $n = 15$

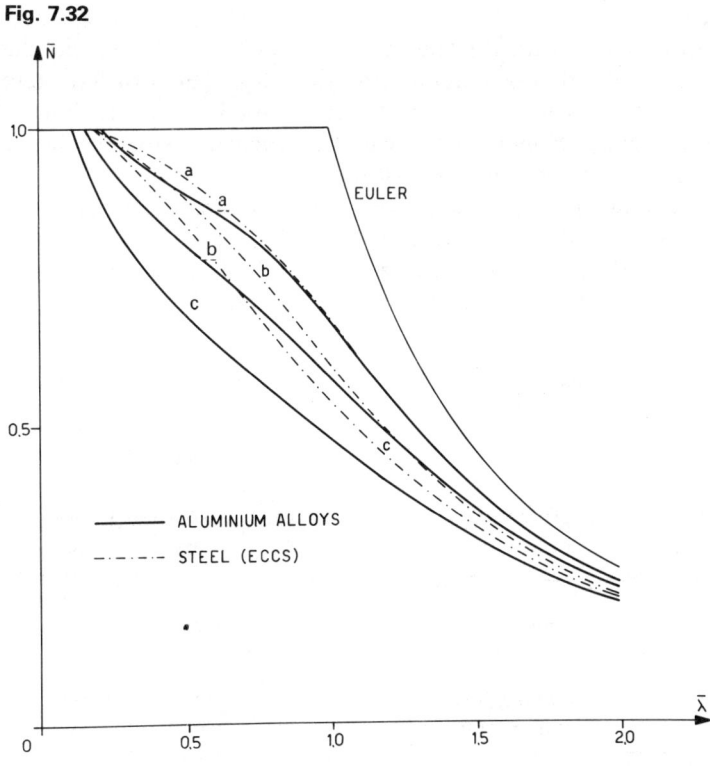

Fig. 7.32

Aluminum alloy structures

Fig. 7.33

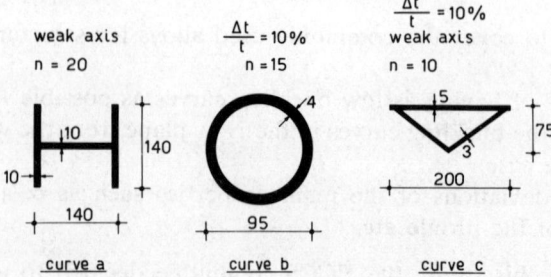

(c) Triangular box section on the minor axis bending with an eccentricity equal to 10 percent and $n = 10$

The cross-sections corresponding to curves (a) and (b) have been used in the experimental research mentioned in Section 7.1.3.1. The cross-section of curve (c) combines both of the unfavorable effects due to asymmetry and eccentricity.

The three curves (a), (b) and (c) drawn in Fig. 7.32 on the classical nondimensionalized $\bar{N}$–$\bar{\lambda}$ diagram are given together with the analogous ECCS curves for steel. These curves are limited by $\bar{N} = 1$, though with small slenderness higher values can be reached owing to the strain-hardening capacity of aluminum alloys.

The buckling curve to be used for a given column has to be chosen on the basis of the indications given in Fig. 7.34.

These criteria have been justified on the basis of comparisons made with experimental results and simulation analysis.

Fig. 7.34 Choice of the buckling curve for aluminum alloy extruded columns

Shape of cross-section	Cross-section with respect to an axis perpendicular to the bending plane	Aluminum alloy with elastic limit $f_{0.2} = 100 \text{ N mm}^{-2}$	
		heat-treated	non-heat-treated
Open Solid	symmetrical	a	b
	asymmetrical	b	c
Hollow	symmetrical	b	c
	asymmetrical	c	c

324

7.2.3.3 Comparison with experimental results

The comparison between the available experimental results and the selected curves (a), (b) and (c) has been carried out in the following way:

The experimental values have been nondimensionalized with respect to the elastic limit $f_{0.2}$ obtained by averaging the results from either stub column tests, or tensile and compression tests on single specimens.
If possible, the statistical analysis used the Student's t test, which is more reliable than the Gaussian distribution for limited data.

Figures 7.35–7.38 show the experimental results represented by the statistical 'fork' when more than eight specimens were tested or by single points in other cases. The main tests considered for the comparison were:

(a) Tests carried out at Liège (ECCS) [2]
(b) Tests carried out in France [29–31]
(c) Tests carried out in Germany [32–34]
(d) Tests carried out in USA [35–37]

American and German tests did not follow the same philosophy adopted by the ECCS and it is therefore difficult to interpret them. In fact these tests were carried out on columns with practically no geometrical imperfections. In general the initial out-of-straightness, the eccentricities of box sections and the geometrical imperfections were not measured.

During the tests carried out in the USA, the load was applied at the center of gravity in order to avoid any initial out-of-straightness and to approximate more closely Euler and Engesser curves. These tests are somewhat old (1938–1940) and are therefore only given to provide qualitative results.

In the tests carried out in Germany (Figs 7.37 and 7.38) the load has been applied with a variable initial eccentricity which was smaller than $L/1000$ in open sections (H and T) and smaller than $e + L/1000$ in tubes, e being equal to the initial eccentricity due to the variations of thickness (eccentricity equal to 10 percent). These results are not conservative since no initial out-of-straightness was considered; in real structures this is usually equal to $L/1000$.

All the comparisons between the experimental results and the curves (a), (b) and (c) are in excellent agreement. Curve (a) can be compared with several tests either from a statistical point of view (Fig. 7.35) or from a deterministic point of view (Fig. 7.36). Curve (b) is also very close to experimental results (Fig. 7.37). Curves (a) and (b) both represent a precise lower bound to the experimental tests of the profiles that they refer to.

On the other hand there are not sufficient test results for the profiles of

Aluminum alloy structures

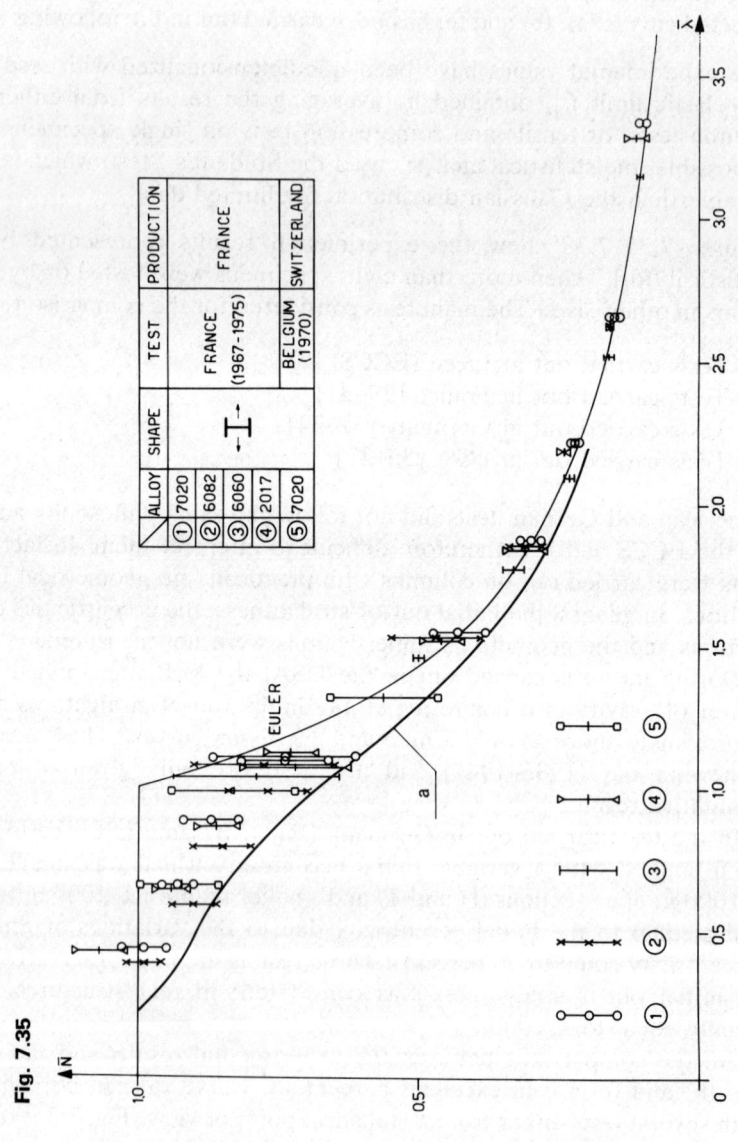

Fig. 7.35

Fig. 7.36

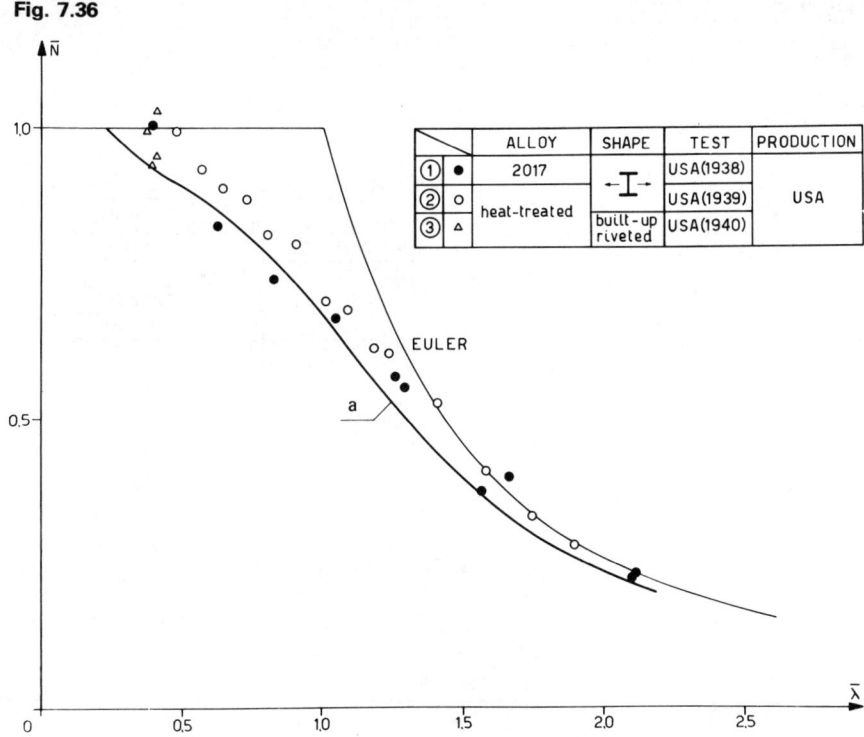

curve (c). The only available results do not lie close to the curve even though they are conservative (Fig. 7.38).

7.2.3.4 Comparison with numerical simulation

Simulation analysis allows the calculation of buckling curves of any profile of a generical material with geometrical and mechanical imperfections (see Section 7.1.2).

Material properties have been taken from the stub column test (the actual material), or simply by taking $f_{0.2} \equiv n$ from the Ramberg–Osgood law (the idealized material). The cross-sections used are shown in Fig. 7.39. They comprise all the extruded profiles in use, both heat-treated (6000 and 7000 series) and non-heat-treated alloys (5000 series).

The comparisons are given in Figs 7.40–7.42. Each simulated case is related to the corresponding buckling curve on the basis of the criteria indicated in Fig. 7.34. The agreement was excellent for all of the curves which represent a lower bound to the cases they refer to. Particularly, curve (c) minimizes the most unfavorable cases influenced by the so-called eccentricity imperfection (Fig. 7.42).

Fig. 7.37

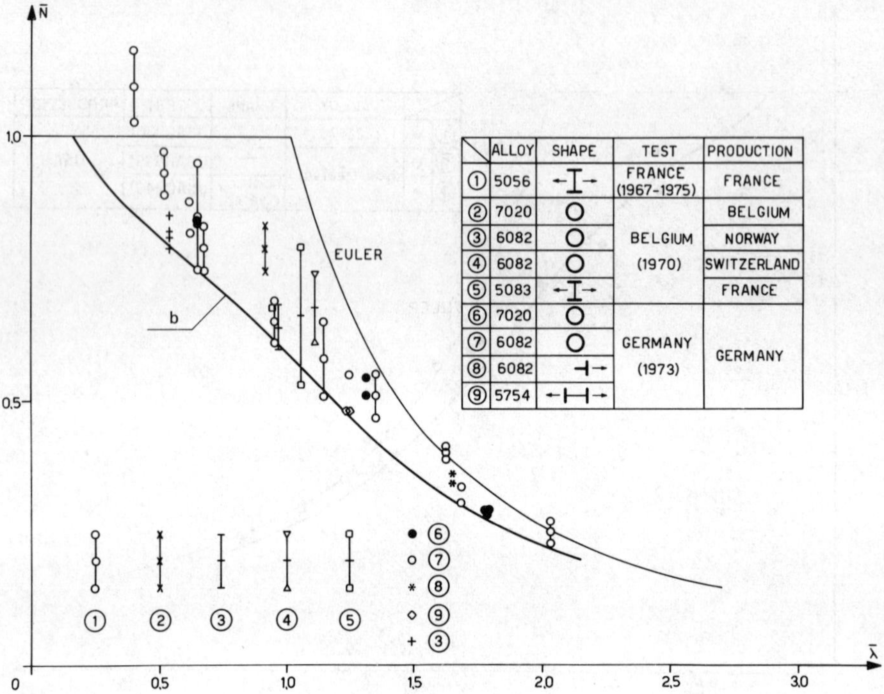

7.2.3.5 Comparison with specifications

Curves (a), (b) and (c) have been compared with those of the following specifications:

BS CP 118   Structural use of aluminum (ref. [7] in Chapter 3)
DIN 4113   The design of aluminum alloy structures (refs. [5, 6] in Chapter 3)
DTU 32.2   Règles de conception et de calcul des charpentes en alliages d'aluminium (ref. [33] in Chapter 3)
ALCAN   The strength of aluminum [63]

The following materials have been considered:

Alloy 5083 H111, $f_{0.2} = 120$ N mm^{-2} (Fig. 7.43)
Alloy 5754 H24, $f_{0.2} = 160$ N mm^{-2} (Fig. 7.44)
Alloy 6060 T6, $f_{0.2} = 170$ N mm^{-2} (Fig. 7.45)
Alloy 6061 T6, $f_{0.2} = 240$ N mm^{-2} (Fig. 7.46)
Alloy 7020 T6, $f_{0.2} = 280$ N mm^{-2} (Fig. 7.47)

Fig. 7.38

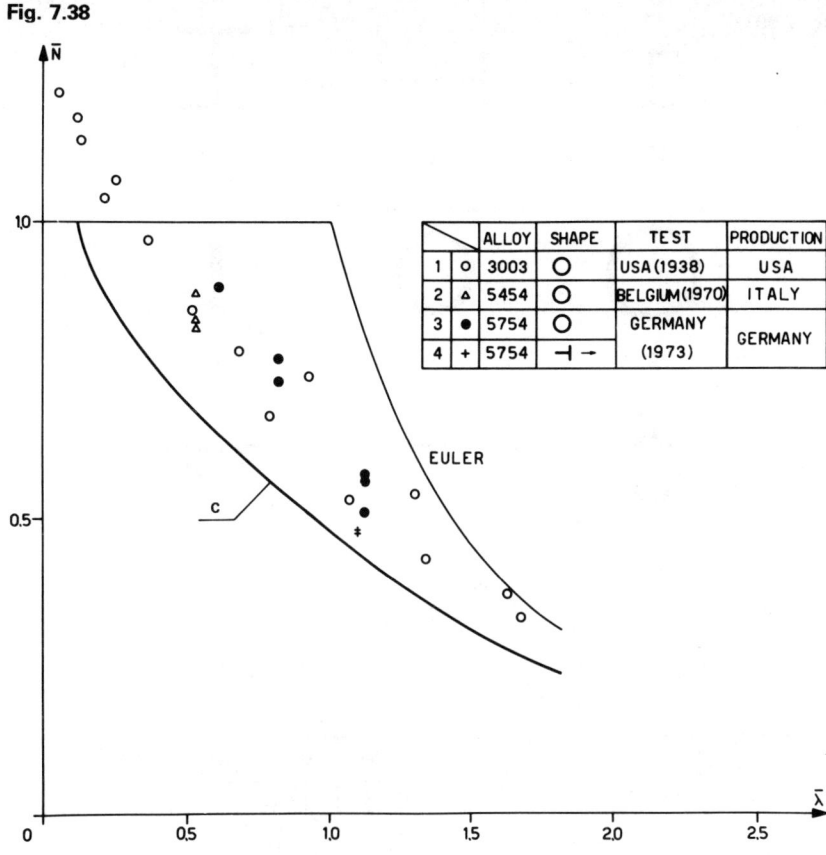

The comparison has been made using the failure curves corresponding to these materials, including the cases in which specifications give the allowable curves (as in BS CP 118). From the analysis of Figs 7.43–7.47 it can be observed that:

(1) The specifications do not differentiate between open and box sections (with the exception of DIN 4113) or between symmetrical and asymmetrical sections.
(2) The curve provided by DIN 4113 for tubes is higher than that for double Ts, in contrast to the ECCS method.
(3) Curves (a) and (b) are approximately located in the region covered by the above specifications.
(4) The shape of curve (c) is completely different from the other available curves and gives the smallest values.

Aluminum alloy structures

Fig. 7.39

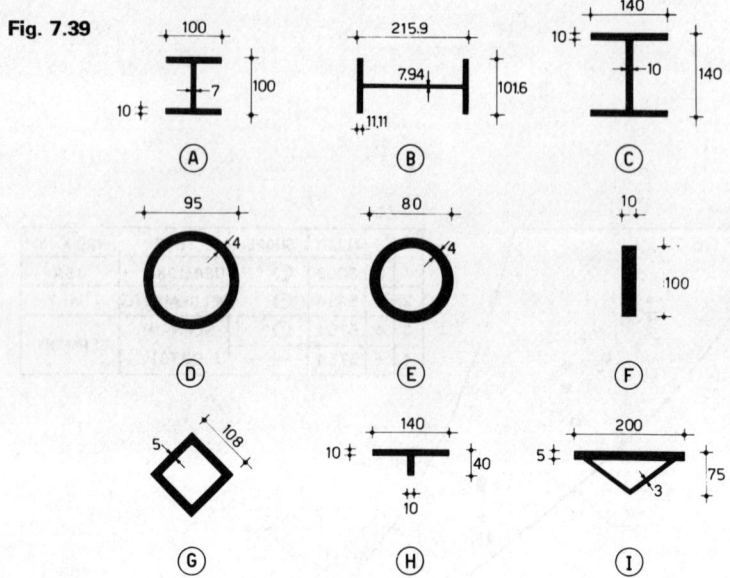

Fig. 7.40

CURVE	SHAPE		ALLOY	Nmm^{-2}		n
				E	$f_{0.2}$	
①	Ⓐ	←I→	6061	72'860	290,5	57
②		←H→				
③	Ⓑ	←I→	7020	76'014	309,5	27,43

Stability of structural elements

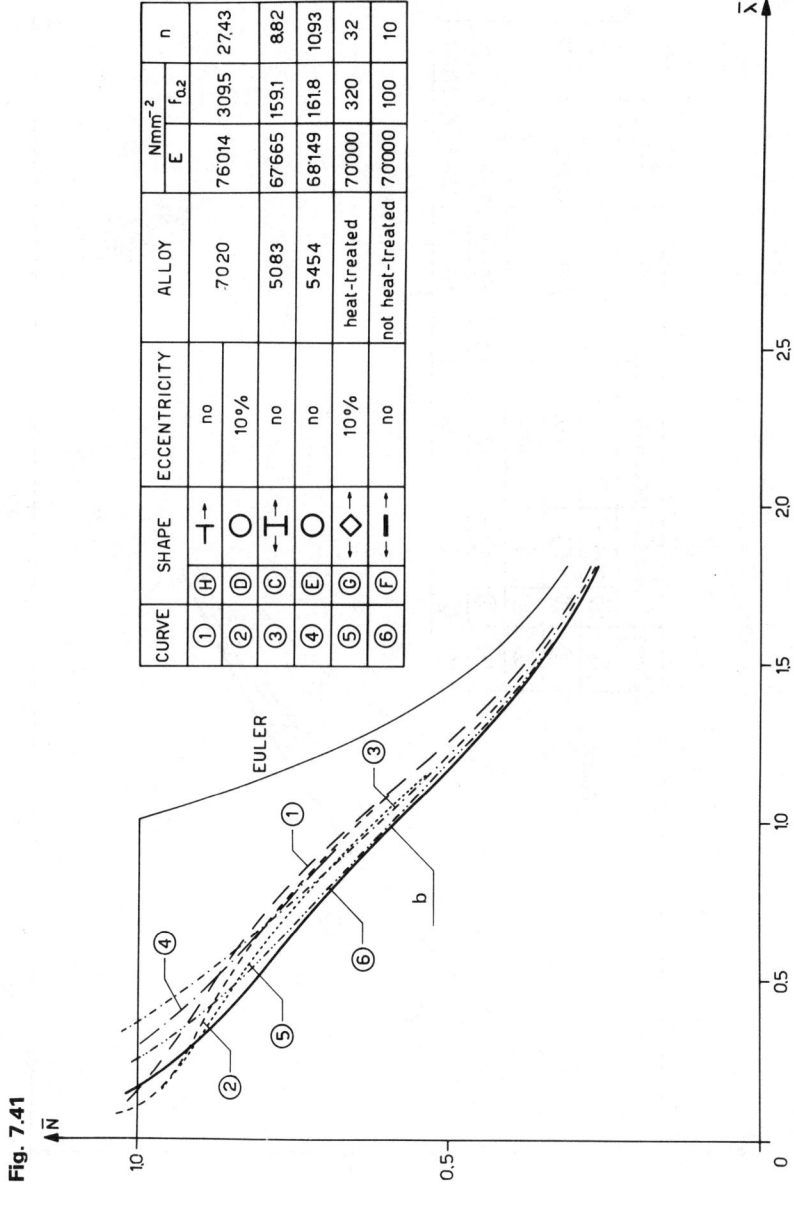

Fig. 7.41

Aluminum alloy structures

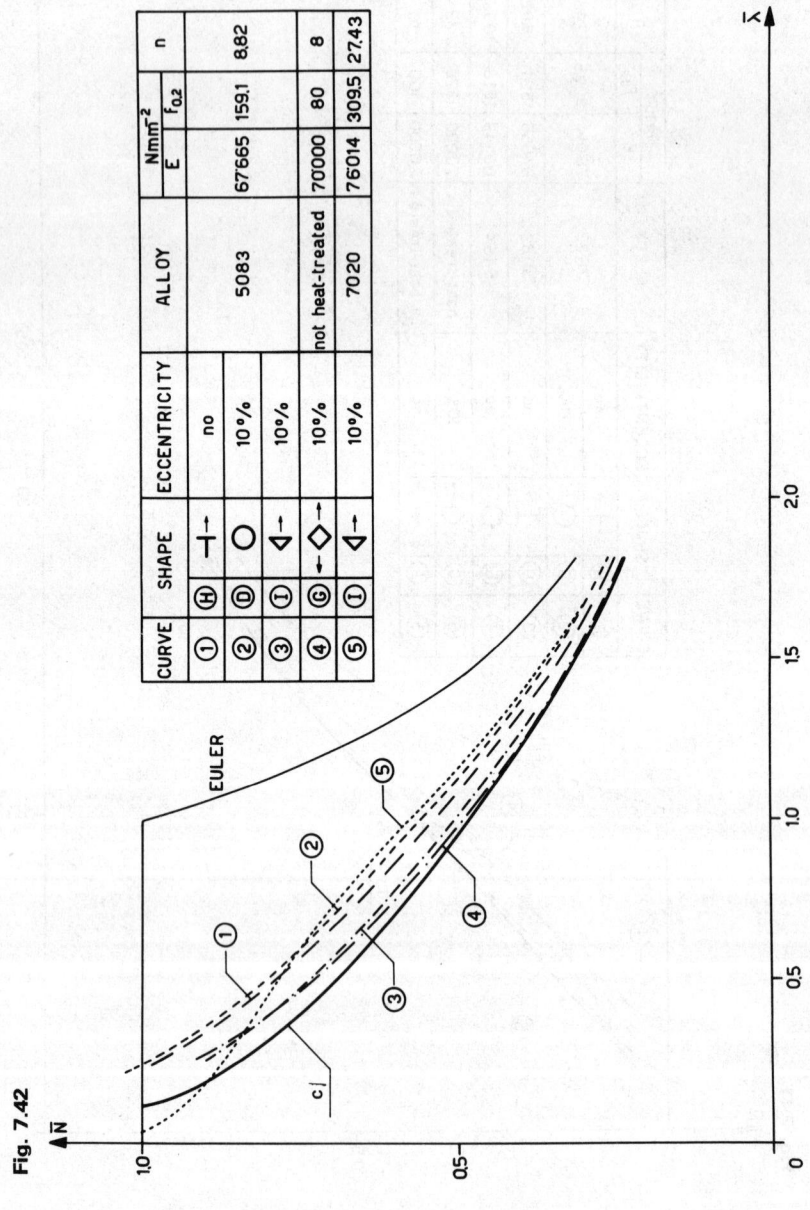

Fig. 7.42

Stability of structural elements

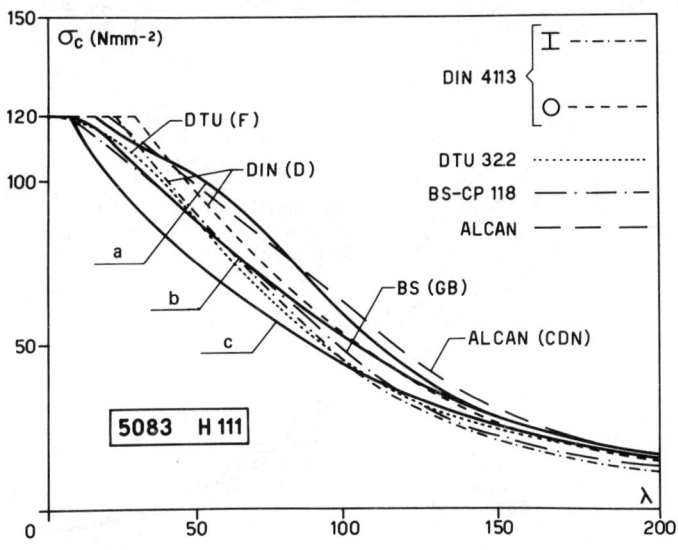

Fig. 7.43

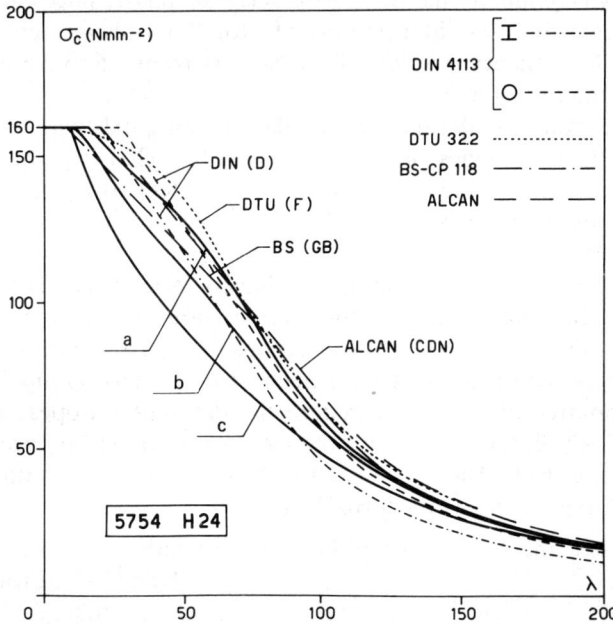

Fig. 7.44

Fig. 7.45

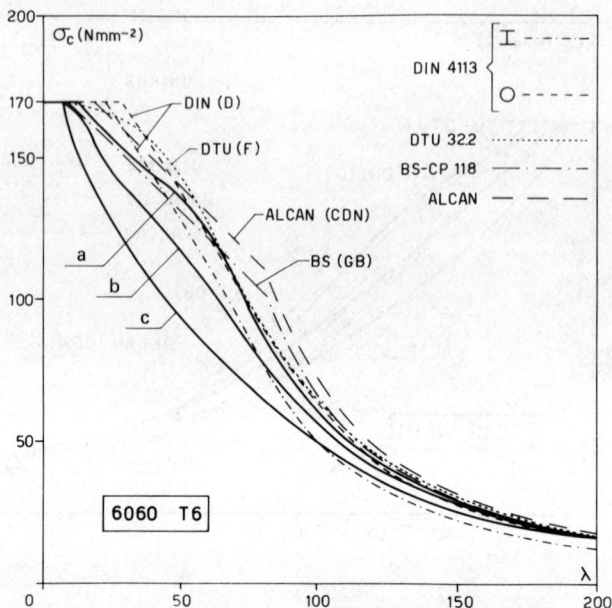

However, this comparison has to be considered only from a qualitative point of view as a nondimensionalized representation of the ultimate behavior. No computational comparisons can be made, since each specification provides its own differing safety factors (or enhancement factors if the probabilistic theory is followed); these vary from 1.5 to 2.5 depending upon the loading conditions.

It is important to underline that, unlike the other existing methods, the ECCS method provides multiple curves for checking buckling of aluminum columns, which is the way in which steel structures are considered. The method is based upon both experimental results and simulation analyses.

Considerations 2 and 4 can be explained if it is assumed that the influence of the imperfection of the cross-section (termed eccentricity) is not taken into account in the specifications of other countries.

Quantitative considerations have been made by task group 2.2 of the ECCS committee in order to compare the different European specifications (A, B, D, F, GB, N, S) with the curves (a), (b) and (c) proposed by the committee itself. They have been used to design three columns (types A, B, C of Fig. 7.48), characterized by:

Two values of the slenderness ratio $\lambda = 60$ and 100
Two shapes   double T and cylindrical tube
Three alloys   5083, 6082, 7020

Fig. 7.46

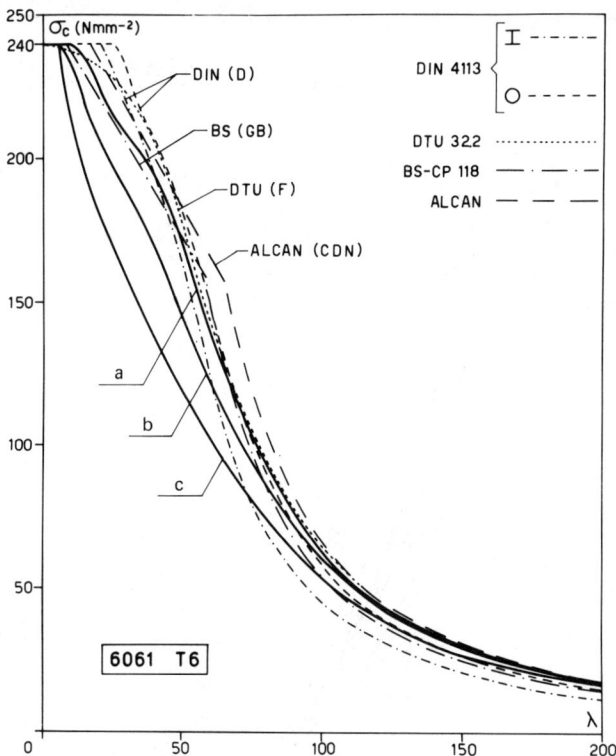

Three loading conditions, with different ratios between live loads and dead loads, the sum being constant (300.000 N).

The minimum thickness for each profile has been established by using the methods of different specifications; the corresponding values are given in Fig. 7.49.

The percentage variations in weight given by the application of different specifications for each profile corresponding to the three curves (a), (b) and (c) are given in Fig. 7.50. It should be noted that the ECCS curves (a) and (b) always provide the minimum weight and thus the most economical design. On the other hand curve (c) seems to be more conservative, as should be expected from the previous comparisons.

The average differences in weight for the three profiles between the different specifications are as follows:

(1) ECCS   0
(2) Sweden   +5   percent
(3) France   +11.5 percent

335

Aluminum alloy structures

Fig. 7.47

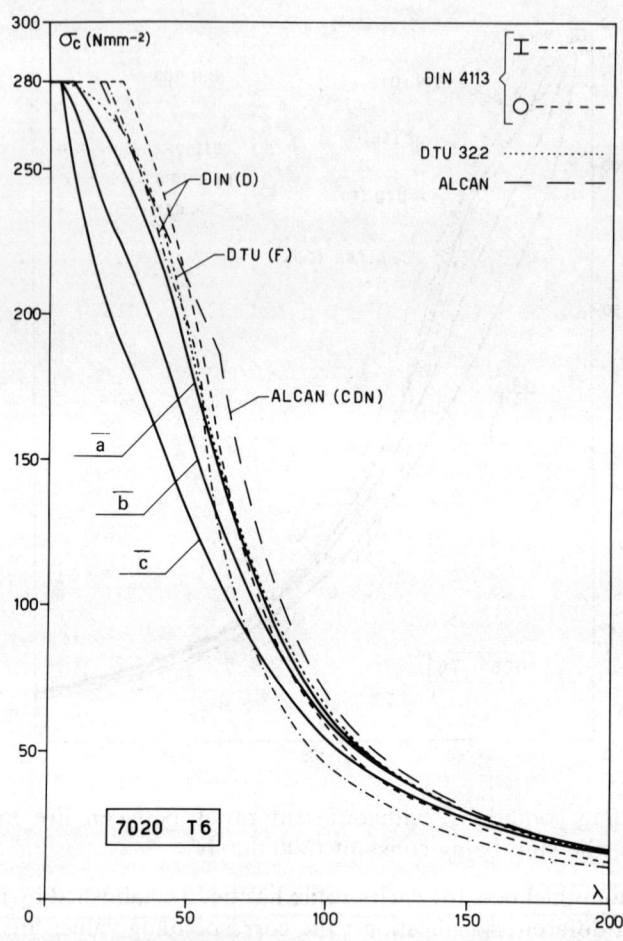

(4) UK	+12	percent
(5) Norway	+12.1	percent
(6) DIN I°	+12.6	percent
(7) Austria	+15	percent
(8) DIN 2°	+20	percent
(9) Belgium	+41	percent

This proves that the ECCS method is the most economical.

7.2.3.6 Codification

From the comparison between curves (a), (b) and (c) and the results of experimental tests and numerical simulations, and on the basis of the

Fig. 7.48

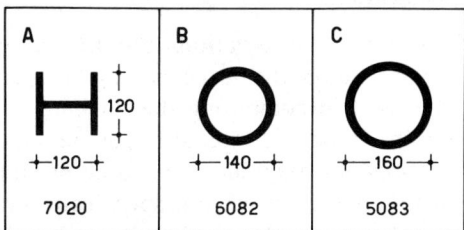

existing specifications it is observed that:

curve (a) has close agreement and represents well the behavior of heat-treated columns with a symmetrical cross-section.

curve (b) provides the same close agreement for non-heat-treated columns with a symmetrical cross-section.

Fig. 7.49 Minimum profile thickness (mm)

Alloy	7020	6082	5083	7020	6082	5083
Shape	A	B	C	A	B	C
λ		60			100	
ECCS	8.1	7.2	11.6	19.4	15.6	18
DIN 4113 (1st)	10.1	8.4	8.2	25.4	20.5	17.4
DIN 4113 (2nd)	11.1	7.7	8.7	30.9	20.7	17.1
UK	8.8	7.2	8.8	24.5	20.1	20.1
Austria	10.1	7	9.3	28.2	18.9	18.2
Norway	9.3	8.4	10.8	23.5	18.2	19.4
Sweden	9.2	7.8	9.5	21.4	17.5	18.1
France	8.9	8	12.7	20.7	17.5	21.3
Belgium	11.5	9.9	11.6	31.8	28.2	19.4

Fig. 7.50 Percentage variations in weight

Section A alloy 7020 curve (a)		Section B alloy 6082 curve (b)		Section C alloy 5083 curve (c)	
1 ECCS		1 ECCS		1 DIN 1st	−14%
2 France	8%	2 Sweden	11%	2 DIN 2nd (ω)	−13%
3 Sweden	11%	3 France	12%	3 Austria	−7%
4 Norway	19%	4 Austria	14%	4 Sweden	−7%
5 UK	21%	5 Norway	17%	5 UK	−2%
6 DIN 1st	29%	6 UK	20%	6 ECCS	
7 Austria	39%	7 DIN 2nd (ω)	25%	7 Norway	2%
8 DIN 2nd (ω)	53%	8 DIN 1st	27%	8 Belgium	5%
9 Belgium	57%	9 Belgium	67%	9 France	15%

Aluminum alloy structures

curve (c) seems to be too conservative from all points of view (tests, simulation and specifications).

On the basis of the above observations the ECCS committee had to devise a method, based upon the three curves (a), (b) and (c), to give practical rules to be used in computing the buckling of columns. These rules had to provide the same approximations as the ECCS recommendations for steel structures and they also had to be as simple as possible.

It has also been noted that curve (c), specifically introduced to take into account asymmetrical profiles, has been deduced considering the eccentricity of box sections. This is neglected in steel box sections even though they have the same manufacturing tolerances.

Therefore, in order to have the same philosophy as adopted in steel structures, the ECCS committee decided:

To use curves (a) and (b) for symmetrical profiles of heat-treated and non-heat-treated alloys respectively.
To cease use of curve (c). So far it is only referred to as a lower bound which takes into account all possible imperfections.
To emphasize that the real behavior of aluminum alloy columns is represented by curves (a), (b) and (c).
To take into account asymmetry by means of a reduction factor and to undertake further research in order to better quantify this effect.

The method currently adopted by the ECCS recommendations [64] on aluminum alloy structures is based upon this philosophy.

Asymmetrical sections are characterized by the following coefficient (Fig. 7.51):

$$\psi = \frac{y_{max} - y_{min}}{h} \qquad (7.21)$$

where y_{max} and y_{min} are the distances of the extreme fibers of the cross-section from the axis perpendicular to the bending plane and passing through the center of gravity ($y_{max} \geq y_{min}$), and $h = y_{max} + y_{min}$.

Fig. 7.51

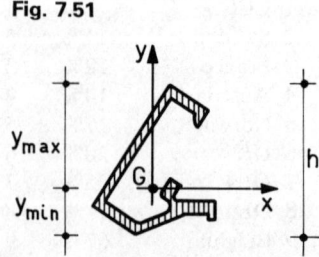

This coefficient can vary from 0 to 1:

$$0 \leq \psi < 1$$

and depends upon the rate of asymmetry, where $\psi = 0$ represents symmetrical sections.

The reduction coefficient which takes into account the asymmetry of the cross-section is given by:

$$k_1 = 1 - \rho\psi^2 \frac{\bar{\lambda}^2}{(1+\bar{\lambda}^2)(1+\bar{\lambda})^2} \qquad (7.22)$$

where

$\rho = 2.4$ in the case of heat-treated alloys (curve a)

$\rho = 3.2$ in the case of non-heat-treated alloys (curve b)

Equation 7.22 yields $k_1 = 1$ for sections symmetrical with respect to the axis perpendicular to the bending plane, and $k_1 < 1$ in the case of asymmetrical sections.

This empirical relationship is based upon a parametric study made through a simulation of 180 columns (T sections). The use of the formula is provisional and needs to be confirmed using further experimental research [22].

7.2.3.7 Calculation method

The load-bearing capacity of an extruded column is given by

$$N_c = k_1 \bar{N} f_d A \qquad (7.23)$$

$\bar{N}$ normalized failure load obtained by the nondimensionalized curves (a) or (b). The values are given in Fig. 7.52 as functions of the slenderness ratio $\bar{\lambda} = \lambda/\pi\sqrt{(E/f_d)}$

k_1 reduction factor given by Eq. 7.22 to approximately take into account the asymmetry of the cross-section

f_d design strength, often considered equal to the elastic limit of the material $f_{0.2}$

A area of the cross-section

It has to be checked that:

$$\frac{N_c}{N} \geq \nu \qquad (7.24)$$

N effective axial load in the column
ν an appropriate safety factor

The values from curves (a) and (b) are used for heat-treated and non-heat-treated alloys, respectively.

Aluminum alloy structures

Fig. 7.52 Values of the normalized stress $\bar{N}$ in relation to the slenderness ratios

Curve (a): Heat-treated alloys

$\bar{\lambda}$	0	0.1	0.2	0.3	0.4	0.5	0.6	0.7	0.8	0.9
0	1	1	1	0.9640	0.9259	0.8929	0.8611	0.8282	0.7863	0.7322
1	0.6716	0.6068	0.5425	0.4822	0.4291	0.3819	0.3407	0.3057	0.2756	0.2495
2	0.2268	0.2069	0.1894	0.1741	0.1606	0.1485	0.1377	0.1280	0.1193	0.1115
3	0.1044	0.0980	0.0921	0.0867	0.0818	0.0773	0.0731	0.0693	0.0658	0.0625
4	0.0595	0.0566	0.0540	0.0516	0.0493	0.0472	0.0452	0.0433	0.0415	0.0398
5	0.0383									

Curve (b): Non-heat-treated alloys

$\bar{\lambda}$	0	0.1	0.2	0.3	0.4	0.5	0.6	0.7	0.8	0.9
0	1	1	0.9693	0.9035	0.8533	0.8084	0.7649	0.7231	0.6776	0.6302
1	0.5808	0.5308	0.4825	0.4355	0.3932	0.3544	0.3202	0.2899	0.2632	0.2397
2	0.2190	0.2008	0.1847	0.1702	0.1572	0.1457	0.1354	0.1261	0.1178	0.1104
3	0.1037	0.0974	0.0916	0.0863	0.0815	0.0771	0.0730	0.0692	0.0657	0.0625
4	0.0595	0.0566	0.0540	0.0516	0.0493	0.0472	0.0452	0.0433	0.0415	0.0398
5	0.0383									

In order to allow the use of calculators when checking buckling, the values of the nondimensionalized curves (a) and (b) can be computed to within 2 percent by the following analytical formulation given by Rondal [13]:

For $0 \leq \bar{\lambda} \leq \bar{\lambda}_0$:

$$\bar{N} = 1,0$$

For $\bar{\lambda} > \bar{\lambda}_0$:

$$\bar{N} = \frac{1}{2\bar{\lambda}^2}\{1 + \alpha\sqrt{(\bar{\lambda}^2 - \bar{\lambda}_0^2)} + \bar{\lambda}^2[1 - 2\beta(\gamma - \bar{\lambda})^\mu\sqrt{(\bar{\lambda}^2 - \bar{\lambda}_0^2)}]$$
$$- \sqrt{[(1 + \alpha\sqrt{(\bar{\lambda}^2 - \bar{\lambda}_0^2)} + \bar{\lambda}^2)^2 - 4\bar{\lambda}^2]}\} \tag{7.25}$$

The values of the parameters $\bar{\lambda}_0$, α, β, γ, μ are given by Fig. 7.53.

Fig. 7.53 Parameter values for buckling curve equation

Parameter	Curve (a)	Curve (b)
$\bar{\lambda}_0$	0.2226	0.1876
α	0.1590	0.2420
β	0.083 if $\bar{\lambda} < 1.1$	0.172 if $\bar{\lambda} < 1.4$
	0.0 if $\bar{\lambda} \geq 1.1$	0.0 if $\bar{\lambda} \geq 1.4$
γ	1.1	1.4
μ	0.966	1.478

Fig. 7.54 Parameter values for simplified buckling curve equation

Parameter	Curve (a)	Curve (b)
$\bar{\lambda}_0$	0.2226	0.1876
α	0.02	0.12
β	9.6	3.6

Equation 7.25 has been introduced into the ECCS recommendations (ref. [19] in Chapter 6).

Later, Rondal provided a simpler formulation (1980) with practically the same approximation as the previous one:

For $0 \leq \bar{\lambda} \leq \bar{\lambda}_0$:

$$\bar{N} = 1$$

For $\bar{\lambda} > \bar{\lambda}_0$:

$$\bar{N} = \frac{1}{2\bar{\lambda}^2} \{[1 + \alpha(\beta - \bar{\lambda})\sqrt{(\bar{\lambda}^2 - \bar{\lambda}_0^2)} + \bar{\lambda}^2]$$
$$- \sqrt{([1 + \alpha(\beta - \bar{\lambda})\sqrt{(\bar{\lambda}^2 - \bar{\lambda}_0^2)} + \bar{\lambda}^2] - 4\bar{\lambda}^2)}\} \quad (7.26)$$

which is a function of the three parameters given in Fig. 7.54.

7.2.4 Welded Bars

7.2.4.1 Computational bases

The method used in the European Recommendations (1978) (ref. [19] in Chapter 6) for checking buckling of welded columns has been derived by ECCS committee 16 from a simulation analysis based upon the available experimental results already defined in Section 7.1.3.2. These results refer to the heat-treated alloy AlMgSi (6082) and to the three profiles T, P, C of Fig. 2.3. The tests allowed the following regions with different mechanical properties to be defined:

A   unaffected base metal for the extruded parts of the P and C sections and for the rolled parts of the T section
A'   unaffected base metal for the rolled parts of the P section
B   partially affected base metal
C   heat-affected zone comprising both base and weld metal

The average values of the mechanical properties E, $f_{0.2}$ and n, which characterize the Ramberg–Osgood law, are given in Fig. 2.49 for the different regions.

Aluminum alloy structures

Compression and tension values have been considered; the values in compression correspond to the actual behavior, whereas those in tension are commonly used in calculations. The principal differences are related to the exponent n, which is always higher when calculated for tension.

The residual stress distributions obtained from tests allowed definitions of the models already shown in Figs 2.41–2.43 for the T, P and C profiles, respectively.

An initial out-of-straightness defined by $v_0/L = 1/1000$ has been used in all of the simulation analyses. These results give sufficient data on the behavior of heat-treated alloys, although extrapolation was necessary for non-heat-treated alloys since there was no experimental data available.

T, P and C profiles, which can be considered representative of welded symmetrical profiles, have also been used to calculate the buckling curves of non-heat-treated alloys. The materials can be characterized by the value of the elastic limit $f_{0.2}$, assumed equal to 10 times the exponent n of the Ramberg–Osgood law. The values 120, 160 and 200 (N mm^{-2}) have been used because they cover the range of non-heat-treated alloys.

It is well known that reduced-strength zones are practically nonexistent in these materials, and residual stresses are therefore the only mechanical imperfection. Tests on simple butt-welded joints showed that the distribution of residual stresses is practically independent of the material. Hence the residual stress pattern in this case is not significantly different from that present in heat-treated alloys. However, maximum residual stress values have to be related to the σ–ε law of the material.

The residual stress distributions of Fig. 7.55 are referred to the maximum tensile stress σ_r, which are given on the basis of the following criteria:

For $n = 20$ the same values obtained from heat-treated profiles have been used

For $n = 16$ the values of σ_r have been derived from previous results on non-heat-treated butt-welded joints

For $n = 12$, σ_r values have been extrapolated from the previous cases by assuming the same ratio with $f_{0.2}$

7.2.4.2 Results of the numerical simulation [12]

Simulation analyses have been carried out by taking account of the different factors which affect the load-bearing capacity of welded members. Both strong (S) and weak (W) axis bending behavior has been studied for each profile. In the case of heat-treated alloys, the presence of the so-called reduced-strength zones has been introduced by varying $f_{0.2}$ and n through the cross-section.

Buckling curves for heat-treated alloys (AlMgSi) are given in Figs

Fig. 7.55

	NON-HEAT-TREATED ALLOYS					
	RESIDUAL STRESSES	AXIS	n	σ_r (N mm^{-2})	CURVE	Nr.
T	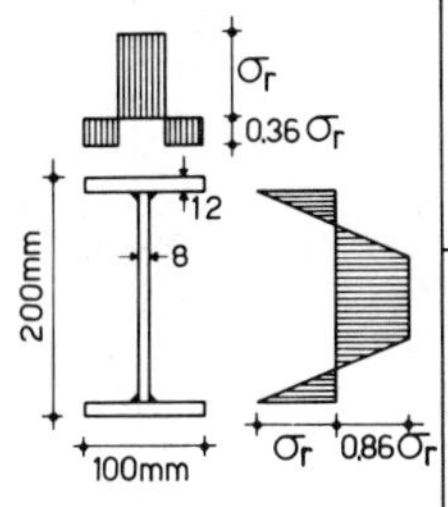 (section diagram: 200mm × 100mm, t=12, t=8; σ_r, $0.36\sigma_r$, $0.86\sigma_r$)	weak	12	90	TW 12	1
			16	120	TW 16	2
			20	140	TW 20	3
		strong	12	90	TS 12	4
			16	120	TS 16	5
			20	140	TS 20	6
P	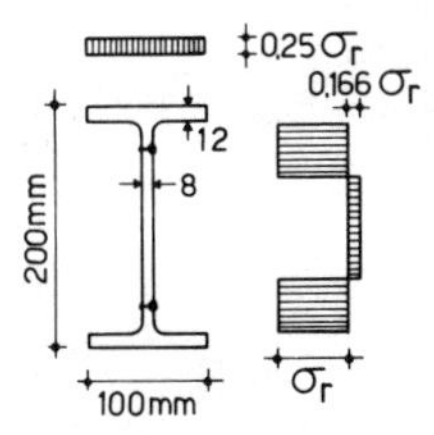 (section diagram: 200mm × 100mm, t=12, t=8; $0.25\sigma_r$, $0.166\sigma_r$, σ_r)	weak	12	90	PW 12	7
			16	120	PW 16	8
			20	120	PW 20	9
		strong	12	90	PS 12	10
			16	120	PS 16	11
			20	120	PS 20	12
C	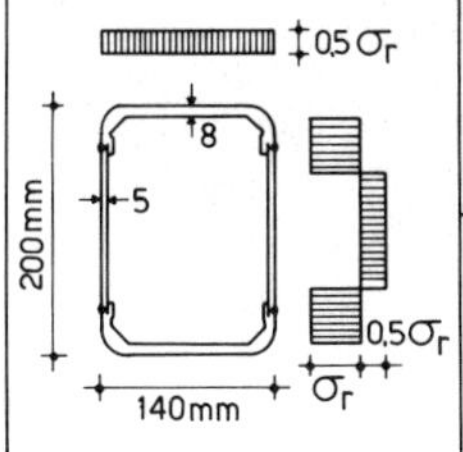 (section diagram: 200mm × 140mm, t=8, t=5; $0.5\sigma_r$, $0.5\sigma_r$, σ_r)	weak	12	60	CW 12	13
			16	80	CW 16	14
			20	80	CW 20	15
		strong	12	60	CS 12	16
			16	80	CS 16	17
			20	80	CS 20	18

7.56–7.59. They are based upon compressive values of the material (subscript c) with the exception of Fig. 7.57 for the T profile, in which tensile values have been considered (subscript t). Forty-eight curves have been computed (three profiles × two materials × two axes × four influence parameters).

For each case (profile, axis, material), the following combinations of influence coefficients have been considered:

Shape effect only (dotted line)
Residual stresses (dashed line)
Effects of variations of $f_{0.2}$ and n (dashed and dotted line)
Cumulative effect of shape, residual stresses and varying $f_{0.2}$ (solid line)

Aluminum alloy structures

Fig. 7.56

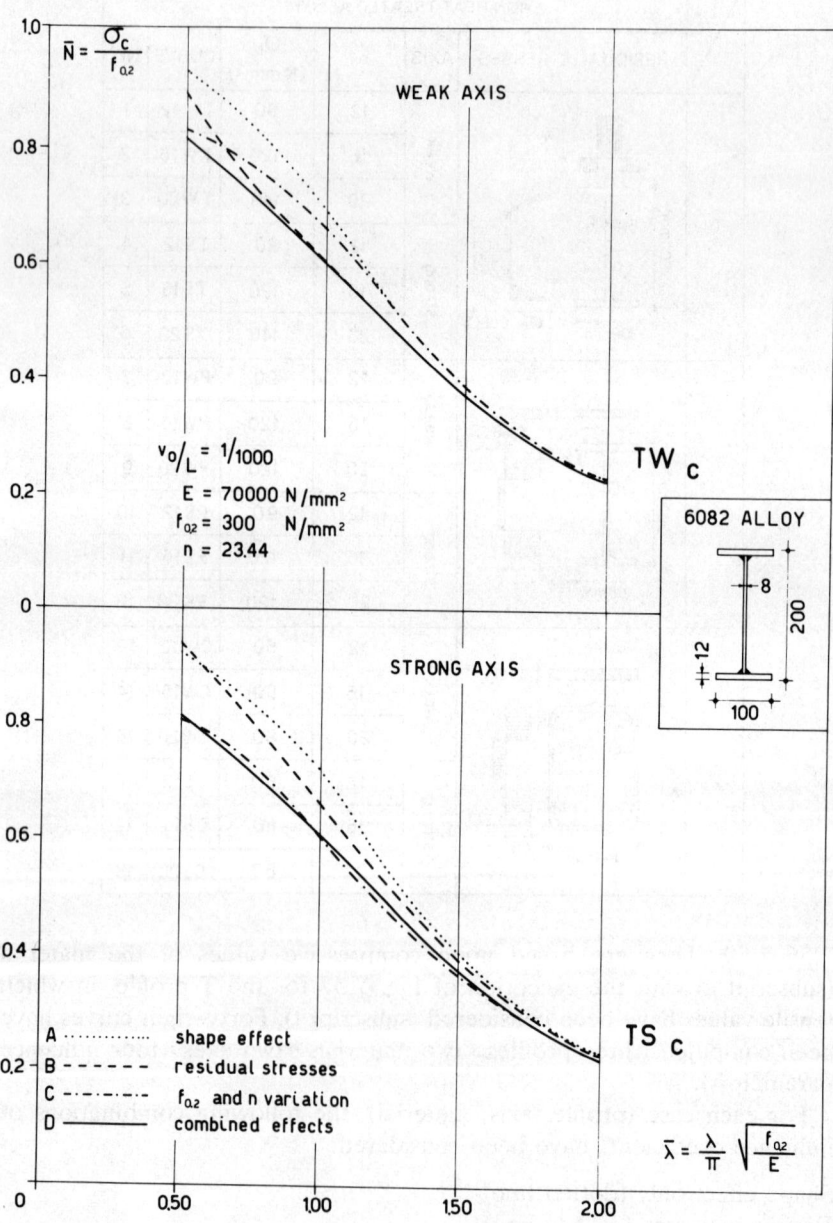

Fig. 7.57

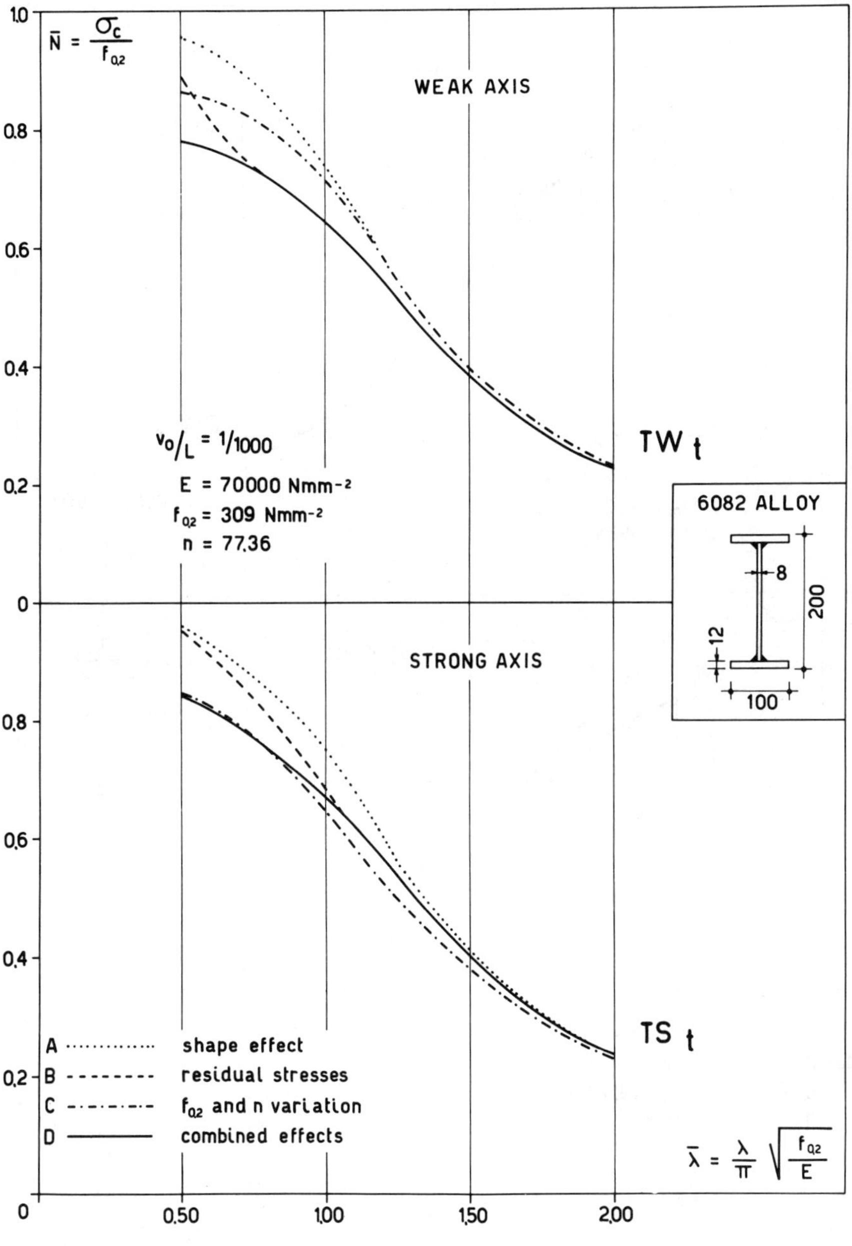

$\bar{N} = \frac{\sigma_c}{f_{0.2}}$
1.0
0.8
0.6
0.4
0.2
0
WEAK AXIS
STRONG AXIS
$v_0/L = 1/1000$
E = 70000 Nmm-2
$f_{0.2}$ = 309 Nmm-2
n = 77.36
TW t
TS t
6082 ALLOY
8
200
12
100
A shape effect
B residual stresses
C $f_{0.2}$ and n variation
D combined effects
$\bar{\lambda} = \frac{\lambda}{\pi}\sqrt{\frac{f_{0.2}}{E}}$
0.50
1.00
1.50
2.00

Aluminum alloy structures

Fig. 7.58

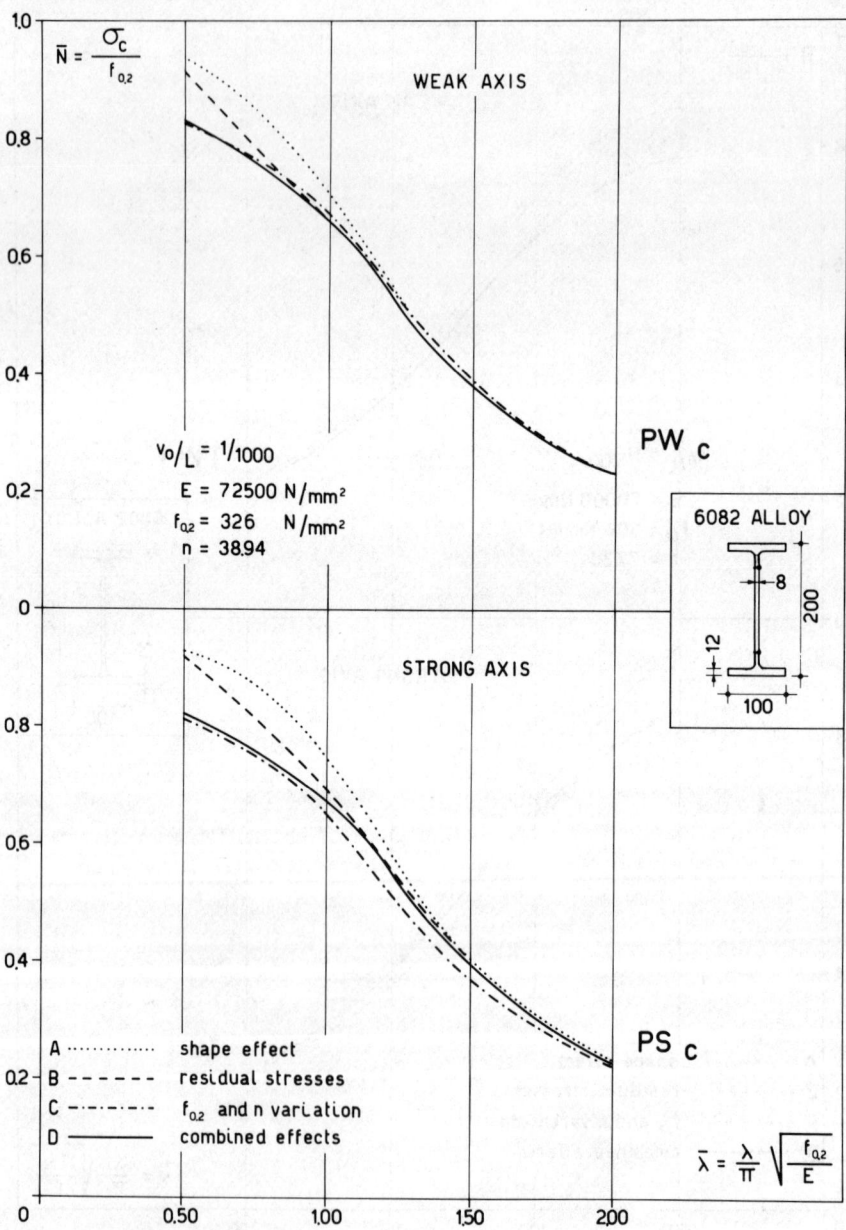

Fig. 7.59

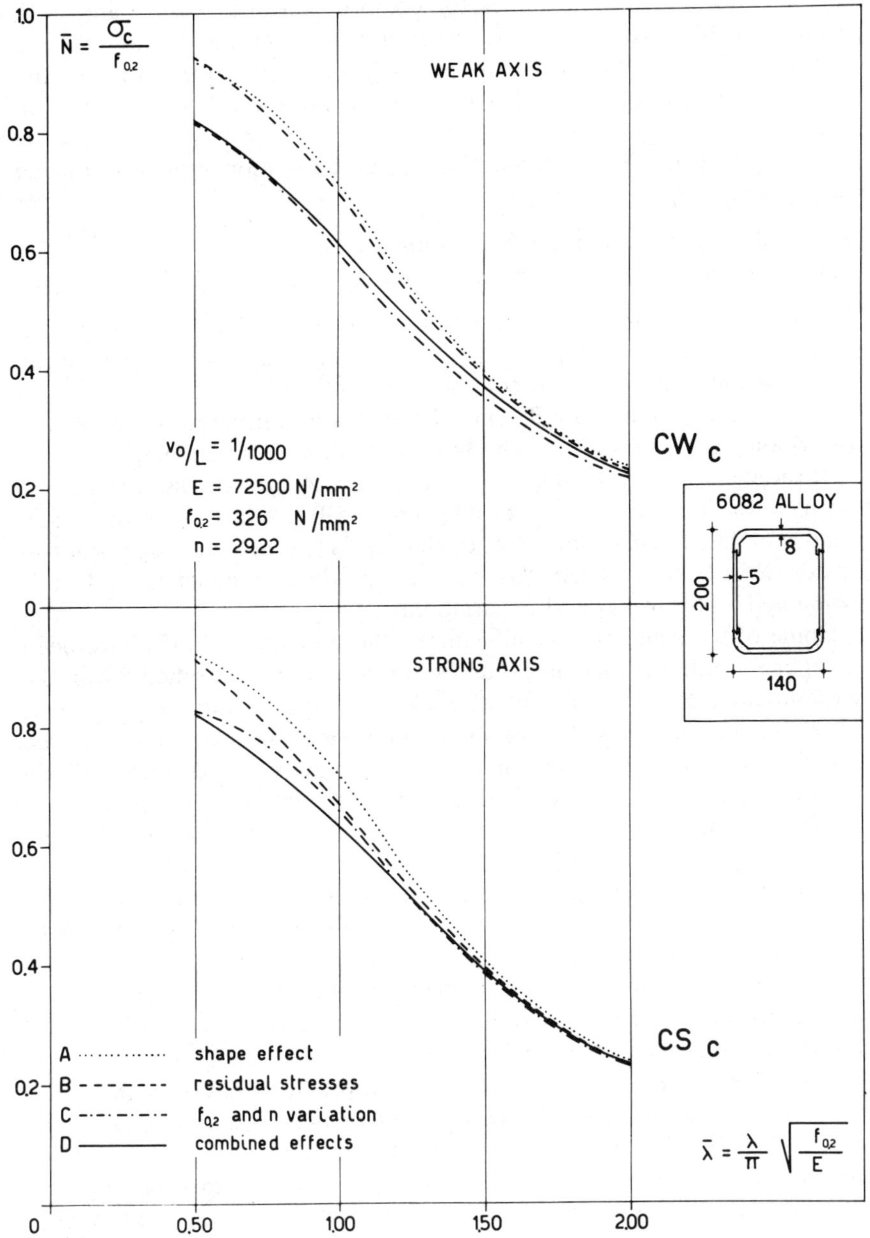

Fig. 7.59 — Column curves for 6082 alloy rectangular hollow section (200 × 140, thickness 8 and 5), showing $\bar{N} = \sigma_c / f_{0.2}$ versus $\bar{\lambda} = \frac{\lambda}{\pi}\sqrt{\frac{f_{0.2}}{E}}$ for weak axis (CW c) and strong axis (CS c).

Parameters: $v_0/L = 1/1000$, $E = 72500$ N/mm², $f_{0.2} = 326$ N/mm², $n = 29.22$.

Curves:
- A .... shape effect
- B − − − − − residual stresses
- C −·−·−·− $f_{0.2}$ and n variation
- D ———— combined effects

Aluminum alloy structures

The comparison between the curves of Figs 7.56 and 7.57 for T profiles shows that the buckling behavior and the effect of the influence factors is qualitatively the same if values for compression and tension are used, even though there is a great difference between the values of the exponent n (23.44 and 77.36 respectively). However, the results of the simulation analysis showed that it is safe to use the calculated values in compression.

In the case of the T profile (Fig. 7.56) the major strength-reducing factors are:

Residual stresses in minor axis bending
$f_{0.2}$ and n variations in major axis bending

In major axis bending it should be observed that residual stresses acting together with the variations of $f_{0.2}$ and n play a beneficial role for the load-bearing capacity of the column.

In the case of the P profile (Fig. 7.58), $f_{0.2}$ and n variations represent the worst imperfection for both major and minor axis bending.

Residual stresses always produce those effects already discussed for T profiles. In the case of the C profile (Fig. 7.59) $f_{0.2}$ and n variations still represent the worst factors. Residual stresses are not very significant in minor axis bending even though they produce a beneficial effect if combined with the $f_{0.2}$ and n variations.

Some of the simulation results among those mentioned [12] have been compared with the buckling curves for minor axis bending from the experimental tests carried out at Liège. The comparison (Fig. 7.60) is excellent for the T and P profiles, despite some small deviations in the input data. Some discrepancy is observed for the C profile, especially in the elastoplastic range of the curves with and without imperfections.

The highest deviation (5 percent) occurs for the dimensional slenderness ratio equal to 1. However, the curve from Liège is lower and is considered in the subsequent computations with the symbol CW'.

In the case of non-heat-treated alloys, for the P, T and C profiles, both major and minor axis bending behavior have been simulated (Figs 7.61–7.63). The effects of mechanical imperfections is emphasized by comparing the buckling curves excluding (dashed line) and including (solid line) the effects of residual stresses.

Also, for non-heat-treated alloys, the T profile behaves in the worst way because of the strength-reducing effect of residual stresses in the region $0.5 < \bar{\lambda} < 1.5$. This effect is also exhibited by P and C profiles. Only in some cases ($\bar{\lambda} < 1$) do residual stresses increase the load-bearing capacity (major axis bending in T sections and minor axis bending in C sections).

Each buckling curve is identified by three parameters which represent

Fig. 7.60

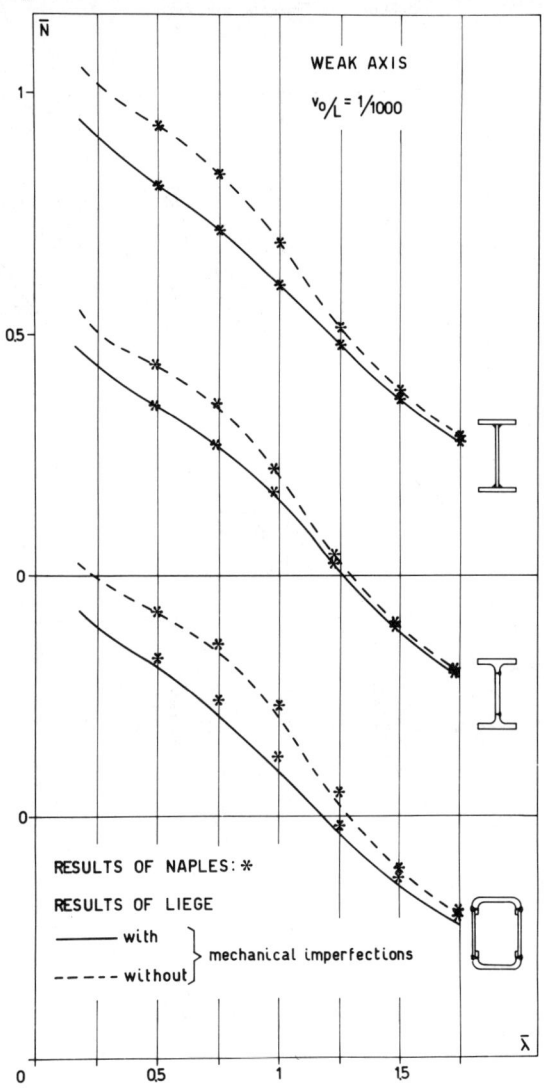

the shape, the axis and the material ($f_{0.2} \equiv n$). Thirty-six buckling curves were calculated (three profiles × three materials × two axes × two imperfections, with and without residual stresses).

7.2.4.3 Exploitation of numerical results [12]

In order to emphasize the strength-reducing effects of mechanical imperfections due to welding, two types of comparison can be made on the

Aluminum alloy structures

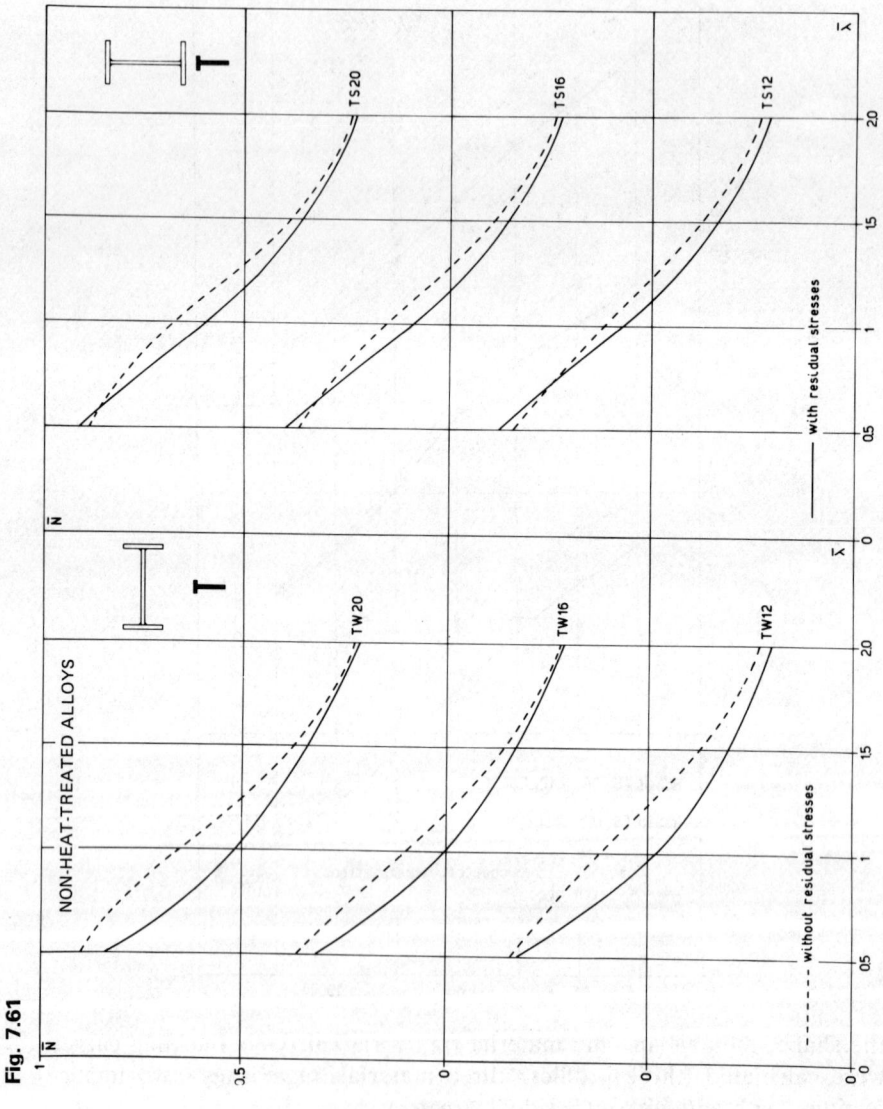

Fig. 7.61

Stability of structural elements

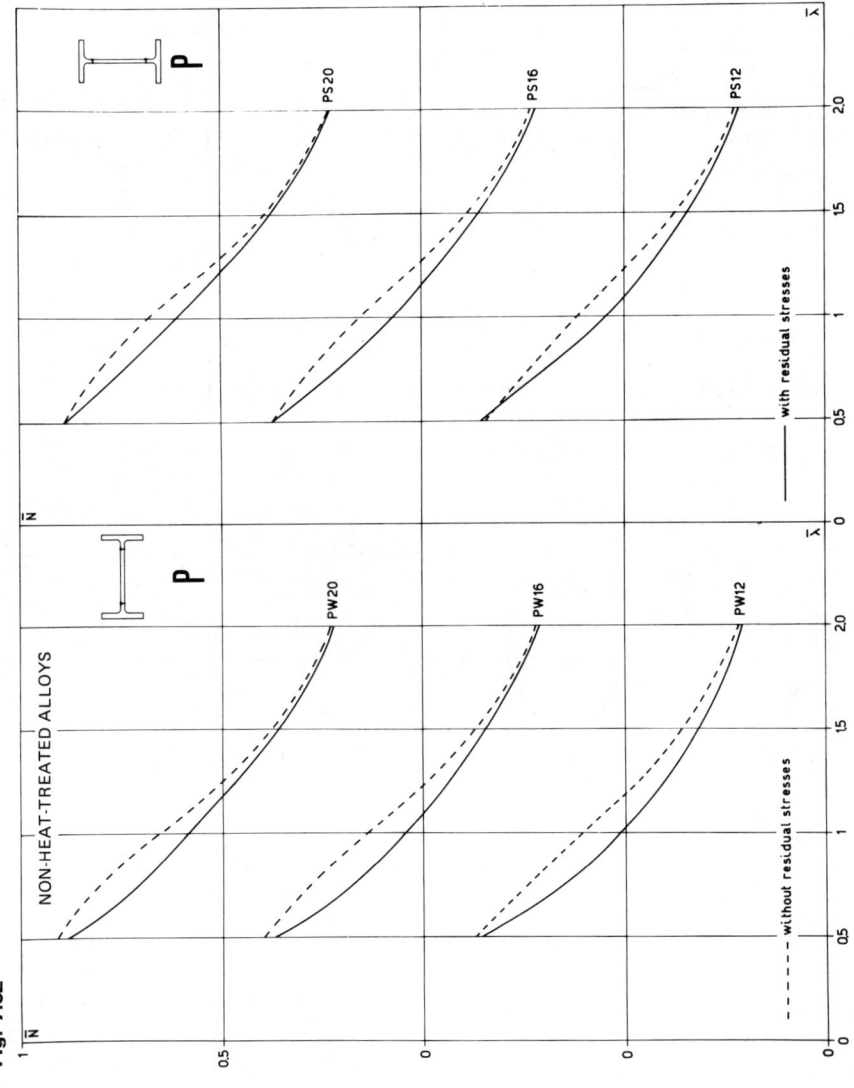

Fig. 7.62

Aluminum alloy structures

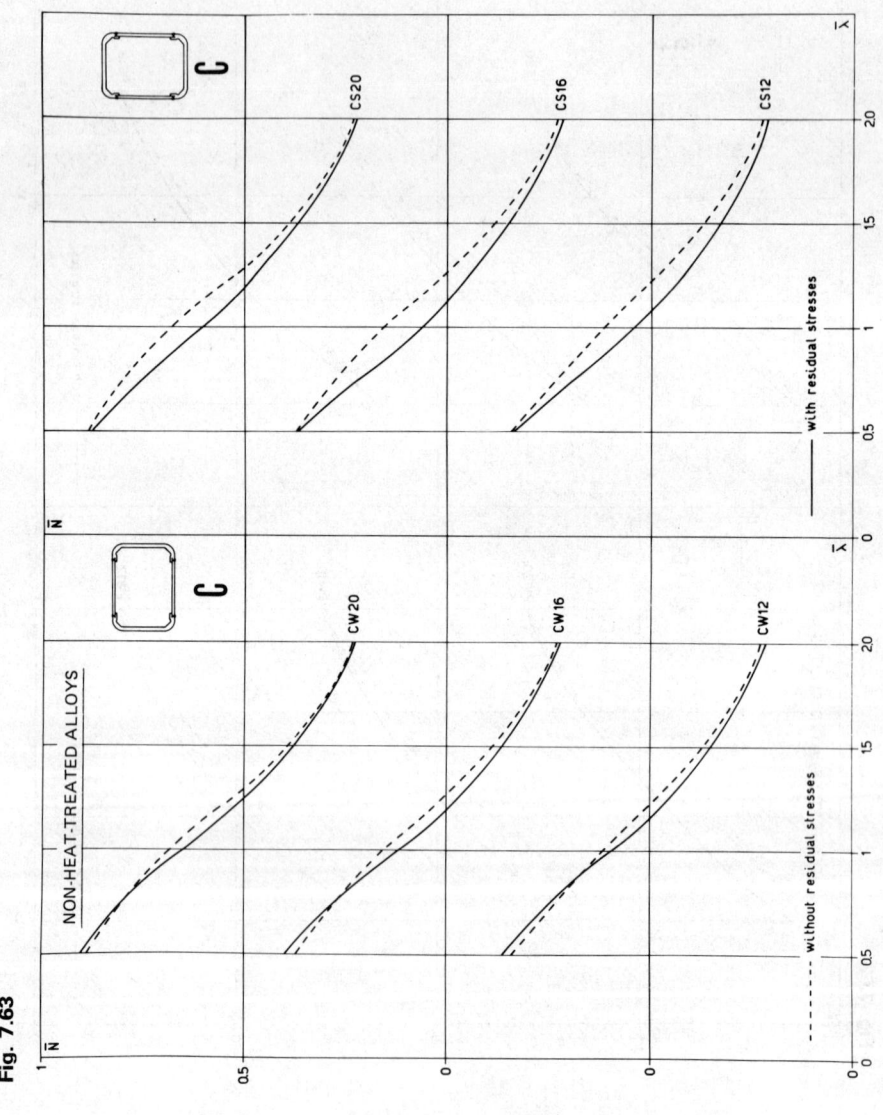

Fig. 7.63

Stability of structural elements

basis of the previous results. For a given cross-section:

N ultimate load when all the imperfections are present
N_0 load-bearing capacity when no mechanical imperfections are present
N/N_0 gives the influence of mechanical imperfections on the load-bearing capacity

The variation of N/N_0 as a function of the slenderness ratio is given in the upper part of Fig. 7.64 for heat-treated alloys. The curves drawn for different cross-sections show that residual stresses have the worst effect in

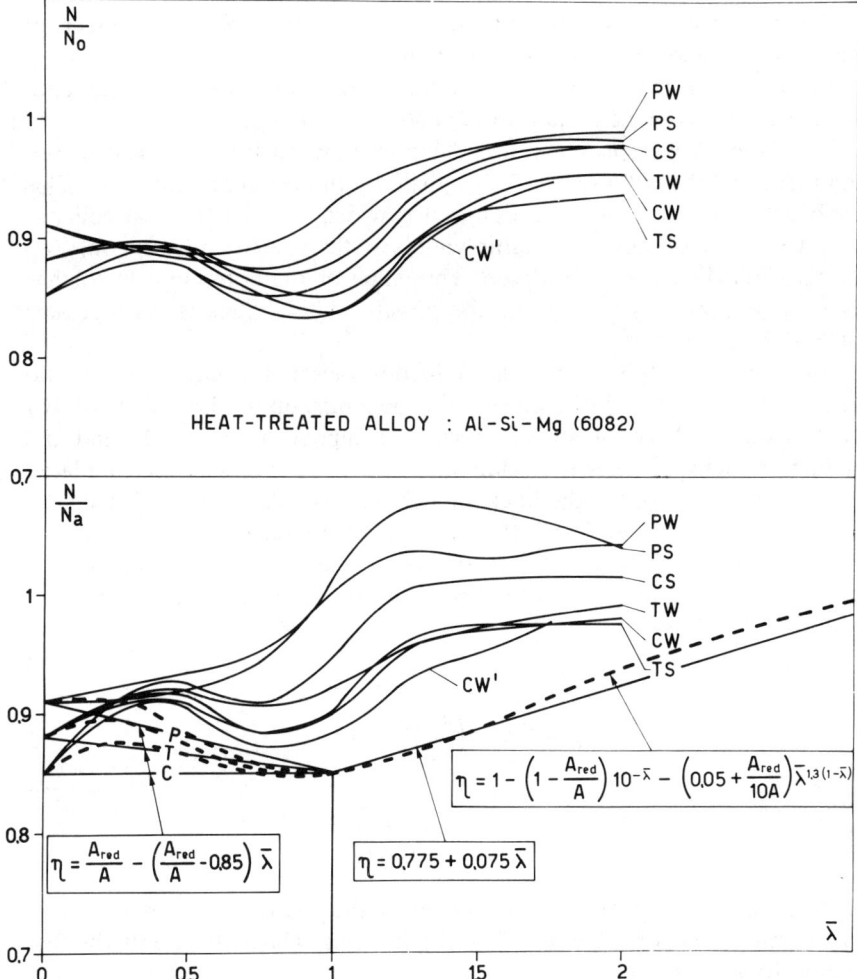

Fig. 7.64

Aluminum alloy structures

the $0.75 < \bar{\lambda} < 1$ region. The N/N_0 ratio is equivalent to the A_{red}/A ratio for $\bar{\lambda} = 0$, where A_{red} is the area of the cross-section taking account of the reduced-strength zones:

$$A_{red} = A - \left[1 - \frac{f_{0.2}^*}{f_{0.2}}\right] \sum_i A_{h,i} \qquad (7.27)$$

This effect, which mostly affects the conventional compression capacity, tends to disappear with increasing slenderness.

This type of comparison is not of practical interest in providing a calculation method to design welded columns. It did not seem advisable to introduce further buckling curves in addition to those already defined for extruded columns (see Section 7.2.3). Hence the same curves are used with appropriate coefficients introduced to take into account the effects of mechanical imperfections due to welding.

In the case of the heat-treated profiles with double symmetry, curve (a) with the collapse loads equal to N_a had to be changed.

The ratio N/N_a represents the deviation between the effective collapse load (N) and the collapse load (N_a) given by the reference curve (a). This ratio also represents a correction factor which permits the real collapse load (N) to be obtained by multiplying it by the value from curve (a) (N_a) in each buckling case considered. The value of the correction factor has been determined by minimizing the possible N/N_a ratios (see lower part of Fig. 7.64).

This ratio, which is termed the reduction factor η, is equal to the ratio A/A_{red} for $\bar{\lambda} = 0$ and therefore only depends upon the shape of the cross-section. Residual stress effects are higher when $\bar{\lambda} = 1$, and the reduction factor reaches the value 0.85 which minimizes all the simulated cases. It is assumed that the buckling behavior is independent of mechanical imperfections when $\bar{\lambda} \geq 3$; the value of η is therefore set equal to 1.

These three points ($\bar{\lambda} = 0, 1, 3$) can be connected through two lines or by an interpolating curve. In the first case the reducing coefficient has the following expressions:

For $0 \leq \bar{\lambda} < 1$:

$$\eta = \frac{A_{red}}{A} - \left(\frac{A_{red}}{A} - 0.85\right)\bar{\lambda} \qquad (7.28)$$

For $1 \leq \bar{\lambda} \leq 3$:

$$\eta = 0.775 + 0.075\bar{\lambda}$$

The second relationship is independent of the shape of the cross-section and of the effects of the reduced-strength zones, which affect only the first region ($0 \leq \bar{\lambda} < 1$).

If a continuous curve is assumed, the relationship is given by:

$$\eta = 1 - \left[1 - \frac{A_{red}}{A}\right]10^{-\bar{\lambda}} - \left[0.05 + 0.1\frac{A_{red}}{A}\right]\bar{\lambda}^{1.3(1-\bar{\lambda})} \quad (7.29)$$

Both Eqs 7.28 and 7.29 for the reduction factor give the values plotted in the lower part of Fig. 7.64.

In the case of non-heat-treated alloys the results of simulation have been analyzed following the same criteria as explained for heat-treated alloys.

The N/N_0 ratios are given in Fig. 7.65. They only show the residual stress effect, which produces the greatest reduction in the case when $\bar{\lambda} \cong 1$.

Since curve (b) is used for non-heat-treated extruded profiles, N/N_b ratios have to be considered in order to define the reduction factor η, which allows N values to be obtained for welded sections from the curve (b) (Fig. 7.66). N/N_b ratios are also given in Fig. 7.67, and in this case the T section represents the worst case, particularly when $\bar{\lambda} \cong 1$. In order to minimize the N/N_b ratios two equations can be used for the reduction factor η, in the same way that η was determined for heat-treated alloys. A bilinear curve can be used to represent the η factor if we assume η equal to 1 for $\bar{\lambda} = 0$ and $\bar{\lambda} = 3$, where the effect of residual stresses seems to be negligible. η is assumed equal to 0.80 for $\bar{\lambda} = 1$. The two lines are given by the following formulas:

For $0 \leq \bar{\lambda} < 1$:

$$\eta = 1 - 0.2\bar{\lambda} \quad (7.30)$$

For $1 \leq \bar{\lambda} \leq 3$:

$$\eta = 0.7 + 0.1\bar{\lambda}$$

Alternatively a continuous curve can be assumed:

$$\eta = 1 + 0.04(4\bar{\lambda})^{(0.5-\bar{\lambda})} - 0.22\bar{\lambda}^{1.4(1-\bar{\lambda})} \quad (7.31)$$

Both Eqs 7.30 and 7.31 are plotted in Fig. 7.67.

7.2.4.4 Reliability of the method

In the case of heat-treated alloys, the reduction factor η permits curve (a') to be obtained from curve (a) (Fig. 7.68). Because of the ratio A_{red}/A, curve (a') depends upon the shape of the cross-section, especially in the region $0 \leq \bar{\lambda} < 1$.

Buckling curves (a'_T), (a'_P) and (a'_C) have been obtained for those sections examined by using both Eqs 7.28 and 7.29. In Fig. 7.68 these curves are compared with all the previous simulation curves (Section

Aluminum alloy structures

Fig. 7.65

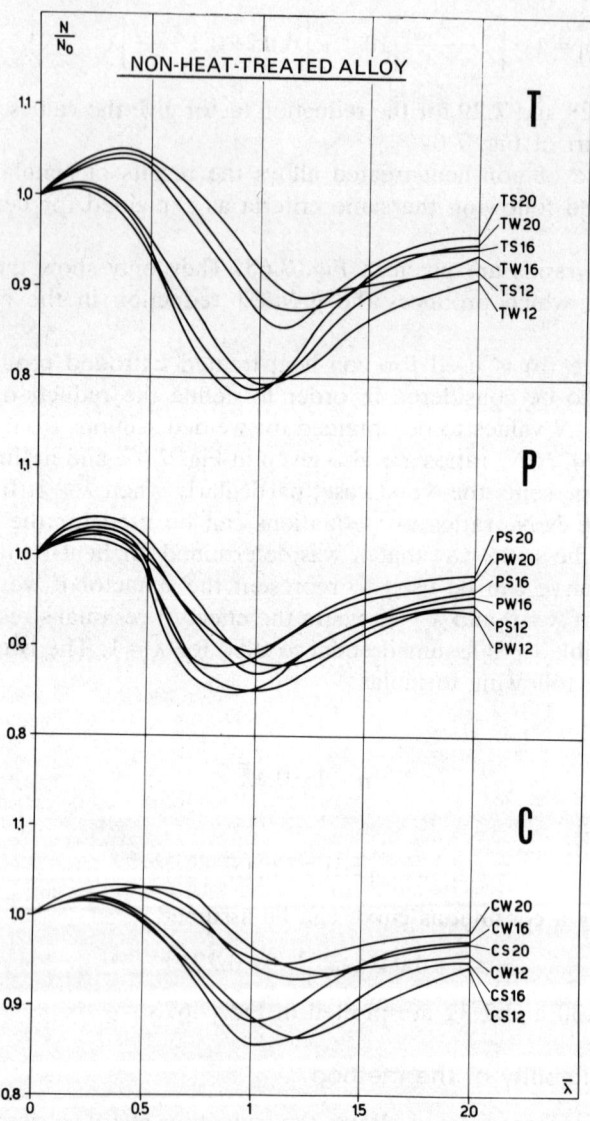

7.2.4.2) and with the experimental results (Section 7.1.3.2). This comparison shows that the T profile represents the worst case, from an experimental viewpoint as well, because of the severe geometrical imperfections of the cross-section. In fact in this case alone the characteristic value of tests fall below curve (a'_T) by about 5 percent.

However, it was decided not to penalize all welded sections owing to

Fig. 7.66

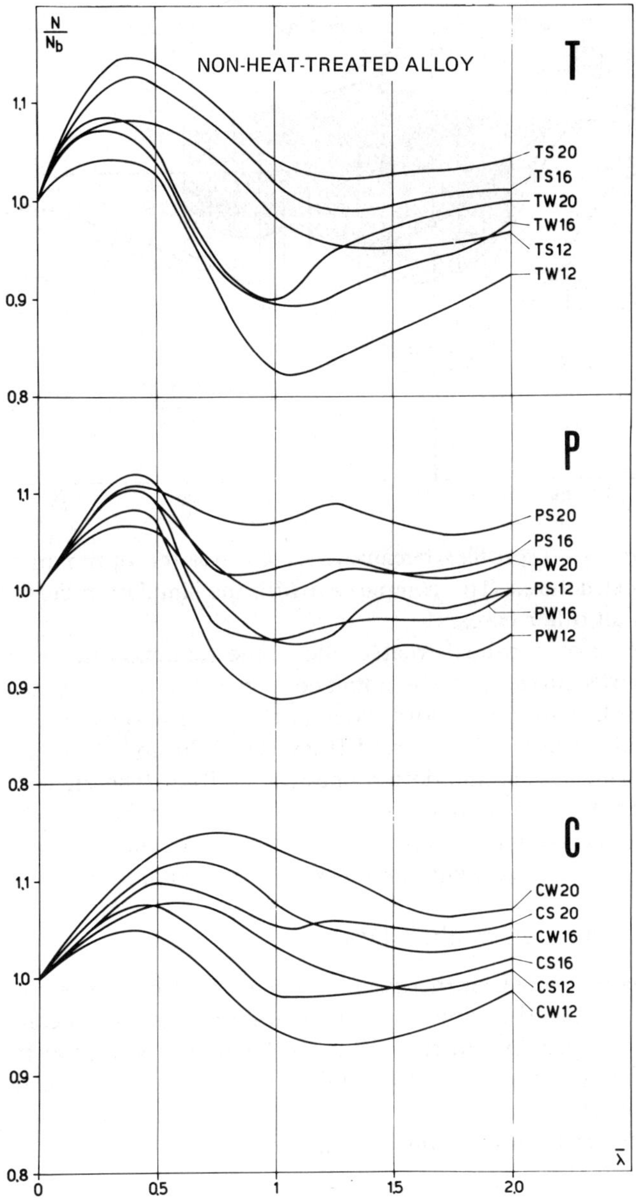

Fig. 7.67

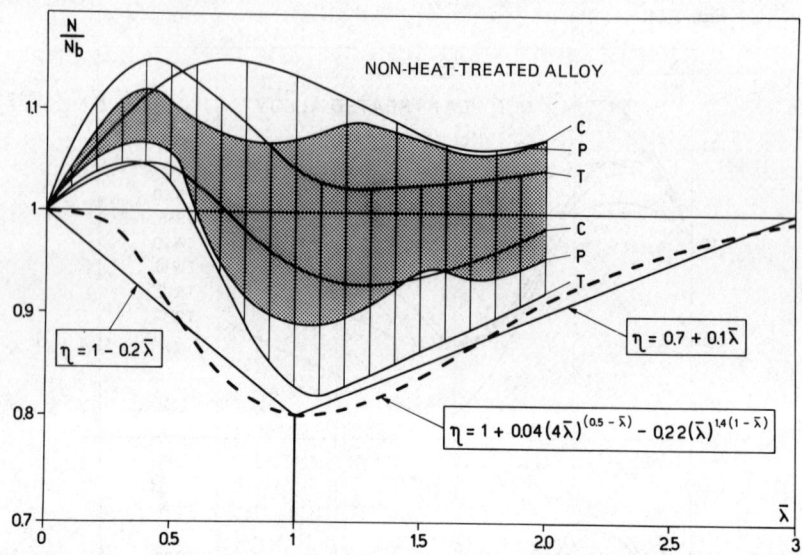

the behavior of T profiles because they are not the optimum choice for aluminum structures. The comparison with the simulation curves is satisfactory in all other cases.

In the case of non-heat-treated alloys, the reduction factor previously defined allows curve (b′) to be obtained from curve (b). This curve (b′), in contrast to heat-treated alloys, does not depend upon the shape of the cross-section. Curves (b′) derived from Eqs 7.30 and 7.31 are given in Fig. 7.69 with dashed and dotted lines respectively. The eighteen simulation curves (Fig. 7.55) which represent the actual behavior of non-heat-treated welded columns have been compared with the (b′) curves in Fig. 7.69, and there is excellent agreement for both representations.

7.2.4.5 Calculation method [15]

The results explained on the buckling behavior of heat-treated and non-heat-treated welded columns have been used by the ECCS committee to deal with this problem in specifications. Since it is not good practice to form asymmetrical sections by welding, only symmetrical sections have been considered.

The load which causes plane buckling of a welded column is given by:

$$N_c = k_2 f_d \bar{N} A \qquad (7.32)$$

f_d design strength, which can be assumed equal to the conventional elastic limit ($f_{0.2}$)

Stability of structural elements

Fig. 7.68

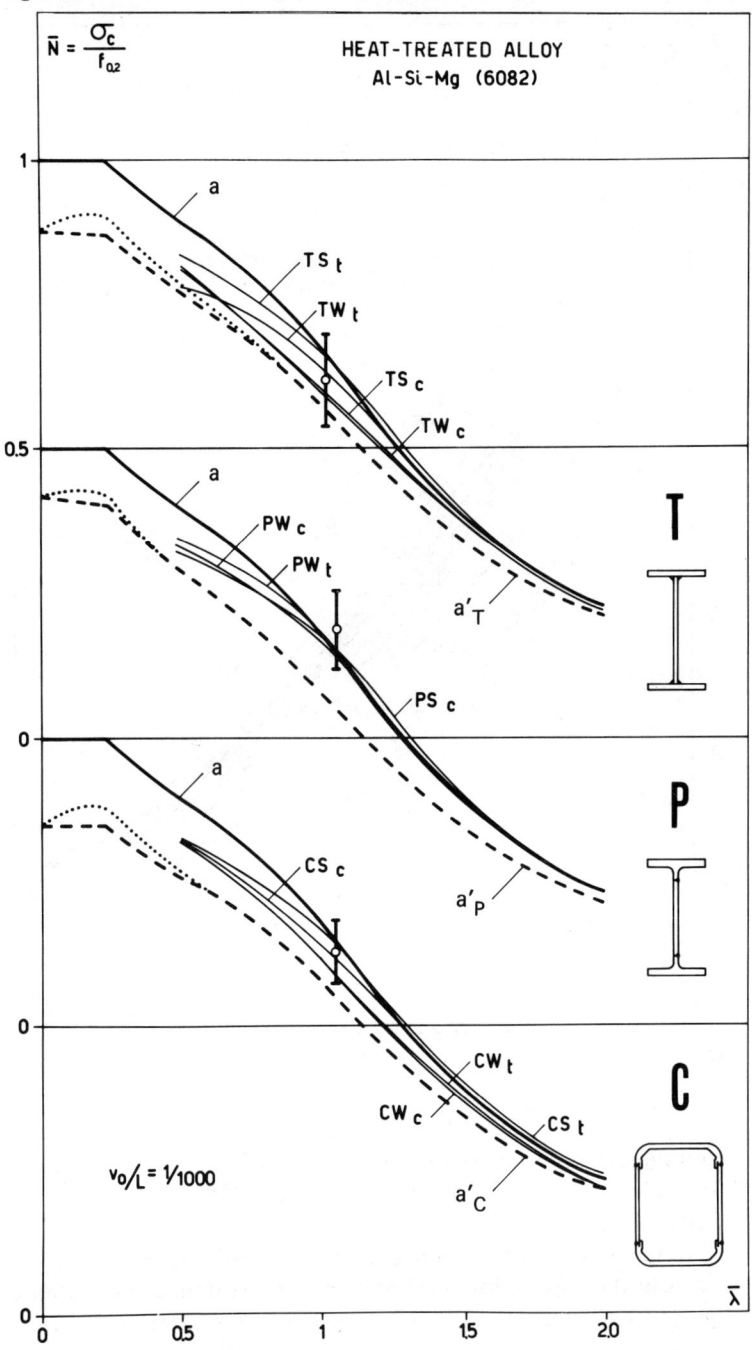

Fig. 7.69

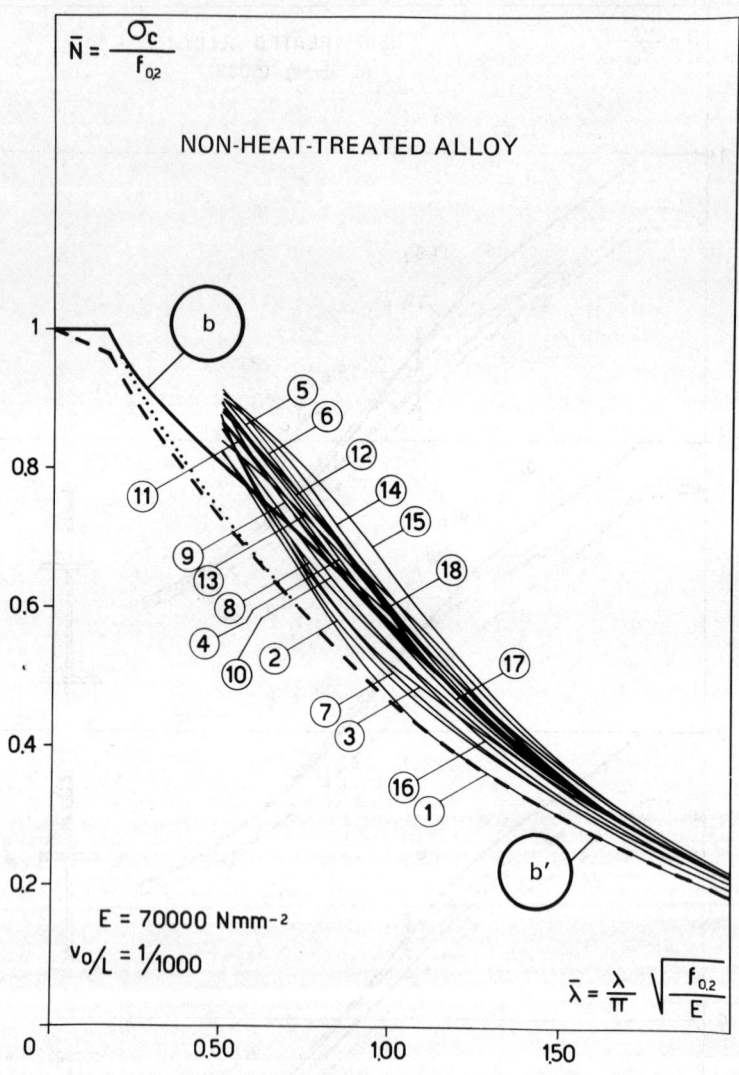

$\bar{N}$ nondimensionalized stress given by buckling curves (a or b) for extruded columns as a function of the slenderness ratio $\bar{\lambda}$

A cross-sectional area of the bar

k_2 a reducing coefficient taking account of welding effects. It is equal to η in the case of longitudinally welded columns and it is equal to $\beta = f_{d,red}/f_d$ in the case of transversely welded columns. $f_{d,red}$

Stability of structural elements

represents the design strength of the base metal in the reduced-strength zone.

Equation 7.24 still has to be checked.

Buckling curves can be selected using Fig. 7.70. In the case of longitudinally welded sections curves (a) and (b) have to be adjusted with the coefficient n previously defined in Section 7.2.4.3.

In the first edition (1978) of the European recommendations for aluminum alloy structures [64] the bilinear equation has been used for the reduction factor η (Eqs 7.28 and 7.30) (see Fig. 7.71). Practical applications of this equation showed that this discontinuous representation is not the most convenient one from the point of view of calculations. The ECCS committee therefore decided to adopt continuous curves (Eqs 7.29 and 7.31) in the second edition of its Recommendations (in preparation) to represent the reduction factor η (see Fig. 7.72).

Fig. 7.70 Selection of buckling curves

Welds	Aluminum alloys		k_2	Area to be used	Design strength to be used
	Heat treated	Non-heat-treated			
Longitudinal	a	b	η	A	f_d
Transversal	a	b	β	A	βf_d

Fig. 7.71 Reduction factor η: bilinear equation

Aluminum alloy	Slenderness range	Factor η
Heat-treated	$0 \leq \bar{\lambda} < 1$	$A_r/A - (A_r/A - 0.85)\bar{\lambda}$
	$1 \leq \bar{\lambda} \leq 3$	$0.775 + 0.075\bar{\lambda}$
	$3 < \bar{\lambda}$	1
Non-heat-treated	$0 \leq \bar{\lambda} < 1$	$1 - 0.2\bar{\lambda}$
	$1 \leq \bar{\lambda} \leq 3$	$0.7 + 0.1\bar{\lambda}$
	$3 < \bar{\lambda}$	1

Fig. 7.72 Reduction factor η: continuous curves

Aluminum alloy	Factor η
Heat-treated	$1 - \left[1 - \dfrac{A_r}{A}\right] 10^{-\bar{\lambda}} - \left[0.05 + 0.1\dfrac{A_r}{A}\right] \bar{\lambda}^{1.3(1-\bar{\lambda})}$
Non-heat-treated	$1 + 0.04(4\bar{\lambda})^{(0.5-\bar{\lambda})} - 0.22\bar{\lambda}^{1.4(1-\bar{\lambda})}$

7.2.5 Built-up Members

7.2.5.1 Calculation methods

Built-up members are usually made of two or more connected chords. They can be classified with respect to the type of connection between chords:

Latticed columns (Fig. 7.73a)
Batten-plated columns (Fig. 7.73b)
Back-to-back columns (Fig. 7.73c)

In the first case the chords can be of different types (Fig. 7.74). Both lattice members and batten plates can be connected to the chords by welding or bolting.

Since the configuration of built-up aluminum members is not significantly different from that for steel, the same analysis as that for steel structures can be used. This analysis is based upon the equivalent slenderness concept. In this way a built-up column can be analyzed like a simple column, with the collapse load obtained from the nondimensionalized

Fig. 7.73

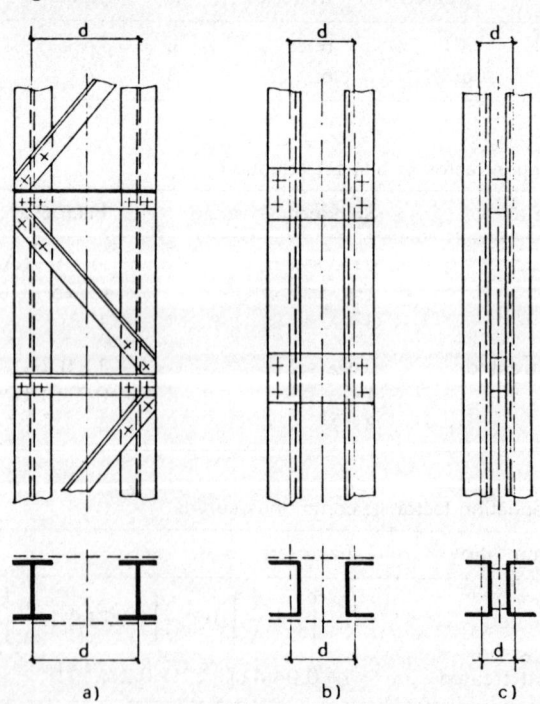

Fig. 7.74

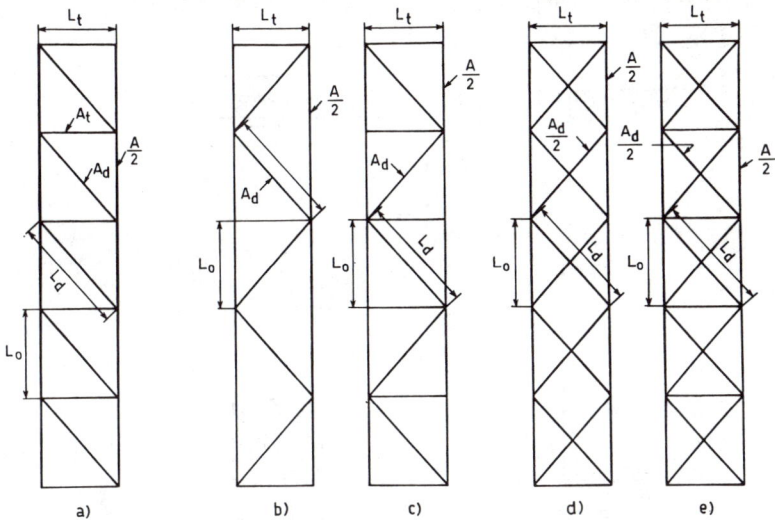

curve (a or b of Section 7.2.3), for a given value of the equivalent slenderness.

The expressions for λ_{eq} have to be evaluated in each case, depending upon the particular type of lacing members, by setting the elastic buckling load of the simple column equal to that of the built-up column.

The analytical steps can be found in books on steel structures [65]. The results reported here have been included for a long time in steel specifications and can now also be found in aluminum specifications – with appropriate modifications.

A typical case occurs when the laced bars are connected by welding to the chords (see Fig. 7.75). In this case the equivalent slenderness can be used, but the reduced-strength zones have to be taken into account by the coefficient:

$$k_2 = 1 - \frac{\sum_{i=1}^{n}(1-\beta)(L_{w,i}+b_{r,i})t_i}{A} \qquad (7.33)$$

A overall cross-section of the chords
$L_{w,i}$ length of the horizontal projection of the welded regions of the chords
$b_{r,i}$ semi-width of the reduced-strength zones, which is assumed equal to 25 mm
t_i thickness of the ith region of the cross-section of the chords
β metallurgical efficiency coefficient defined in Section 4.2.1, and given by $\beta = f_{d,red}/f_d$

Aluminum alloy structures

Fig. 7.75

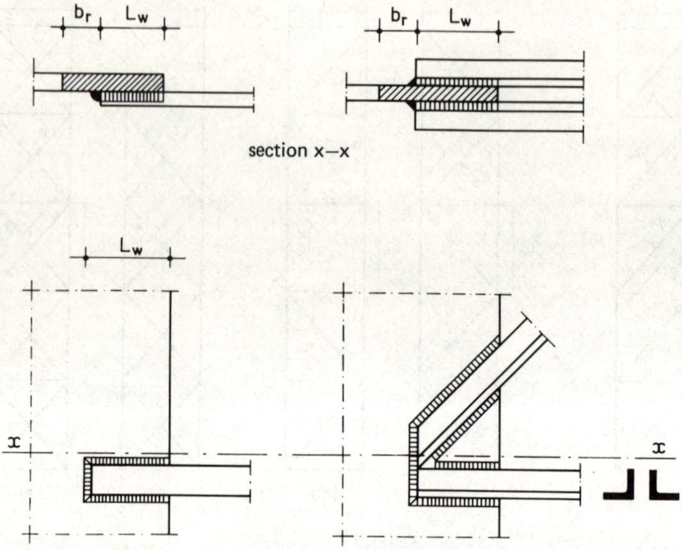

The summation is that of all the rectangular parts which make up the total cross-section of the chords.

The columns made up of two or more elements connected by lacings or batten plates usually have the cross-section of Fig. 7.76.

In order to check the buckling of these bars and to calculate the equivalent slenderness to be used when choosing a buckling curve from

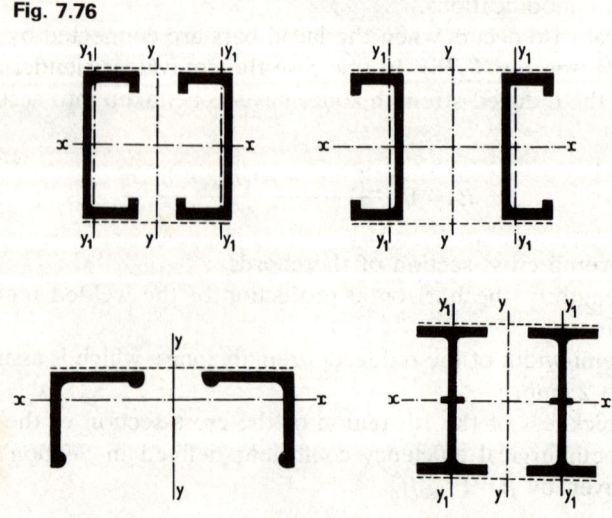

Fig. 7.76

specifications, two different cases are considered:

(a) The case in which lateral bending occurs in a direction normal to a principal axis of inertia which passes through all the cross-sections of the elements of the built-up column (axis x–x of Fig. 7.76). In this case the slenderness λ_x is calculated in the same way as for simple columns.
(b) The case in which lateral bending occurs in a direction normal to a principal axis of inertia which does not pass through all the cross-sections of the elements of the built-up column (y–y axis of Fig. 7.76). In this case the equivalent slenderness $\lambda_{y,eq}$ is calculated as shown in the following sections.

7.2.5.2 Latticed columns

In the case of built-up columns having the configuration of Fig. 7.74 the equivalent slenderness can be computed by the following expressions:

Configuration (a):

$$\lambda_{y,eq} = \sqrt{\left[\lambda_y^2 + \frac{10A}{L_0 L_t^2}\left(\frac{L_d^3}{A_d} + \frac{L_t^3}{A_t}\right)\right]} \qquad (7.34)$$

Configurations (b), (c), (d), (e):

$$\lambda_{y,eq} = \sqrt{\lambda_y^2 + \frac{10 A L_d^3}{L_0 L_t^2 A_d}} \qquad (7.34')$$

λ_y slenderness of the entire column in the plane of bending of the laced bars ($= \beta L / i_y$)
A entire cross-sectional area of both chords
A_d cross-sectional area of a diagonal for configurations (a), (b) and (c), and of the two diagonals in one bay for configurations (d) and (e)
A_t cross-sectional area of the tie
L_d length of the diagonal
L_t spacing of the chords
L_0 length of the diagonal projected onto the axis of the column

Overall buckling of the column has to be checked in addition to local buckling of each chord subjected to the load it carries.

Laced connections of built-up columns have to be checked against a fictitious force equal to

$$V_{eq} = \frac{f_d}{\sigma_c} \frac{N}{100} \qquad (7.35)$$

Aluminum alloy structures

N axial load applied at the end of the column
f_d design strength of the material
σ_c critical stress corresponding to the equivalent slenderness λ_{eq}

7.2.5.3 Batten-plated columns

Batten plates can be considered as a connection having infinite flexural rigidity with respect to the chords. Under this hypothesis the equivalent slenderness can be expressed by the following formula:

$$\lambda_{y,eq} = \sqrt{(\lambda_y^2 + \lambda_1^2)} \tag{7.36}$$

$\lambda_y = \beta L/i_y$, i_y being the radius of gyration of the entire cross-section with respect to the y–y axis. $\lambda_1 = L_1/i_{1,\min}$, L_1 being the spacing of the batten plates and $i_{1,\min}$ the minimum radius of gyration of the single element.

The formula is applicable if

$$\begin{array}{ll} \dfrac{L_1}{i_{1,y}} \leq \dfrac{\lambda_x}{2}\left(4 - 3\dfrac{\nu N}{A\sigma_c}\right) & \text{when } \dfrac{\lambda_x}{2}\left(4 - 3\dfrac{\nu N}{A\sigma_c}\right) > 50 \\[2mm] \text{or} & \\[2mm] \dfrac{L_1}{i_{1,y}} \leq 50 & \text{when } \dfrac{\lambda_x}{2}\left(4 - 3\dfrac{\nu N}{A\sigma_c}\right) \leq 50 \end{array} \tag{7.37}$$

where N is the total load acting on the built-up column, A is the entire cross-sectional area, σ_c is the critical stress corresponding to the equivalent slenderness λ_{eq}, and ν is the assumed safety factor.

Among batten-plated columns are those which are made of two or four aluminum shapes spaced at a distance equal to the thickness of the connection plates. If the spacing L_1 of the connections is smaller than:

$$L_1 \leq \frac{450}{\sqrt{f_d}} i_{\min} \tag{7.38}$$

where f_d is the design strength of the material (in $N\,mm^{-2}$), the column can be checked as a single bar (see Section 7.2.5.4). In particular, in the case of built-up members made of two angles in the relative positions of Fig. 7.77, the connections are made with ties orthogonal to each other. These ties are connected to the elements by welding, riveting or using high-strength steel bolts. At least two rivets or bolts have to be used.

Fig. 7.77

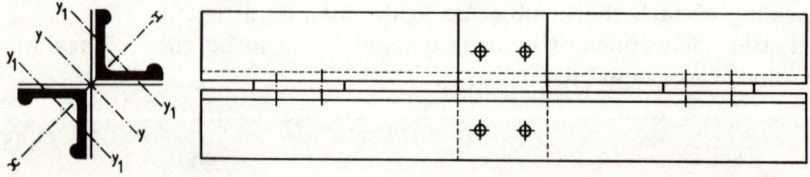

Fig. 7.78

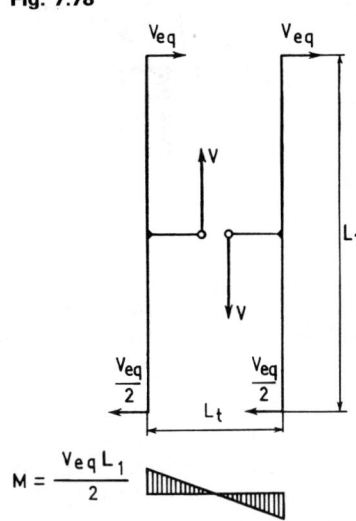

In order to check locally the chords over the length L_1 equal to their spacing, the effective force V_{eq} (Eq. 7.35) can be used. This force has to be increased by the amount $5\,(L_t/i_{1,min}-20)$ percent in those cases in which the maximum spacing between the axes of the chords is higher than $20i_{1,min}$.

The batten plates are dimensioned with the hypothesis that the effective shear V_{eq} is resisted in equal parts by the chords (see Fig. 7.78); that is,

$$V = \frac{V_{eq}L_1}{nL_t} \qquad (7.39)$$

in which n (equal to 1 or 2) is the number of batten plates in the connected cross-section of the column.

In order to have an efficient connection, it is good practice to divide the bar into at least three regions with the batten plates equally spaced. Their connection has to be made by welding, riveting or bolting with at least two rivets or high-strength steel bolts.

The connection between the batten plates and the chord has to be checked against a shear V (Eq. 7.39) and a bending moment equal to:

$$M = \frac{V_{eq}L_1}{2}$$

7.2.5.4 Back-to-back columns

These built-up columns are made of elements joined together by means of a thin connecting plate (Fig. 7.73c). The profiles are at a distance

Fig. 7.79

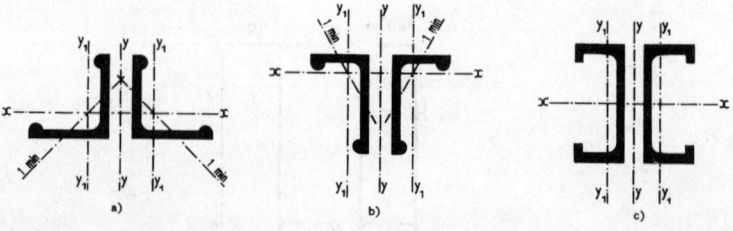

equal to the thickness of the connection plate which must be always less than three times the thickness of the profile (Fig. 7.79).

When the connection has only to resist buckling of a single profile (e.g. the deflection in the plane normal to the minor axis bending of the profiles of Fig. 7.79a and b) the slenderness of the built-up member is calculated using a different radius of gyration to that of the single profile (e.g. the x–x axis of Fig. 7.79 cases a and b). In these cases the built-up member can be considered as a simple column.

The connections are usually made of plates welded, riveted or bolted to the angles with a spacing equal to:

$$L_1 \leq \frac{450}{\sqrt{f_d}} i_{\min} \quad \text{(with } f_d \text{ in N mm}^{-2}\text{)}$$

When the connection has to prevent buckling in a direction normal to a principal axis of the built-up column which does not pass through all of the cross-sections of the elements of the built-up column (e.g. the y–y axis of Fig. 7.79), the slenderness is calculated with respect to a radius of gyration which is greater than that of the single profile in the same direction. In this case the built-up column has to be checked like a single column using the equivalent slenderness:

$$\lambda_{eq} = \sqrt{(\lambda^2 + \lambda_1^2)} \tag{7.40}$$

λ is the slenderness of the column. $\lambda_1 = (L_1/i_{\min})$, L_1 being the spacing of the connections and $i_{\min}$ the minimum radius of gyration of the single profile.

When

$$\lambda_1 > \frac{450}{\sqrt{f_d}} \tag{7.41}$$

(f_d being the design strength of the material) more refined calculations are required.

However, it is good practice to divide the column into at least three panels and to weld, rivet or bolt the connecting plates. At least two

Fig. 7.80

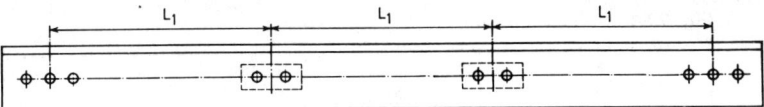

high-strength steel bolts or two calibrated bolts on the axis of the member are required (see Fig. 7.80).

7.2.6 Flexural–torsional Instability

Flexural–torsional buckling is likely to occur, even in a bar subjected to simple compression, with an open cross-section made of thin plates. The criterion of equivalent slenderness can also be used in order to consider the phenomenon through a simple plane problem (see Section 7.2.3).

In the case of symmetrical sections with respect to the y–y axis (Fig. 7.81) the following expression can be assumed:

$$\lambda_{eq} = \frac{L\beta}{i_y}\sqrt{\left[\frac{c_0^2+i_0^2}{2c_0^2}\left\{1+\sqrt{\left(1-\frac{4c_0^2\left[i_p^2+0.093\left(\frac{\beta^2}{k^2}-1\right)e_0^2\right]}{(c_0^2+i_0^2)^2}\right)}\right\}\right]} \quad (7.42)$$

β buckling coefficient defined in Section 7.2.1
k warping coefficient equal to 0.5 and 1.0 respectively for ends prevented from warping and free to warp

$$c_0 = \sqrt{\left[\frac{(\beta/k)^2 I_\omega + 0.039(\beta L)^2 I_T}{I_y}\right]} \quad (7.43)$$

e_0 coordinate of the shear center C with respect to the principal axes of inertia

$i_p^2 = (I_x + I_y)/\Delta$

    I_x and I_y being the moments of inertia with respect to the principal axes

$i_0^2 = e_0^2 + i_p^2$

I_T polar moment of inertia of the cross-section. If the cross-section is made of n rectangles, each of length h_i and thickness t_i, it can be assumed:

$$I_T = \tfrac{1}{3}\sum_{i=1}^{n} h_i t_i^3 \quad (7.44)$$

I_ω warping moment of inertia. This is equal to zero if the cross-section is made of several plates intersecting on one axis (e.g. L

Aluminum alloy structures

Fig. 7.81

1	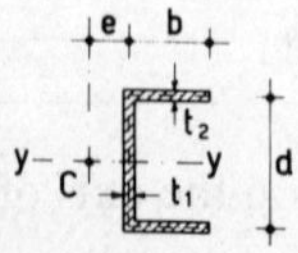 C-section with t_1, t_2, dims e, b, d	$e = \dfrac{3b}{\psi+6}$; $I_\omega = \dfrac{d^2 b^3 t_2}{12}\left(\dfrac{2\psi+3}{\psi+6}\right)$ con $\psi = \dfrac{d\,t_1}{b\,t_2}$
2	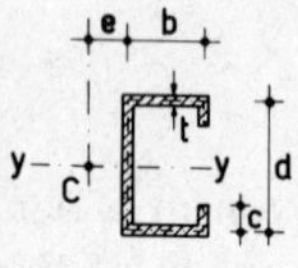 C-section with lips (lips down), dims e, b, d, c, t	$e = \dfrac{d^2 b^2 t}{I_y}\left[\dfrac{1}{4} + \dfrac{c}{2b} - \dfrac{2c^3}{3d^2 b}\right]$; $I_\omega = \dfrac{b^2 t}{6}\left[4c^3 + 6dc^2 + 3d^2 c + d^2 b\right] - e^2 I_y$
3	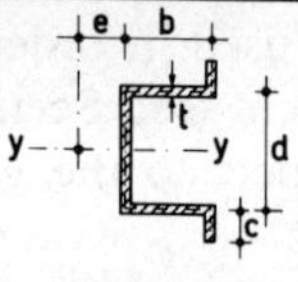 C-section with lips (one up, one down)	$e = \dfrac{d^2 b^2 t}{I_y}\left[\dfrac{1}{4} + \dfrac{c}{2b} - \dfrac{2c^3}{3d^2 b}\right]$; $I_\omega = \dfrac{b^2 t}{6}\left[4c^3 - 6dc^2 + 3d^2 c + d^2 b\right] - e^2 I_y$
4	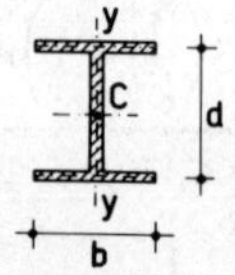 I-section, dims b, d	$I_\omega = \dfrac{d^2}{4} I_y$
5	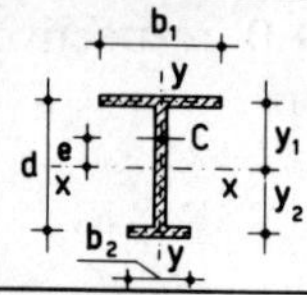 Unequal-flange I-section, b_1, b_2, d	$e = \dfrac{1}{I_y}\left[y_1 I_1 - y_2 I_2\right]$; $I_\omega = \dfrac{d^2 I_1 I_2}{I_y}$ I_1 and I_2 being the moment of inertia of the flanges respect to y-y axis
6	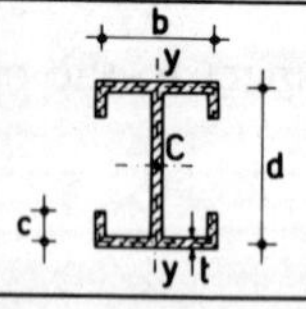 T-with lip, dims b, c, d, t	$I_\omega = \dfrac{d^2 I_y}{4} + c^2 b^2 t\left[\dfrac{d}{2} + \dfrac{c}{3}\right]$
7	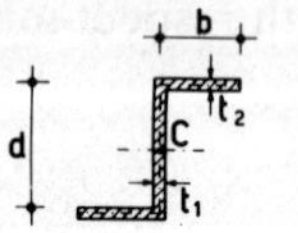 Z-section, t_1, t_2, b, d	$I_\omega = \dfrac{d^2 b^3 t_2}{12}\left[\dfrac{2d+b}{d+2b}\right]$
8	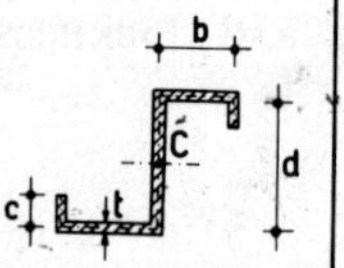 Z-section with lips, b, c, d, t	$I_\omega = \dfrac{b^2 t}{12(2b+d+2c)}\left[d^2(b^2+2bd+4bc+6dc) + 4c^2(3db+3d^2+4bc+2dc+c^2)\right]$

Stability of structural elements

and T profiles), whereas in the case of C, Z, Ω and double T sections it is equal to the values given in Fig. 7.81

The presence of roots or bulbs, typical of aluminum alloy extruded sections, leads to an increase in I_T which can be approximately calculated in the following way (see Fig. 7.82):

$$\Delta I_T = [(\chi_1 + \alpha\chi_2)t]^4 \qquad (7.45)$$

t average thickness of the elements connected by the root or of the element reinforced by the bulb

α ratio between the height of the root (radiused transition) or of the bulb and the thickness of the connected elements

χ_1, χ_2 empirical constants whose values are given in Fig. 7.82 for the more common cases

Fig. 7.82

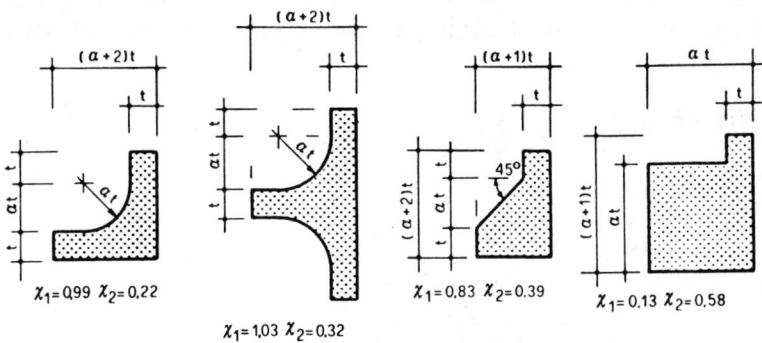

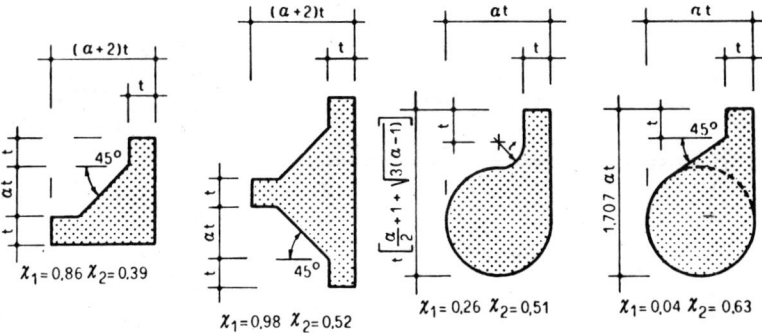

7.3 Members under bending

7.3.1 Physical Behavior

Buckling behavior of beams is usually referred to as flexural–torsional buckling.

Consider a deep plate girder with two end moments. As the applied load increases, the vertical displacements increase and the deflection of the beam remains in the same plane. For a given value of the load, defined as the critical load, the beam buckles laterally and twists.

As an example we can consider a double T section. In this case the compressed flange reaches its critical load and buckles in the lateral plane with lowest inertia, whereas the flange under tension does not buckle. As a result of this behavior there is a torsional rotation of the sections of the beam (see Fig. 7.83).

In real structures the presence of the geometrical and mechanical imperfections of the industrial bar make the process more gradual because there is no bifurcation of equilibrium. The deformations develop gradually with the increase of loads beyond the elastic range up to collapse conditions, these identifying the maximum load-bearing capacity of the beam.

Fig. 7.83

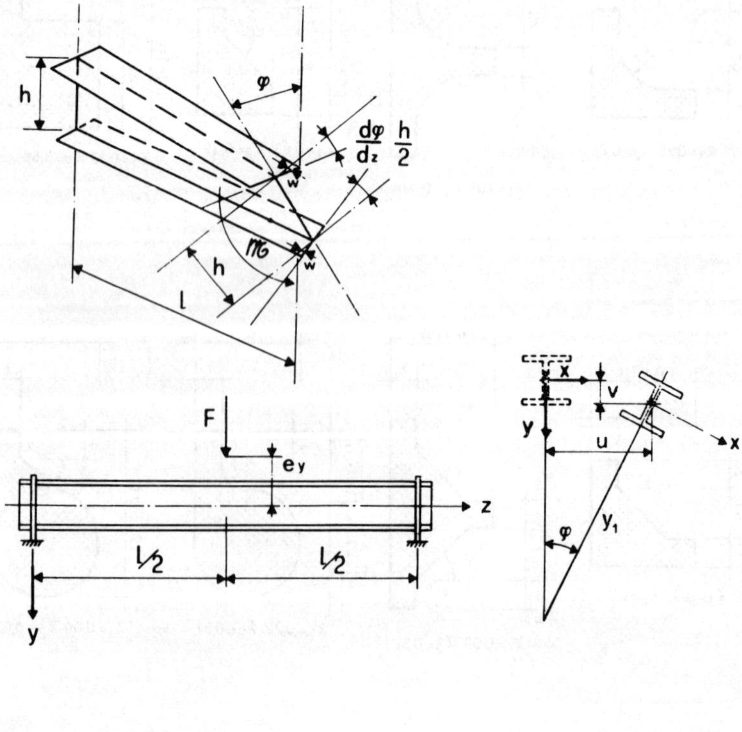

Flexural–torsional buckling has been studied extensively both in the elastic and in the elastoplastic range, especially with respect to steel structures [49–53]. The corresponding literature can be consulted in order to find more details of the experimental research and of the analytical studies undertaken.

The research (Section 7.3.2) and the calculation method (Section 7.3.3) for aluminum alloy structures will be reported here, and in particular the results elaborated and used by the ECCS committee in its recommendations are given [65].

7.3.2 Experimental and Theoretical Results

Elastic bifurcation theory leads – as is well known – to the following expression for the critical bending moment of double T sections:

$$M_{cr,D} = \psi_1 \frac{\pi}{L_{c,h}} \sqrt{(EI_y GI_T)} \sqrt{\left(1 + \frac{\pi^2}{k^2}\right)} \qquad (7.46)$$

where

$$k = L_{c,h}\sqrt{(GI_T/EI_\omega)} \qquad (7.47)$$

ψ_1 a coefficient depending upon the load distribution and boundary conditions

$L_{c,h}$ effective length of the column, which is represented by the distance between points of inflection

EI_y lateral bending rigidity

GI_T Saint Venant's torsional rigidity

EI_ω warping torsional rigidity

Equation 7.46, which was first developed for steel structures, can also represent the behavior of aluminum alloy beams given these assumptions. More details on Eq. 7.46 are given in texts on steel [65].

Complex simulation programs were required in order to investigate the inelastic range including the presence of imperfections. Even though the aspects of the problem were analyzed, no practical design rules were developed in this manner. A method which was useful for this purpose is that proposed by Linder [53] for steel structures. The bending moment which causes buckling of a beam is given by:

$$M_D = M_{pl}\left[\frac{1}{1+\bar{\lambda}_M^{2n}}\right]^{1/n} \qquad (7.48)$$

M_{pl} plastic moment

n 'system factor'

$\bar{\lambda}_M = \sqrt{\dfrac{M_{pl}}{M_{cr,D}}}$ is the bending slenderness ratio, where $M_{cr,D}$ is the critical elastic moment which can be calculated from Eq. 7.46.

Aluminum alloy structures

This approach can also be used in aluminum alloys provided that the appropriate value of n is used.

The ECCS recommendations (1978) assumed $n = 2.5$ in steel structures, even though this value was later criticized on the basis of several Japanese test results.

In order to use Eq. 7.48 for aluminum alloys, it was necessary to calibrate the n coefficient on the basis of the available experimental results, which were neither recent nor numerous. In the past, several tests were carried out by ALCOA in the USA, and in particular:

Dumont and Hill undertook tests on rectangular rolled sections [54] and on extruded double Ts [55]
Hill undertook tests on asymmetrical double T profiles [56]
Clark and Jombock undertook tests on double T 2014-T6 profiles with a linearly varying bending moment [57]
Clark and Rolf undertook tests on 2000 series rectangular sections, the alloy being heat treated and non-heat treated [58]

More recently some tests have been carried out in Germany by Klöppel and Barsch [34] on double T AlMgMn and AlZnMg1 profiles. The results of these tests are given in Figs 7.84–7.89 in the monodimensional plane.

Fig. 7.84

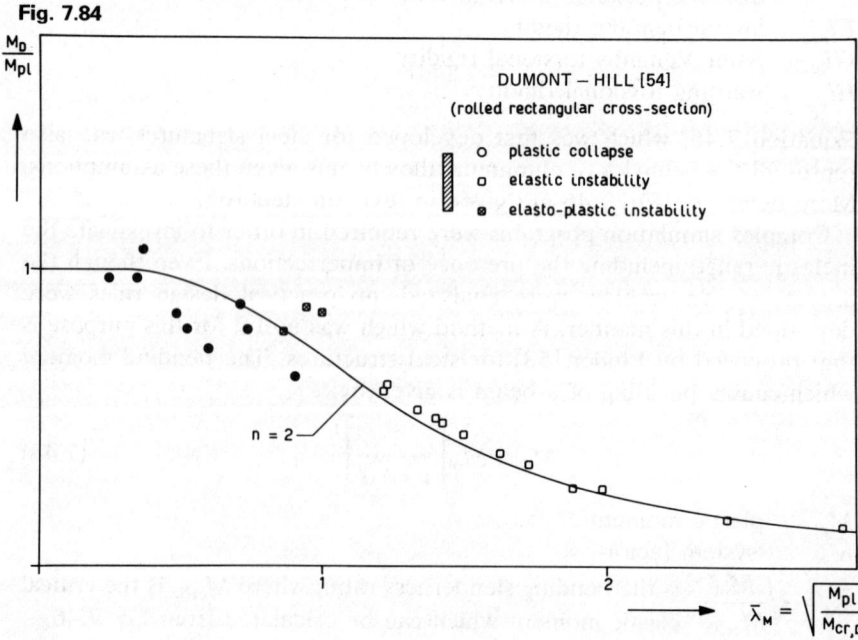

Stability of structural elements

Fig. 7.85

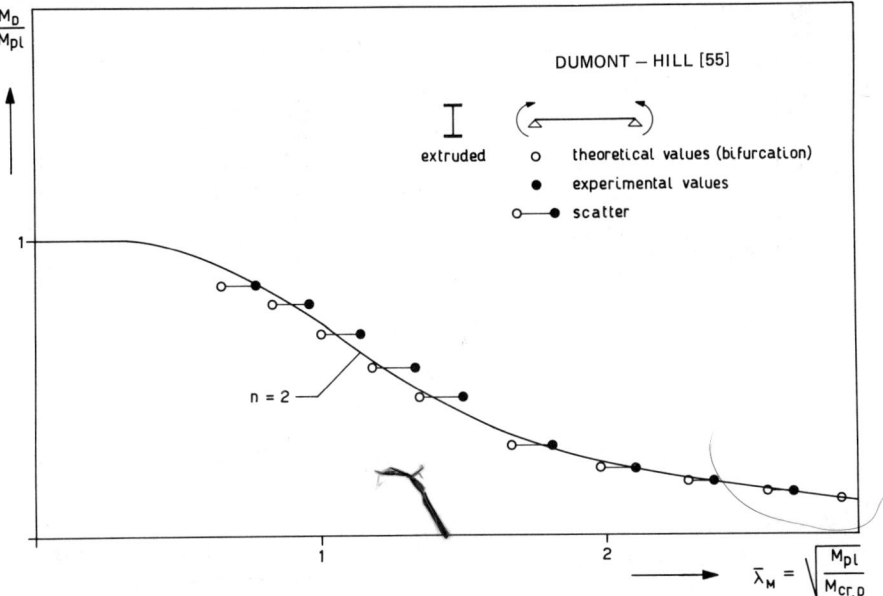

Fig. 7.86

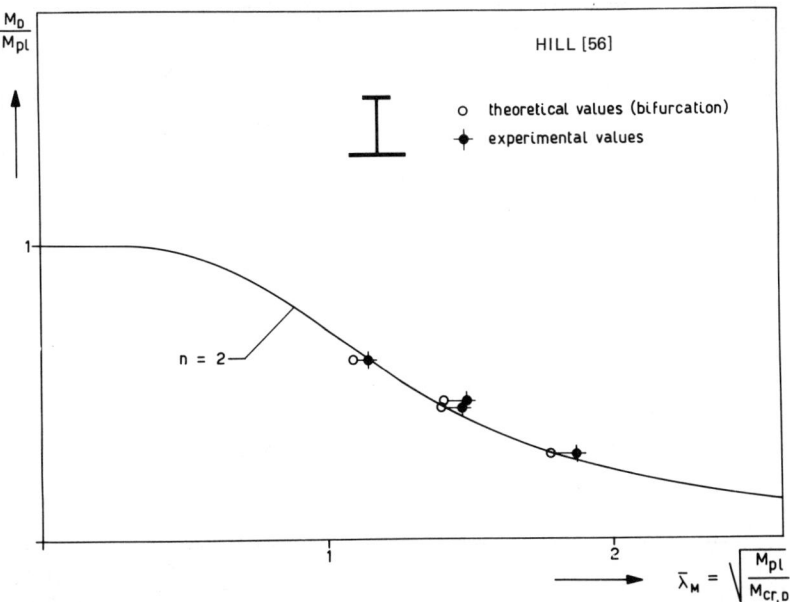

Aluminum alloy structures

Fig. 7.87

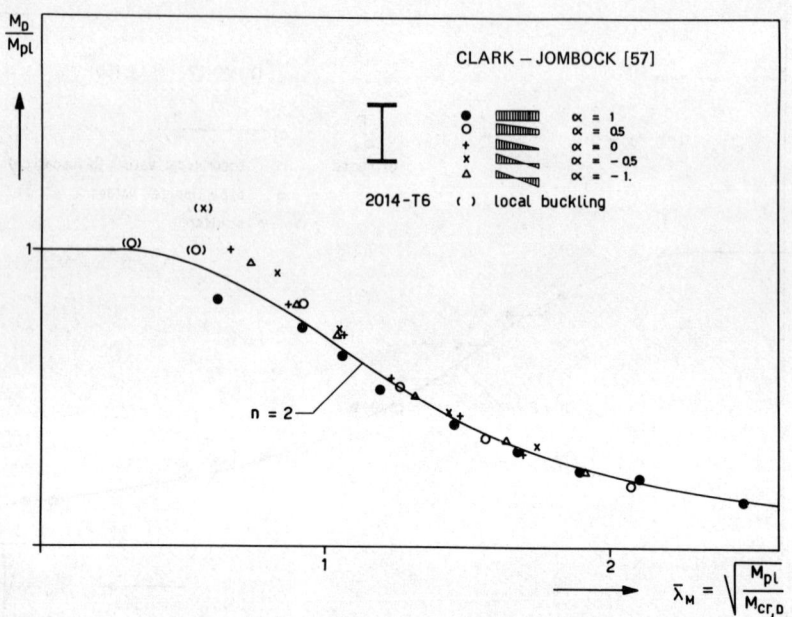

Fig. 7.88

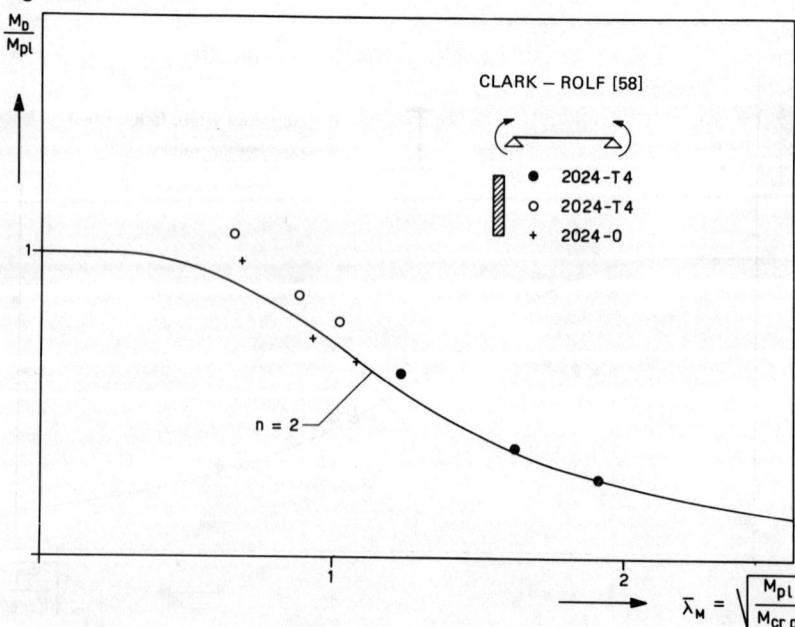

Fig. 7.89

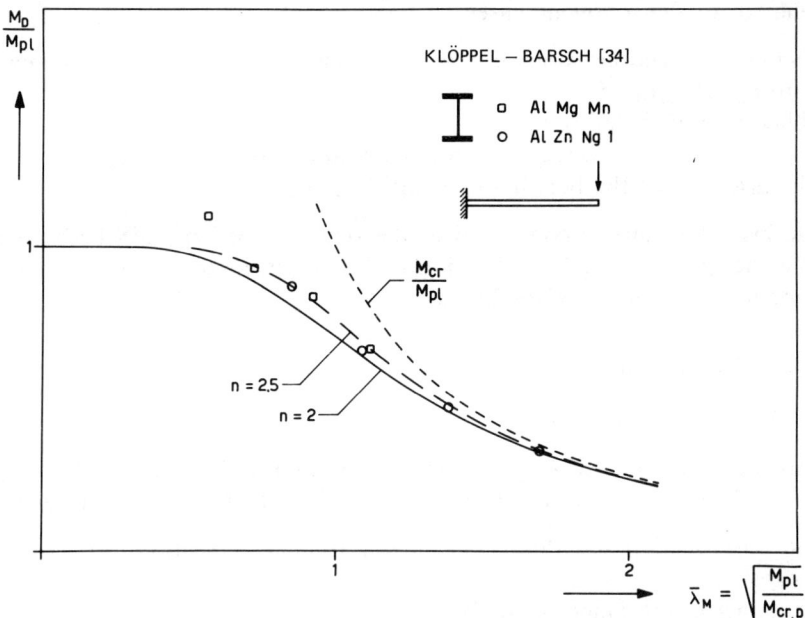

The curve represented by Eq. 7.48 is also given in these diagrams. The n coefficient which seems to interpret the experimental results conservatively is $n = 2$ [11]. The ECCS committee decided to provisionally adopt this value in the first edition of the recommendations on aluminum alloy structures while waiting for further experimental results. The committee also decided to simplify M_{pl} computations assuming (as in steel structures)

$$M_{pl} = \alpha_p W f_{0.2} \qquad (7.49)$$

α_p geometrical shape factor of the cross-section
W bending section modulus
$f_{0.2}$ conventional elastic limit of the material

The highest deviations of the formula from the experimental values are in the range

$$0 < \bar{\lambda}_M < 0.75$$

in which local buckling phenomena are of extreme importance since they cause local buckling of compressed parts.

Even though some experimental results fall below the analytical curve, this is not of major concern since actual beams usually have support conditions more restrained than those of the experimental tests. The

Aluminum alloy structures

proposed curve, with $n = 2$, seems appropriate to represent the averaged behavior of the following cases:

Rectangular sections, double T (symmetrical and asymmetrical) sections (with $b_{max}/b_{min} \leq 1.5$)
Rolled and extruded sections
Heat-treated and non-heat-treated aluminum alloys
All variations of the bending moment diagram

The last possibility is connected to the structure of Eq. 7.48 because it takes account indirectly of loading and support conditions through the computation of $M_{cr,D}$, given by Eq. 7.46.

7.3.3 Calculation Methods

7.3.3.1 ECCS method

This method is based upon the application of Eq. 7.48 with $n = 2$. The range of applicability of this formula is based on the following conditions:

The cross-section has double symmetry (double T sections with equal flanges)
Loads act in the plane of the web
Distortion of the cross-section and local buckling phenomena are not taken into account
Supports do not allow lateral displacements and rotations outside the bending plane

It has to be checked that:
$$M_{max} \leq M_D$$
$$M_{max} \leq W f_d \qquad (7.50)$$

M_{max} maximum bending moment caused by design loads
M_D ultimate flexural–torsional buckling moment
W section modulus of the cross-section
f_d design strength of the material which is assumed equal to the conventional elastic limit

The M_D value is given by the relationship

$$M_D = \frac{M_{pl}}{\sqrt{(1 + \bar{\lambda}_M^4)}} \qquad (7.51)$$

where $\bar{\lambda}_M = \sqrt{(M_{pl}/M_{cr,D})}$ is the slenderness ratio, and M_{pl} and $M_{cr,D}$ are given by Eqs 7.49 and 7.46, respectively.

The values of the M_D/M_{pl} ratio are given in Fig. 7.90. When the value of $M_{cr,D}$ cannot be calculated exactly, the approximate method given in Section 7.3.3.2 can be used.

Fig. 7.90 M_D/M_{pl} ratio

$\bar{\lambda}_M$	0	0.1	0.2	0.3	0.4	0.5	0.6	0.7	0.8	0.9
0	1	1	0.9992	0.9960	0.9874	0.9701	0.9409	0.8980	0.8423	0.7771
1	0.7071	0.6370	0.5704	0.5092	0.4545	0.4061	0.3639	0.3270	0.2949	0.2670
2	0.2424	0.2211	0.2023	0.1857	0.1711	0.1580	0.1463	0.1359	0.1265	0.1181
3	0.1104	0.1035	0.0972	0.0914	0.0862	0.0814	0.0769	0.0729	0.0691	0.0656
4	0.0624	0.0594	0.0566	0.0540	0.0516	0.0493	0.0472	0.0452	0.0434	0.0416
5	0.0400	0.0384	0.0370	0.0356	0.0343	0.0330	0.0319	0.0308	0.0297	0.0287

7.3.3.2 Approximate methods

When a double T symmetrical or asymmetrical section deflects in the plane of the web an approximate analysis involves checking the buckling of the compressed flange assuming it to be isolated from the web. In fact it can be simplistically assumed that buckling is due to the compressed flange. The problem is then reduced to that of a bar under compression, and the method of Section 7.2.3.7 can be used. This method gives a critical value of N for any given value of the slenderness of the flange.

It has to be checked that:

$$\sigma = \frac{N_f}{A_f} \leq f_d \bar{N}$$

f_d design strength of the material
A_f area of the compression flange

$$N_f = \frac{M_{eq}}{I_x} S_x$$

     is the axial force acting on the flange assumed independent from the web, under design loads
I_x moment of inertia of the entire cross-section with respect to the axis of gravity x
S_x static moment of the compression flange with respect to the x axis
M_{eq} reference equivalent moment

The reference equivalent moment is computed from the maximum and the average bending moment in the portion of beam considered in the following way:

$M_{eq} = 1.3 M_M$ in the case of simply supported or continuous beams, with the limitation $0.75 M_{max} \leq M_{eq} \leq M_{max}$

$M_{eq} = M_m$ in the case of cantilever beams, with the limitation $0.5 M_{max} \leq M_{eq} \leq M_{max}$

This approximate method, first used for steel structures, is physically immediate and allows the limitation of double symmetry to be overcome.

7.3.3.3 Welded members

The methods explained in Sections 7.3.3.1 and 7.3.3.2 can also be used in the case of built-up welded members provided that the effects of the heat-affected zones are accounted for. In particular if the cross-section of the beam is built up by longitudinal welds, the geometrical properties of the reduced effective cross-section have to be used:

The reduced area:

$$A_{red} = A - \sum_{i=1}^{n} b_{r,i} t_i \left[1 - \frac{f_{d,red}}{f_d}\right] \quad (7.52)$$

The reduced moment of inertia:

$$I_{red} = I - \sum_{i=1}^{n} b_{r,i} t_i y_i^2 \left[1 - \frac{f_{d,red}}{f_d}\right] \quad (7.53)$$

The reduced section modulus:

$$W_{red} = \frac{I_{red}}{y_{max}} \quad (7.54)$$

A nominal area of the cross-section normal to the axis of the weld (base metal and weld metal)
I moment of inertia of the nominal cross-section of area A
$b_{r,i}$ semiwidth of the single reduced-strength zone
t_i average thickness of the base metal in that zone
$f_{d,red}$ reduced design strength in the reduced-strength zone
f_d design strength of the base metal
y_i distance between the center of gravity of the ith reduced-strength zone and the center of gravity of the entire cross-section.

If the cross-section is built up through transverse welds, it is sufficient to use the reduced design strength $f_{d,red}$ instead of the design strength of the base metal.

When transverse welds in a simply supported beam are at a distance from the ends less than $0.1L$, no reduction of strength is necessary.

7.4 Members under compression and bending

7.4.1 Physical Behavior

Columns can be subjected to bending caused by:

Eccentricity of the axial load
Transverse loads

With respect to the shape of the cross-section of the bar, buckling phenomena corresponding to bending and compression can be of two types:

Plane buckling
Flexural–torsional buckling

Plane buckling, which occurs in the same plane as the load eccentricity, is usually of influence in the case of open profiles with double symmetry (double T) and in the case of box sections with high torsional rigidity. However, in all these sections the shear center practically coincides with the center of gravity.

Flexural–torsional buckling is influential in the case of sections with single symmetry or without symmetry (T, C, Ω etc. or L, Z etc.) in which the position of the center of gravity differs from that of the shear center. The noncoincidence of the shear center with the center of gravity causes torsion, even in the case of axial load, which results in lateral deflection of the bar and twisting of its cross-section. However, it is also of interest to study plane buckling in these cases because there are often restraints which prevent torsional rotation of the cross-section.

7.4.2 Interaction Domains

Failure conditions of a beam column are expressed by the interaction M–N curves. The first interaction formula for M–N, proposed to correctly interpret the behavior of aluminum alloy beam columns, was that of Clark and Hill [59, 60]. Under this approach it has to be checked that:

$$\frac{N}{N_c} + \frac{M}{M_u[1 - N/N_{cr}]} \leq 1 \qquad (7.55)$$

N axial load
M bending moment in the case of constant moment diagram, or equivalent bending moment in the case of different distributions
N_{cr} Euler critical load
N_c collapse load of the bar under simple compression
M_u ultimate moment of the bar. This corresponds to the moment which causes flexural–torsional buckling M_D (e.g. Eq. 7.51) when lateral buckling is allowed; otherwise it corresponds to the plastic moment of the cross-section M_{pl} (e.g. Eq. 7.49)

Under this twofold definition of M_u, Eq. 7.55 interprets both plane buckling and flexural–torsional buckling.

In order to define the value of M_{pl} to be used in Eq. 7.56 in the case of aluminum profiles, three alternatives related to the different definitions of

Aluminum alloy structures

the conventional ultimate state of the cross-section are possible (Chapter 6, Section 6.3):

$$M_{pl} \begin{cases} Wf_{0.2} & \text{(elastic limit state)} \\ \alpha_p Wf_{0.2} & \text{(plastic limit state)} \\ \alpha Wf_{0.2} & \text{(inelastic limit state)} \end{cases}$$

which gives increasing values of M_{pl}, since α_p and α are the geometric and effective shape factor of the cross-section, respectively (see Section 6.3.4.6). An analogous formula has been proposed by Massonnet [65] for steel structures and it is used in several specifications.

Later in the Netherlands, TNO (1974) [65] proposed a modified equation (Eq. 7.56) in order to make the design of beam columns more economic. This was based on recent experimental results on steel columns.

The following analysis was then derived and used in the ECCS steel recommendations:

plane buckling:

$$\frac{N}{N_{pl}} + \frac{\mu}{\mu - 1} \frac{M + Ne^*}{M_{pl}} \leq 1 \qquad (7.56)$$

flexural torsional buckling:

$$\frac{N}{N_{pl}} + \frac{\mu_x}{\mu_x - 1} \frac{kM_x + Ne_x^*}{M_{pl,x}} \leq 1$$

$$\frac{N}{N_{pl}} + \frac{\mu_x}{\mu_x - 1} \frac{kM_x}{M_{pl,x}} + \frac{\mu_y}{\mu_y - 1} \frac{Ne_y^*}{M_{pl,y}} \leq 1 \qquad (7.57)$$

N axial load
M bending moment or the equivalent moment
N_{pl} load which causes complete plasticity of the cross-section
M_{pl} plastic moment of the cross-section
$\mu/(\mu - 1)$ amplifying coefficient, with $\mu = N/N_{cr}$
k $= M_{pl}/M_D$; M_D is the buckling moment (Eq. 7.51)
e^* a conventional eccentricity given by

$$e^* = \left[\frac{N_{pl}}{N_c} - 1\right]\left[1 - \frac{N_c}{N_{cr}}\right]\frac{M_{pl}}{N_{pl}} \qquad (7.58)$$

N_c collapse load for plane buckling of a bar under simple compression
x, y identify the bending planes related to the corresponding variables

The two formulations have been examined by the ECCS committee which checked these analyses against the available experimental results for aluminum beam columns (see Section 7.4.3). On this basis the ECCS committee developed its recommendations (see Section 7.4.4).

7.4.3 Analysis of the Experimental Results

The experimental results on aluminum alloy bars are the result of research carried out at ALCOA (USA) in the 1950s and more recently, on a larger scale, in Germany. In particular:

Hill and Clark [59, 60]: forty tests on the lateral torsional buckling of heat-treated I and H profiles

Holt (see ref. [62]): nine tests on heat-treated (2000 series) beam column tubes

Clark [61]: twelve tests on rectangular box beam columns and sixteen tests on heat-treated (6000 series) rectangular solid sections

Klöppel and Barsch [34]: thirty-two tests on heat-treated double T extruded sections (AlMgSi1 and AlZnMg1); thirty-one tests on non-heat-treated (AlMgMn) double T and cylindrical sections; twenty-seven tests on T sections made of the same alloys

All the tests, except the first group which considered flexural–torsional buckling, are referred to plane buckling.

These experimental results have been used to check the validity of the interaction formulas of Clark–Massonnet (Eq. 7.55) and of TNO (Eqs 7.56 and 7.57) [65]. The nondimensional N–M planes of Fig. 7.91 have been constructed with the following meanings:

Curve A corresponds to Eq. 7.55 in which the conventional elastic limit $M_{0.2}$ is assumed for M_{pl} ($\alpha = 1$)
Curve A' corresponds to Eq. 7.55 in which the conventional plastic limit is assumed for M_{pl} ($\alpha = \alpha_p$)
Curve B corresponds to Eq. 7.56 or 7.57 in which the conventional elastic limit $M_{0.2}$ is assumed for M_{pl} ($\alpha = 1$)
Curve B' corresponds to the application of Eq. 7.56 or 7.57 in which the conventional plastic limit is assumed for M_{pl} ($\alpha = \alpha_p$).

Following the layout of Fig. 7.91 the experimental results are given in Figs 7.92 (flexural–torsional buckling) and 7.93–7.98 (plane buckling). The N_c values used in representing the experimental results in the N–M plane have been calculated using the nondimensional curves (a) and (b). The rules given in Section 7.2.3 have been followed. From considering the comparison between the analytical and experimental results the following can be said:

(a) Flexural–torsional buckling cases (Fig. 7.92)
    (i) The experimental results which caused flexural–torsional buckling

Aluminum alloy structures

Fig. 7.91

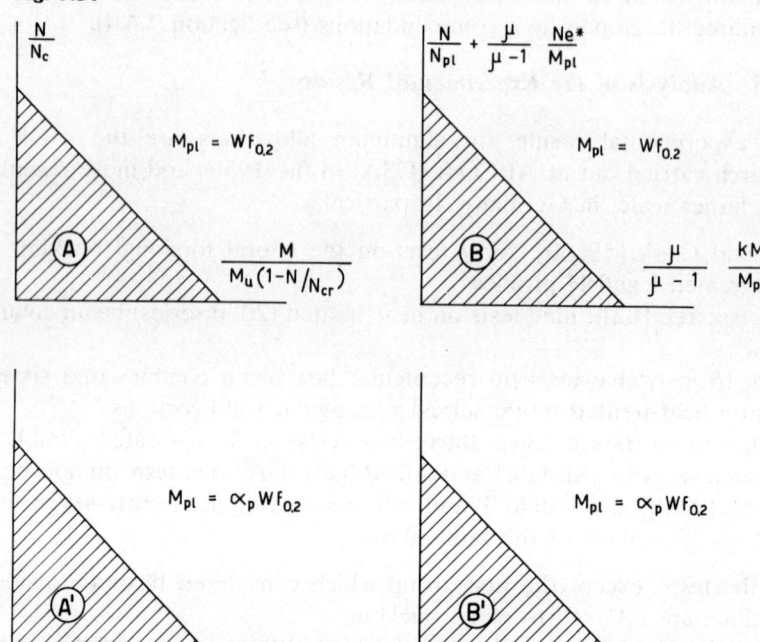

proved that the TNO formulation (Eq. 7.57) cannot be accepted in the case of aluminum alloy beam columns since many experimental results fall below curve B (elastic range) and B' (plastic range). The calculated values were up to 25 percent smaller than the experimental ones. It was generally observed that the discrepancy is higher with increasing slenderness and bending moment M.

(ii) Clark and Massonnet's formula (Eq. 7.55) always gives conservative results with the exception of some points relative to the third alloy. The plastic range, in particular, is on the safe side. This discrepancy between calculated and experimental values is within a few percent.

(b) Plane buckling cases (Figs 7.93–7.98)

(i) Equation 7.55 always gives conservative results. These results are sometimes too conservative, especially if the elastic curve is adopted (curve A). In the case of profiles calculated in the plastic range the opposite is true (A' curve of Fig. 7.98).

(ii) Equation 7.56 gives results which are very close to the experimental results. The deviations are smaller than those obtained by Eq. 7.55 (see Figs 7.94–7.96). Even though some points fall below the

Fig. 7.92

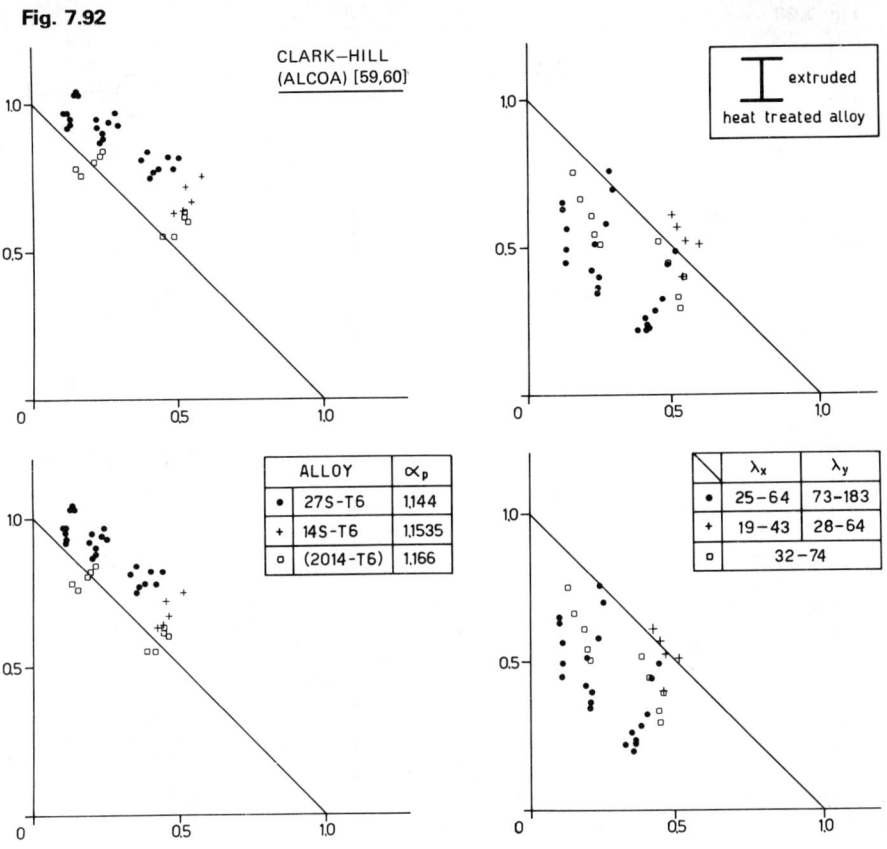

curve (also in the elastic calculations) the deviations between the experimental and the calculated values are always less than 5 percent.

From the above considerations the following conclusions can be made:

In the case of flexural–torsional buckling, the TNO formulation (Eq. 7.57) is not conservative, whereas the Clark–Massonnet equation (Eq. 7.55) provides satisfactory results.
In the case of plane buckling, both approaches are conservative. The TNO formulation, however, is closer to the experimental results.

A summary of the comparison between the experimental results and the application of the TNO formula is given by the histograms of Figs 7.99, 7.100 which compare the agreement among the experimental loads F_{exp} and the computed loads F_{calc} with respect to flexural–torsional buckling and plane buckling respectively. The TNO formulation results are conser-

Aluminum alloy structures

Fig. 7.93

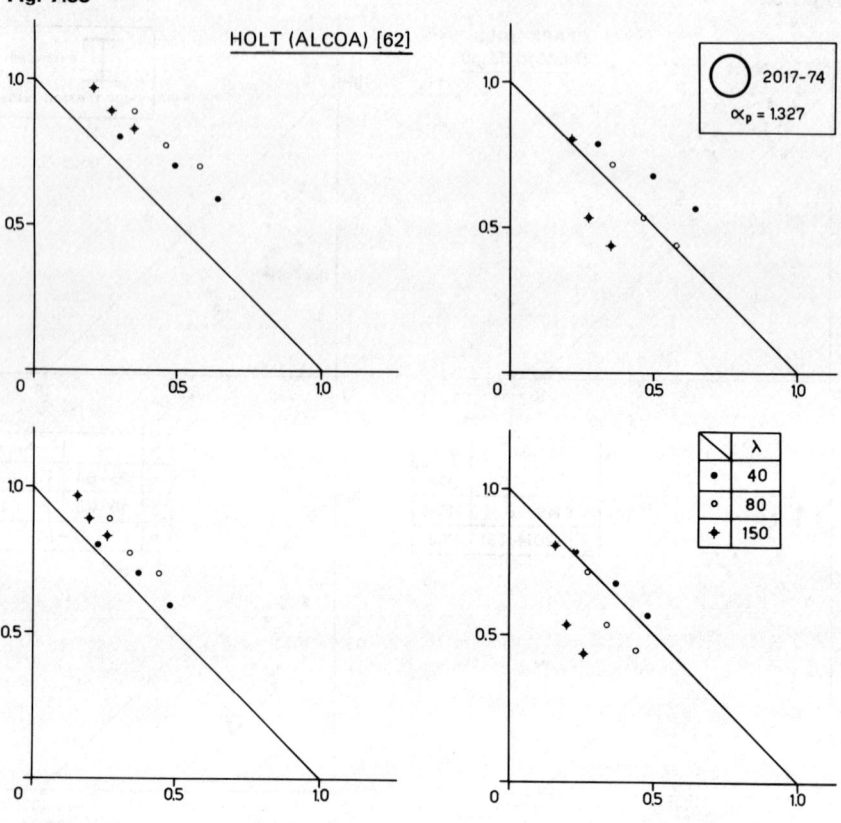

vative by the following percentages [11]:

Flexural–torsional buckling (40 tests, see Fig. 7.99):
(a) Elastic case 47.5 percent
(b) Plastic case 37.5 percent
Plane buckling (130 tests, see Fig. 7.100):
(a) Elastic case 99 percent
(b) Plastic case 93 percent

7.4.4 Calculation Methods

7.4.4.1 ECCS method

On the basis of the experimental results (Section 7.4.3), the ECCS committee decided to adopt the most conservative formulation, which was the one proposed by Clark–Massonnet, for both the plane buckling

Fig. 7.94

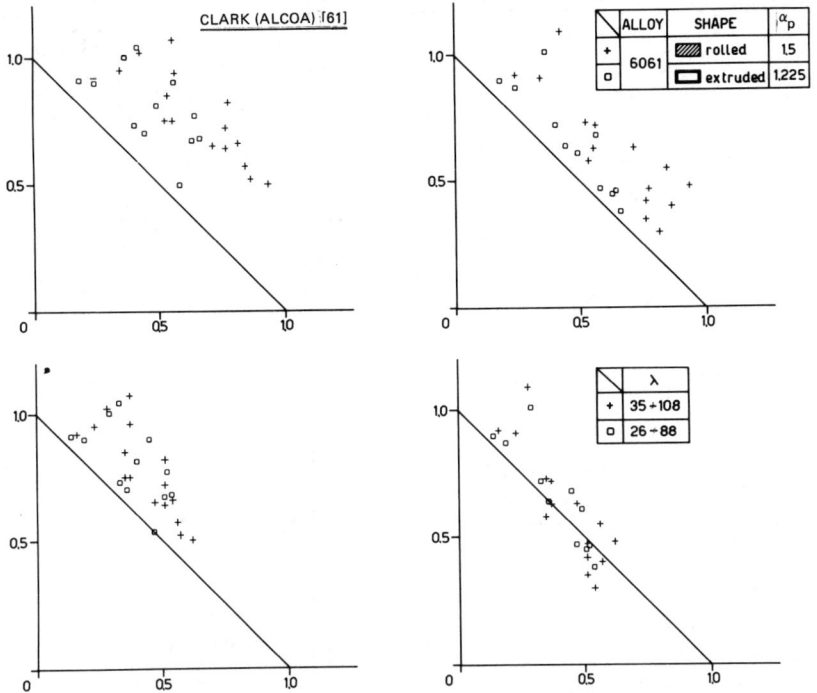

case and for the flexural–torsional case. The method was also extended to biaxial bending. The following equations were derived:

Plane bending
It has to be checked that:

$$\frac{N}{N_c} + \frac{\mu}{\mu - 1}\frac{M_{eq}}{M_u} \leq 1$$
$$\frac{N}{A} + \frac{M_{max}}{W} \leq f_d \qquad (7.59)$$

where the meaning of the symbols has been previously defined. It should be noted that the ultimate moment M_u has the twofold meaning:

$$M_u = \begin{cases} M_D & \text{in the case of flexural–torsional buckling} \\ M_{pl} & \text{in the case of plane buckling} \end{cases}$$

However the equivalent moment is calculated as follows. Assume:

M_a and M_b are the end moments of the bar (with $|M_b| \geq |M_a|$)
$M_1 = \beta |M_b|$, with $\beta = 0.6 + 0.4(M_a/M_b) \geq 0.4$

Aluminum alloy structures

Fig. 7.95

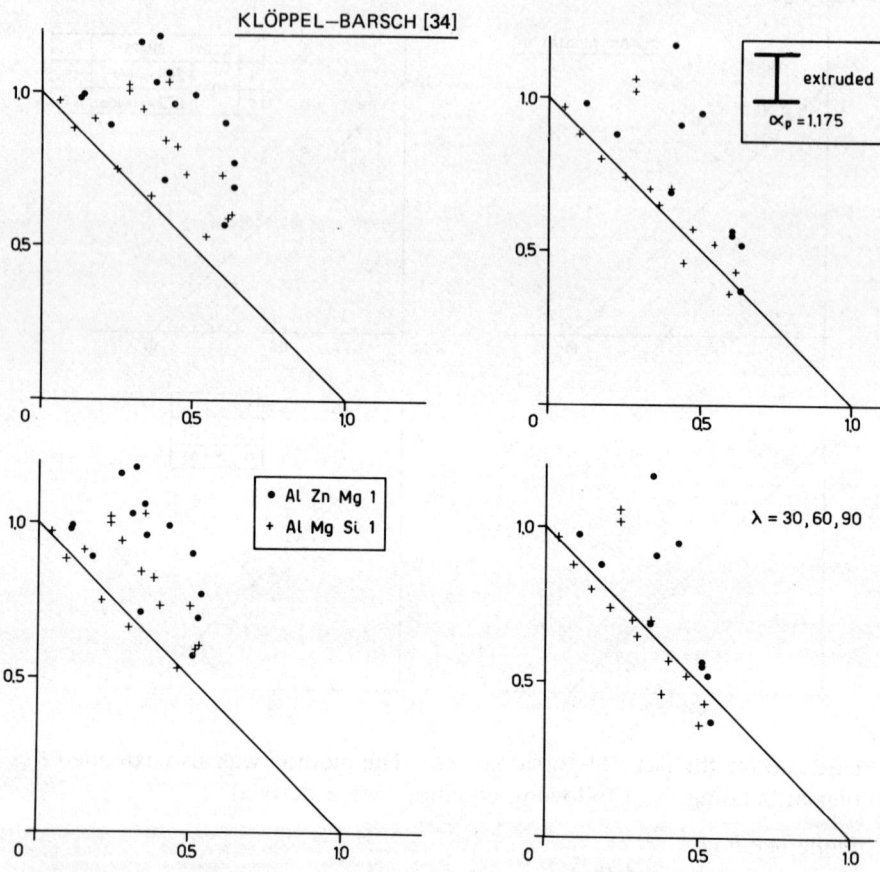

M_2 is the absolute value of the bending moment on the axis of the bar with transverse loads applied (when the moment diagram has no peaks between the ends, $M_2 = |M_b|$). If there are no transverse loads it is assumed that $M_2 = 0$

The value to be used as M_{eq} is the greater of M_1 and M_2 defined in this way. The value of M_{max} to be used in the second of Eq. 7.59 is the greater of $|M_b|$ and M_2.

Biaxial bending
It has to be checked that:

$$\frac{N}{N_c} + \frac{\mu_x}{\mu_x - 1} \frac{M_{eq,x}}{M_{u,x}} + \frac{\mu_y}{\mu_y - 1} \frac{M_{eq,y}}{W_y f_d} \leq 1$$

$$\frac{N}{A} + \frac{M_x}{W_x} + \frac{M_y}{W_y} \leq f_d$$

(7.60)

Fig. 7.96

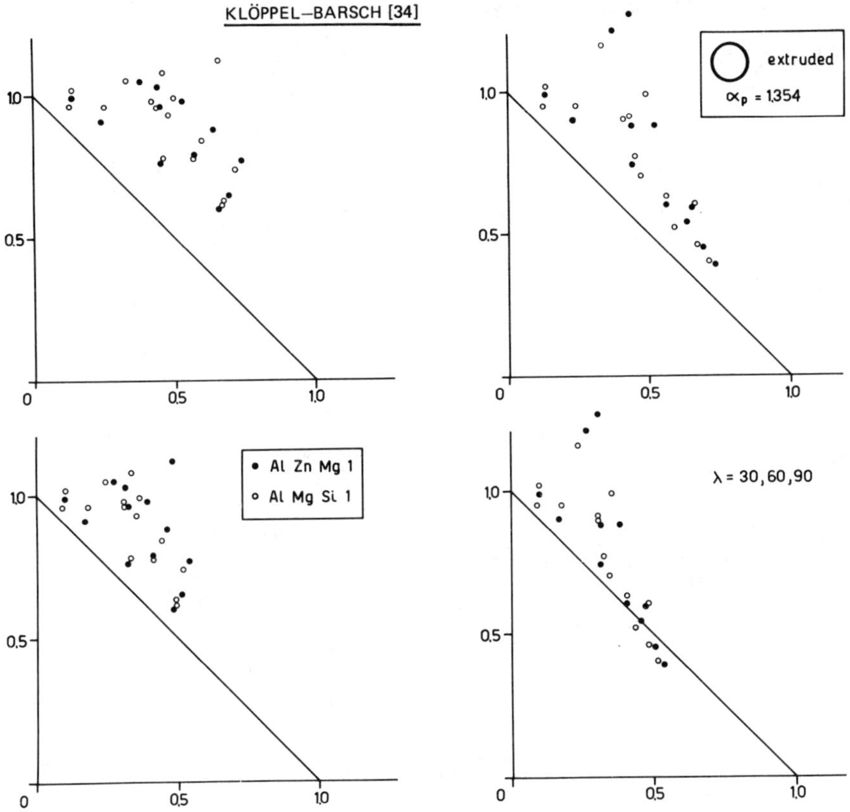

where the subscript x signifies major axis properties and the subscript y signifies minor axis properties.

The $\mu = \sigma_{cr} A/N$ ratio, which characterizes the amplifying coefficient, can be calculated by using the slenderness ratio corresponding to the bending plane to obtain the value of Euler critical stress from Fig. 7.101.

Equations 7.59 and 7.60 have also been used in Italian specifications (UNI 8634), where Massonnet's expression [65] is used for the equivalent moment:

$$M_{eq} = \sqrt{[0.3(M_a^2 + M_b^2) - 0.4 M_a M_b]} \qquad (7.61)$$

The two definitions of equivalent moment practically lead to the same results:

Load distribution:	ECCS	Massonnet
Bitriangular	0.4M	0.45M
Constant	M	M
Triangular	0.55M	0.60M

Aluminum alloy structures

Fig. 7.97

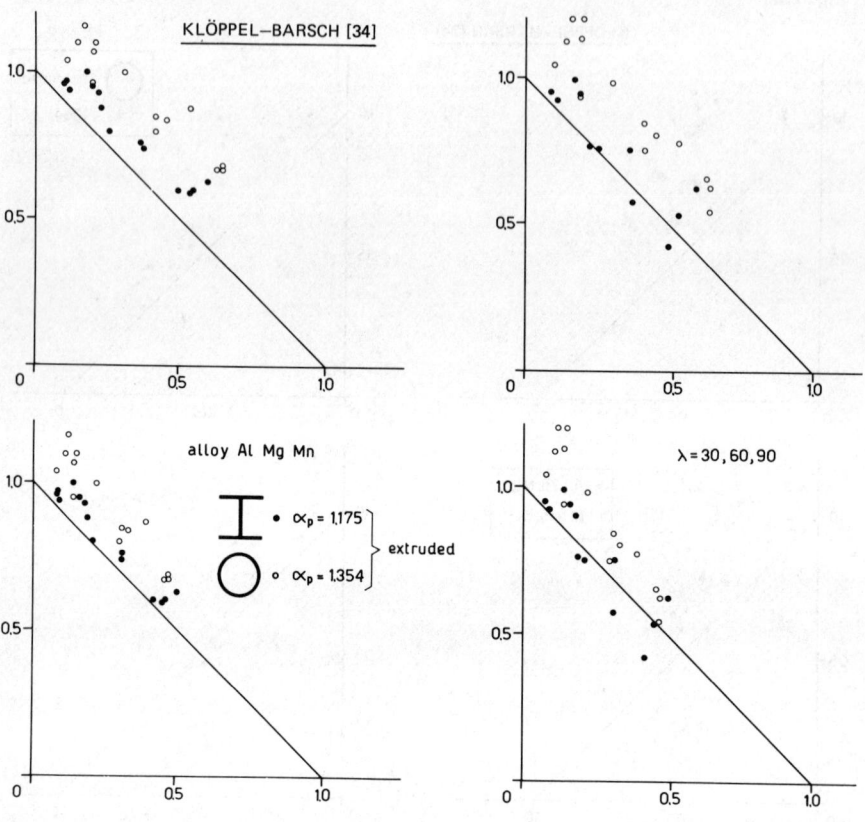

where M is the bending moment which characterizes the distribution under consideration.

7.4.4.2 DIN method

The most recent edition of the DIN 4113 specifications (1980) suggests two different analytical models for checking the buckling of columns and beam columns. These methods reflect the different philosophies of the most advanced schools in metal construction: Karlsruhe [45] and Darmstadt [34].

The first method [45] requires two steps, beginning with a comparison of the axial and reference loads, the latter given by:

$$N_{\text{ref}} = \frac{N_{\text{cr}}}{\gamma_1 \nu} \tag{7.62}$$

Fig. 7.98

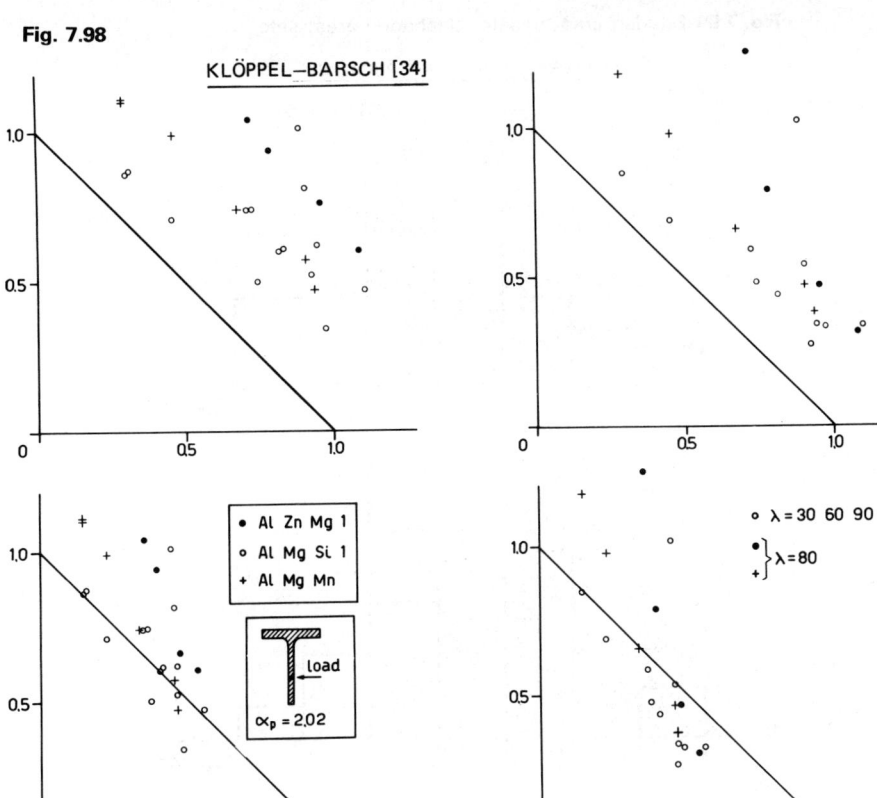

N_{cr}	Euler critical load
ν	safety factor
γ_1	a numerical coefficient varying from 2.5 to 3.0 according to the particular alloy

The following values are assumed for ν:

Loading condition:	Profiles	Tubes
Primary	1.5	1.7
Secondary	1.33	1.5

Dependent upon

$$N > N_{ref} \quad \text{or} \quad N < N_{ref} \quad (7.63)$$

Aluminum alloy structures

Fig. 7.99 Shaded area: unsafe. Unshaded area: safe

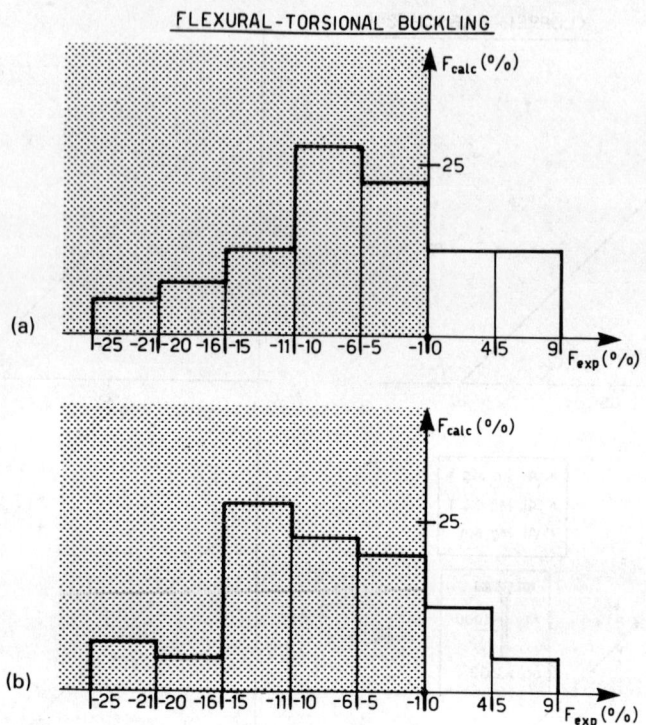

Eqs 7.64 or 7.65 will be used:

$$1 + \frac{M_{eq} + Ne^*}{M_{pl}\left(1 - \dfrac{\nu N}{N_{cr}}\right)} \leq \frac{\gamma_2}{\nu} \tag{7.64}$$

where γ_2 is a numerical coefficient varying from 0.75 to 0.85, depending upon the alloy

$$\psi + \frac{M_{eq} + Ne^*}{M_{pl}\left(1 - \dfrac{\nu N}{N_{cr}}\right)} \leq \frac{1}{\nu} \tag{7.65}$$

where

$$\psi = 1 + \gamma_3(1 - \gamma_2)\frac{M_{pl}}{N_{cr} - \nu N}$$

where γ_3 is a numerical coefficient varying from 1.5 to 2.0 depending upon the alloy.

Fig. 7.100 Shaded area: unsafe. Unshaded area: safe

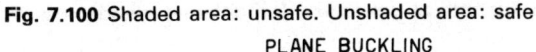

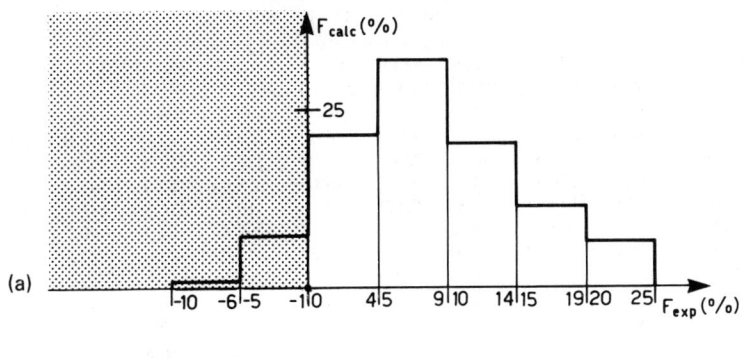

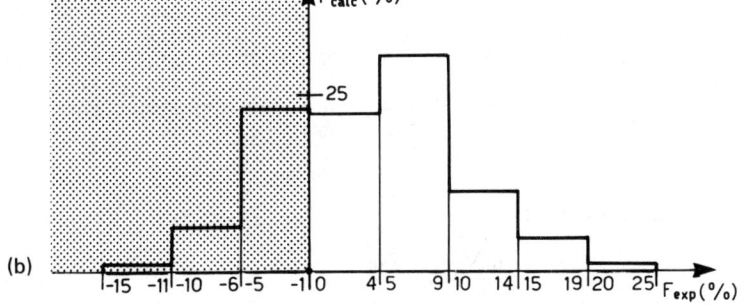

The initial eccentricity e^* is given by specifications as a function of the slenderness ratio, the type of alloy and the shape of the cross-section.

The expression for the equivalent moment M_{eq} is different from those examined in Section 7.4.4.1 and is similar to that of the ECCS:

$$\beta = 0.8 + 0.3 \frac{M_a}{M_b}$$

In the second method proposed by DIN 4113 [34] two different cases are considered: pure compression and compression with an eccentricity. In the first case it is required that:

$$\frac{\omega N}{A} \leq \sigma_{adm} \qquad (7.66)$$

ω values are given in tables as a function of slenderness ratio for each type of alloy.

In the case of beam columns, the centre of compression has to be taken into account (Fig. 7.102). If $y_t = y_c$ or $y_t < y_c$ (cases a and b of Fig. 7.102)

Aluminum alloy structures

Fig. 7.101 Values of the Euler critical stress σ_{cr} for different slenderness λ (in N mm^{-2})

λ	0	1	2	3	4	5	6	7	8	9
10	6908.7	5709.7	4797.7	4088.0	3524.9	3070.5	2.698.7	2390.6	2132.3	1913.3
20	1727.2	1566.6	1427.4	1306.0	1199.4	1105.4	1022.0	947.7	881.2	821.5
30	767.6	718.9	674.7	634.4	597.6	564.0	533.1	504.7	478.4	454.2
40	431.8	411.0	391.7	373.6	356.9	341.2	326.5	312.8	299.9	287.7
50	276.3	265.6	255.5	245.9	236.9	228.4	220.3	212.6	205.4	198.5
60	191.9	185.7	179.7	174.1	168.7	163.5	158.6	153.9	149.4	145.1
70	141.0	137.1	133.3	129.6	126.2	122.8	119.6	116.5	113.6	110.7
80	107.9	105.3	102.7	100.3	97.9	95.6	93.4	91.3	89.2	87.2
90	85.3	83.4	81.6	79.9	78.2	76.6	75.0	73.4	71.9	70.5
100	69.1	67.7	66.4	65.1	63.9	62.7	61.5	60.3	59.2	58.1
110	57.1	56.1	55.1	54.1	53.2	52.2	51.3	50.5	49.6	48.8
120	48.0	47.2	46.4	45.7	44.9	44.2	43.5	42.8	42.2	41.5
130	40.9	40.3	39.7	39.1	38.5	37.9	37.4	36.8	36.3	35.8
140	35.2	34.8	34.3	33.8	33.3	32.9	32.4	32.0	31.5	31.1
150	30.7	30.3	29.9	29.5	29.1	28.8	28.4	28.0	27.7	27.3
160	27.0	26.7	26.3	26.0	25.7	25.4	25.1	24.8	24.5	24.2
170	23.9	23.6	23.4	23.1	22.8	22.6	22.3	22.1	21.8	21.6
180	21.3	21.1	20.9	20.6	20.4	20.2	20.0	19.8	19.5	19.3
190	19.1	18.9	18.7	18.5	18.4	18.2	18.0	17.8	17.6	17.4
200	17.3	17.1	16.9	16.8	16.6	16.4	16.3	16.1	16.0	15.8

it has to be checked that:

$$\frac{\omega N}{A} + 0.9 \frac{Ne}{W_c} \leq \sigma_{adm} \qquad (7.67)$$

In the case in which $y_t > y_c$ (case c of Fig. 7.102), the following condition is checked (in addition to that of Eq. 7.67):

$$\frac{\omega N}{A} + \frac{300 + 2\lambda}{1000} \frac{Ne}{W_t} \leq \sigma_{adm} \qquad (7.68)$$

Fig. 7.102

W_c and W_t are the section moduli with respect to compression and tension respectively due to the bending moment Ne.

7.4.4.3 Welded members

The methods previously explained, particularly that of ECCS, can be adapted to the case of welded members provided that the reduced cross-sectional properties are used.

In the case of longitudinally welded members, the expressions for the reduced cross-sectional properties A_{red}, I_{red} and W_{red} to be used when calculating N_c, M_u are given by Eqs 7.52–7.54. Transversely welded built-up members can be approximated as consisting of a material which has a reduced design strength $f_{d,red}$ corresponding to the material of the heat-affected zone.

7.5 Plates

7.5.1 Physical Behavior

Plates are defined as those structural elements which are characterized by having two dimensions greater than the thickness. The stress field in these elements lies in their midplane.

These structural elements can be found in several structural systems (Figs 7.103–7.105), such as those of built-up sections, hollow extruded

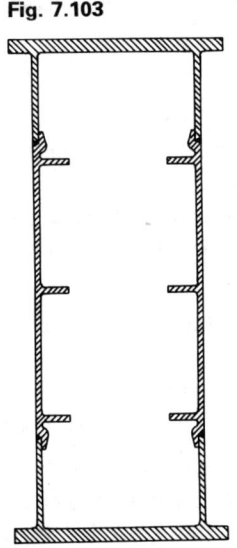

Fig. 7.103

Aluminum alloy structures

Fig. 7.104

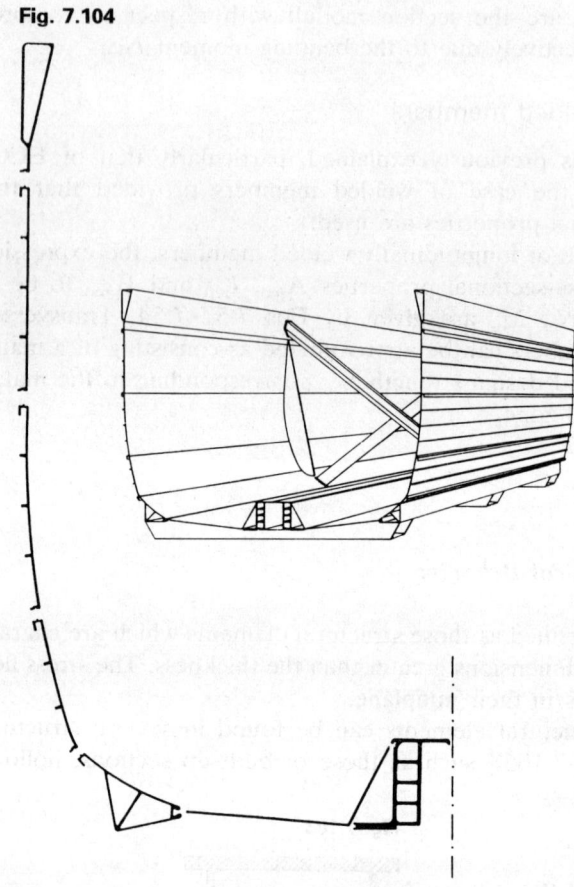

profiles, stiffened plates and shells. The plate is often stiffened by longitudinal or transverse stiffeners (stiffened or orthotropic plates). The individual parts of a section (e.g. the web) can also be considered as plates. Local and global buckling of these parts has to be analyzed (Fig. 7.106). Plates subjected to compression have to be checked against different modes of buckling:

(a) Buckling of the individual parts of a section, or the single panel between stiffeners
(b) Local buckling of the stiffeners
(c) Buckling of the stiffeners and panels together outside their plane

Even though there are many cases, which are difficult to list fully, two main cases can be identified: local buckling of the single plate element (case a) or of the single stiffener (case b), as well as global buckling of the

Stability of structural elements

Fig. 7.105

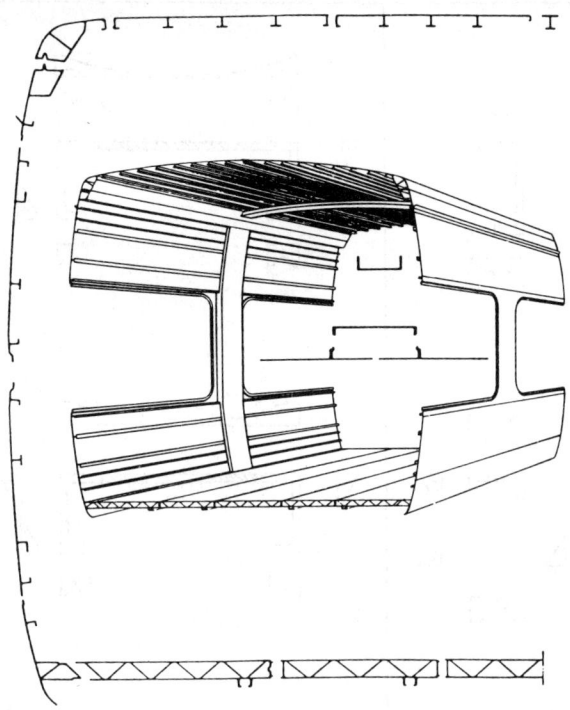

Fig. 7.106

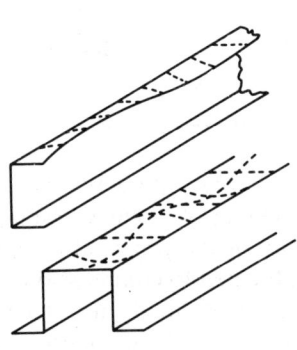

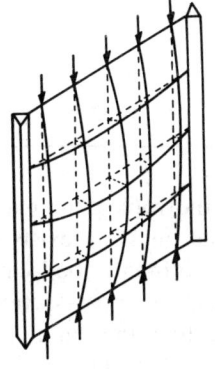

a)   b)

Fig. 7.107

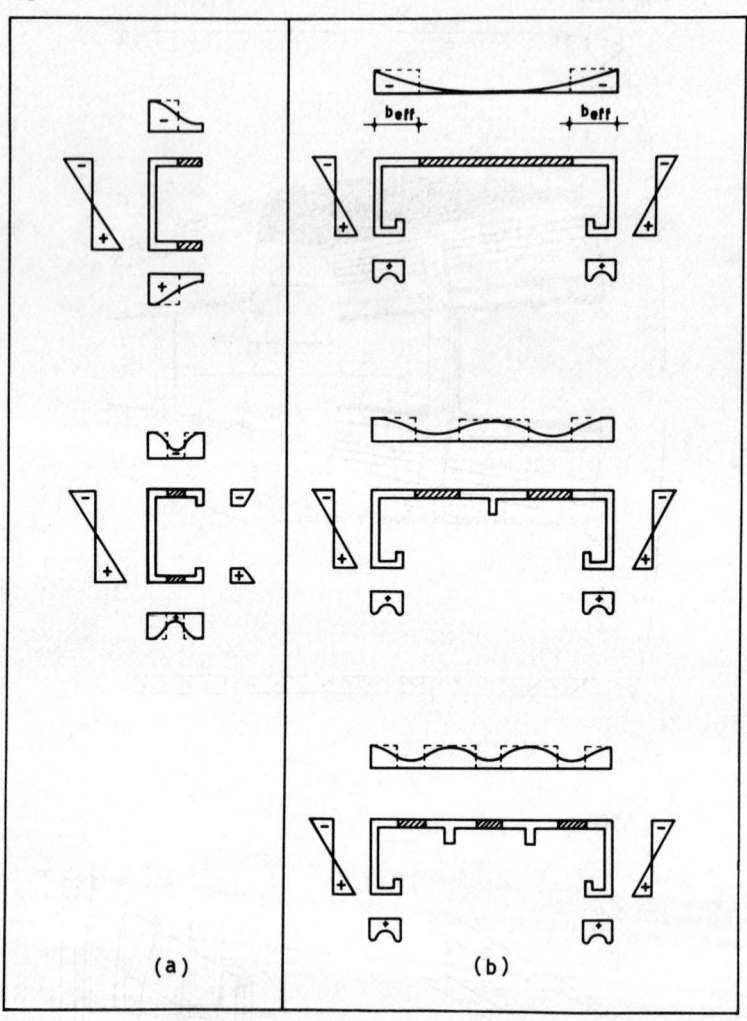

entire orthotropic plate (case c). However, some interaction between local and global buckling exists and local buckling of a single component can affect the ultimate load-carrying capacity of the principal element.

If part of the cross-section of a bar subjected to compression undergoes local buckling, the bar cannot reach its collapse load. In order to avoid a decrease of load-bearing capacity because of this interaction, the values of the ratio b/t which correspond to local and global buckling together can be identified by using the equal risk criterion with respect to both phenomena. In the case of bent members the local effects are not always

Stability of structural elements

negative. With respect to the ratios b/t of the compressed parts, two cases can be identified:

Buckling occurs after the structure achieves its maximum load-bearing capacity

Buckling occurs before the structure reaches its maximum load-bearing capacity

In the first case there is no reduction in load-bearing capacity. However, depending upon the values of the b/t ratio, local buckling can decrease ductility of the structure because of the prevention of the redistribution of stresses due to plasticity.

In contrast, for the second case, with very high values of b/t (thin profiles), local buckling occurs in some regions of the cross-section in which it is no longer possible to reach the elastic limit stress $f_{0.2}$ (Fig. 7.107). Thus, local buckling decreases the section modulus and hence the ultimate capacity of the structure.

The ultimate capacity of stiffened plates also depends upon the way they are loaded. In the case of plates under compression, collapse is reached through an unstable, increasing phenomenon because no stabilizing mechanism arises in the element. The post-critical capacity of the compressed plate is therefore limited and cannot be relied upon.

Plates under bending, however, exhibit a large post-critical range which cannot be neglected because of a stabilizing mechanism which occurs after the buckling phase.

The effects of longitudinal stiffeners in aluminum alloy structures are very important because of the ease of obtaining them by the extrusion process.

7.5.2 Compressed Parts of Bars

7.5.2.1 Width–thickness ratios

The problem of interaction between global buckling of a bar and local buckling of an element can be overcome by stipulating that the local buckling load is greater than that corresponding to global buckling of the bar:

$$\sigma_{c,loc} \geq \sigma_{c,glob} \qquad (7.69)$$

If the first term of Eq. 7.69 is set equal to the second term, values of b/t are obtained which correspond to the maximum load-bearing capacity of the bar and therefore to the most economical solution.

When applying Eq. 7.69, some theory has to be adopted to calculate the stress fields. The $\sigma_{c,glob}$ term should correspond to the collapse stress for a given slenderness λ of the buckling curve (see Section 7.2.3).

Aluminum alloy structures

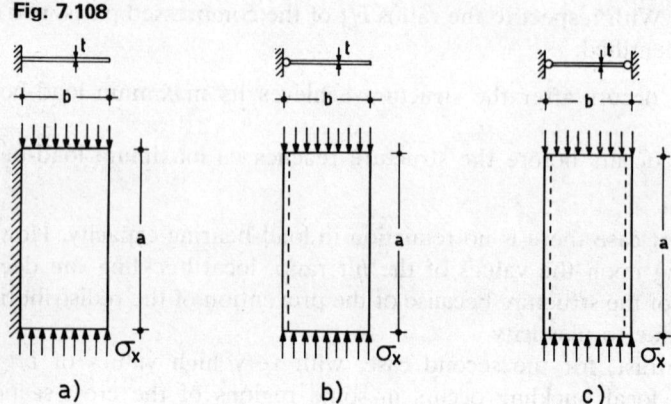

Fig. 7.108

However, in order to simplify the problem, specifications usually consider two main cases which correspond to the separate physical behavior:

$$\sigma_{c,glob} = f_{0.2} \quad \text{for} \quad \lambda < \lambda_0 \quad \text{(plastic collapse)} \tag{7.70}$$

$$\sigma_{c,glob} = \sigma_{cr} \quad \text{for} \quad \lambda > \lambda_0 \quad \text{(elastic buckling)} \tag{7.71}$$

where $f_{0.2}$ is the conventional elastic limit of the material, σ_{cr} is the critical Euler stress, and $\lambda_0 = \pi\sqrt{(E/f_{0.2})}$. In this way the bars are set into two categories: slender columns which are subjected to Euler buckling, and nonslender columns whose ultimate capacity is governed by the material strength.

The $\sigma_{c,loc}$ term can also be obtained by the elastic theory of the simply compressed panel (Fig. 7.108):

$$\sigma_{c,loc} = \sigma_{cr} = k \frac{\pi^2 E}{12(1-\nu^2)} \left(\frac{t}{b}\right)^2 \tag{7.72}$$

where k is a coefficient which depends upon the support conditions. In the cases of Fig. 7.108, it is equal to:

(a) $k = 1.28$
(b) $k = 0.425$
(c) $k = 4.00$

If Eq. 7.70 or Eq. 7.71 is made equal to Eq. 7.72, the following expression is obtained for b/t:

$$\frac{b}{t} = \pi \sqrt{\left[\frac{kE}{12(1-\nu^2)\sigma_{c,glob}}\right]}$$

which becomes

$$\frac{b}{t} = 251.53\sqrt{k}\sqrt{\left(\frac{1}{\sigma_{c,glob}}\right)}$$

Fig. 7.109 Values of b/t

$f_{0.2}$ (N mm^{-2})	b/t (a)	b/t (b)	b/t (c)
80	31.8	18.3	56.2
125	25.5	14.7	45.0
160	22.5	13.0	39.8
200	20.1	11.6	35.6
260	17.6	10.2	31.2
280	17.0	9.8	30.1

or

$$\frac{b}{t} = 1.533\sqrt{k}\sqrt{\left(\frac{G}{\sigma_{c,\text{glob}}}\right)}$$

for $E = 70\,000$ N mm^{-2} and $\nu = 0.3$, since:

$$G = \frac{E}{2(1+\nu^2)}$$

is the modulus of elasticity in shear. If $\sigma_{c,\text{glob}} = f_{0.2}$, the values of Fig. 7.109 result.

The elastic analysis can be extended to the inelastic range if it is assumed that:

$$\sigma_{c,\text{loc}} = \left(\frac{t}{b}\right)^2 G_t \qquad (7.73)$$

where G_t is the tangent modulus value of G in the inelastic range. If it is assumed that

$$G_t = \frac{2G}{1+[E/E_t/4(1+\nu)]} \qquad (7.74)$$

which was proposed by Lay [66], the tangent modulus E_t can be obtained by the Ramberg–Osgood law (Eq. 2.32), assuming ultimate conditions at $\varepsilon = 2$ percent and setting $E = 70\,000$ N mm^{-2} and $10n = f_{0.2}$ N mm^{-2}. In this way the b/t values given in Fig. 7.110 [67] are obtained.

Another expression for b/t derived by Lay from torsional buckling theory, which is valid for a flange restrained by the web, is the following:

$$\frac{b}{t} = \sqrt{\left\{\frac{3.15}{3+f_t/f_{0.2}}\frac{1}{\varepsilon_{0.2}}\frac{1}{1+[E/E_t/4(1+\nu)]}\right\}} \qquad (7.75)$$

where $f_{0.2}$ and f_t are the elastic limit and the ultimate strength of the

Aluminum alloy structures

Fig. 7.110 Values of b/t

$f_{0.2}$ (N mm^{-2})	E_t (N mm^{-2})	G_t (N mm^{-2})	b/t
80	504	1940	4.29
125	442	1710	3.39
160	428	1660	3.01
200	418	1620	2.70
220	418	1620	2.59
260	422	1640	2.41
280	419	1630	2.32

material, respectively, and $\varepsilon_{0.2}$ is the deformation corresponding to the stress $f_{0.2}(f_{0.2}/E)$.

If the same values of the tangent modulus E_t of Fig. 7.110 are used in Eq. 7.75, the b/t values given in Fig. 7.111 are obtained. These values correspond to a support condition which falls between cases (a) and (b) of Fig. 7.108.

Generally it is observed that there is a large deviation of the elastic analysis from the inelastic one. However, it seems better to refer to inelastic results which are more conservative and closer to experimental evidence.

7.5.2.2 Equivalent slenderness method

An alternative approach to the determination of b/t ratios is the equivalent slenderness method. This approach, already used in French specifications (DTU régles Al) (ref. [33] in Chapter 3) and British specifications (BS CP 118) (ref. [7] in Chapter 3), has also been used by the ECCS committee in its European recommendations [64].

In order to avoid local buckling of the compressed regions of columns,

Fig. 7.111 Values of b/t

$f_{0.2}$ (N mm^{-2})	f_t (N mm^{-2})	b/t
80	180	4.4
125	280	3.3
160	220	3.1
200	280	2.8
220	300	2.6
260	320	2.5
280	360	2.4

beam columns or beams, the same approach as used for checking buckling of columns is applicable provided that the critical stress is calculated for a conventional slenderness ratio given by:

$$\lambda_{eq} = \gamma b/t \qquad (7.76)$$

where b and t are the width and thickness respectively of the element of the cross-section in which local buckling is likely to occur. γ is the local buckling coefficient, which can be derived from the condition that the critical load of the entire bar is equal to that of a component of the bar itself (the panel having b and t dimensions):

$$\frac{\pi^2 E}{\lambda_{eq}^2} = \rho k \frac{\pi^2 E}{12(1-\nu^2)} \left(\frac{t}{b}\right)^2 \qquad (7.77)$$

By comparing this formula with Eq. 7.76:

$$\gamma = \sqrt{\left[\frac{12(1-\nu^2)}{\rho k}\right]} \qquad (7.78)$$

where k depends upon the support conditions of the panel (see Fig. 7.108), and ρ depends upon the loading condition on the panel (see Fig. 7.112) and is equal to 1 in the case of uniform load.

Actually, webs can be considered as pin-ended elements (case a of Fig. 7.112, $k = 4$), whereas flanges can be considered as pin-free elements (case b of Fig. 7.112, $k = 0.425$). In the case of cross-sections built up from these components, the value assumed for the conventional slenderness ratio λ_{eq} will be the largest value obtained by considering separately each component which is partially or completely compressed.

The coefficient ρ, which affects the expression for γ (Eq. 7.78), can be calculated in the following way. After bifurcation, the same deformed shape is assumed for all of the cases of Fig. 7.113, which correspond to groups (a) and (b) of Fig. 7.112. Therefore the same elastic deformation energy is associated with this deformed shape. The work of external loads, on the other hand, varies with respect to the load distribution. The values of the coefficient ρ can be calculated as the ratio:

$$\rho_i = \frac{E_0}{E_i} \qquad (7.79)$$

between the work done by external loads E_0 for a uniformly distributed load and that for a given distribution E_i for the same support condition.

If a buckled shape is assumed which results in a parabolic curve in the plane of the panel along the loaded edges, the values of the work of the external loads E_i and of the coefficient ρ_i are as given in Fig. 7.113 (columns 3 and 4). The values of γ given in column 5 are obtained by

Aluminum alloy structures

Fig. 7.112

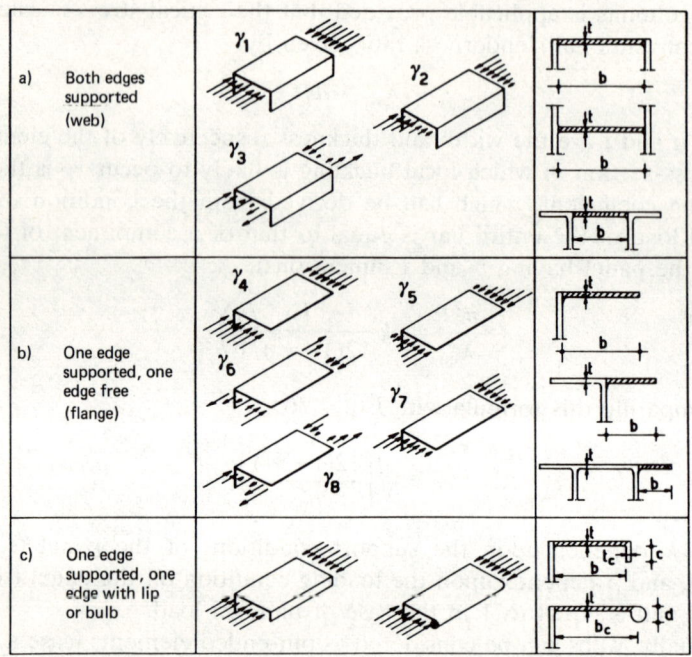

Fig. 7.113 Gamma coefficients

Boundary conditions	Case	E_i	ρ_i	Coefficient γ		ECCS	
				Theoretical	CP118	HT	NHT
(a) Both edges are simply supported (web) $k = 4$	1	1/6	1	1.652	1.60	1.50	1.20
	2	1/12	2	1.168	1.20	1.00	0.85
	3	1/32	5.33	0.715	0.70	0.45	0.45
(b) One edge supported and one edge free (flanges) $k = 0.425$	4	1/3	1	5.069	5.10	5.10	4.70
	5	1/4	1.333	4.390	4.40	4.40	4.20
	6	17/96	1.883	3.693	3.60	3.70	3.40
	7	1/12	4	2.534	2.50	2.60	2.40
	8	1/96	32	0.896	1.80	1.80	1.30

using Eq. 7.78. These values are then compared with those given in British specifications (BS CP 118) and the ECCS European recommendations. These specifications give different values of γ for non-heat-treated (NHT) and heat-treated (HT) alloys.

With the exception of case 8, the agreement between the theoretical results and the values given in specifications can be considered satisfactory. However, it is not understood why in some cases recommended values are coincident with theory and in other cases they differ.

When the edges of a profile are reinforced by lips or stiffeners (case (c) of Fig. 7.112), the following procedure can be adopted. When the flanges are reinforced by stiffeners, the value to be used for the effective slenderness ratio λ_{eq} is the highest from those calculated in the following ways:

(1) Considering the flange as a web element with a uniform stress distribution $(\gamma = \gamma_1)$
(2) Considering the stiffener as a flange element $(\gamma = \gamma_4)$
(3) Considering the interaction between web and stiffener, with

$$\gamma = \gamma_4[1 - c^2/(85 t_c^2)] \qquad (7.80)$$

where c and t_c are the width and the thickness of the stiffener, respectively

In the case of a flange reinforced by a lip, the value to be used for the effective slenderness ratio λ_{eq} is the highest from those calculated as follows:

(1) Considering the flange as a web element with uniform stress distribution $(\gamma = \gamma_1)$ having width $b = b_c + t/2$
(2) Considering the interaction between flange and lip with

$$\gamma = \gamma_4[1 - d^2/(\delta t^2)] \qquad (7.81)$$

where d is the diameter of the lip, t is the thickness of the web, and δ is a coefficient equal to 275 in heat-treated alloys and 310 in non-heat-treated alloys.

A more refined evaluation of the γ coefficient is given by British specifications (BS CP 118) (ref. [7] in Chapter 3) for some cases of C and double T sections with reinforced and nonreinforced flanges. γ values for the entire section are given in Figs 7.114–7.117 as functions of the ratios between the geometrical dimensions of the cross-section (a, b) and the thicknesses of the different parts $(t_1$ and $t_2)$. For lipped sections (Figs 7.116 and 7.117) the γ curve is selected by also considering the value of the radius of gyration i of the lip about the axis through its centroid and parallel to the parent flange. In each case the value of the equivalent

Aluminum alloy structures

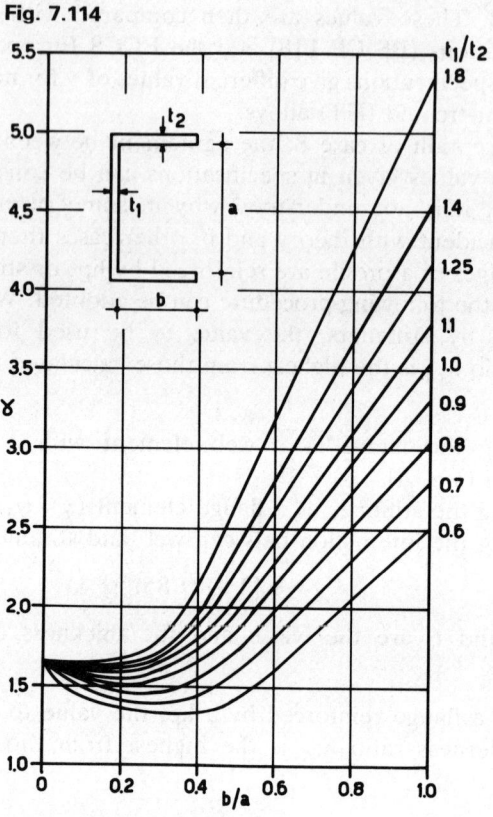

Fig. 7.114

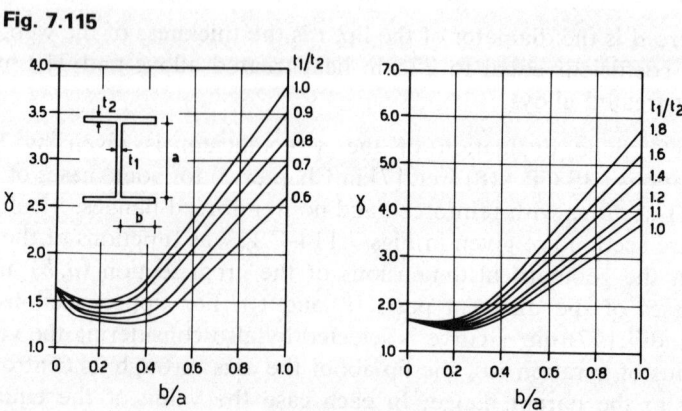

Fig. 7.115

Fig. 7.116

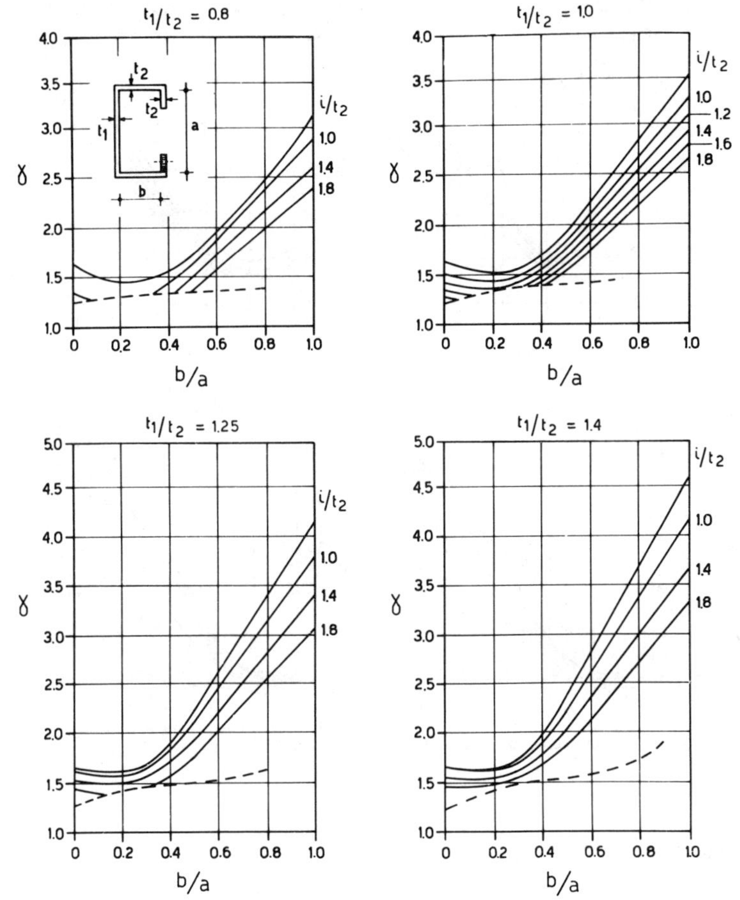

slenderness is given by

$$\lambda_{eq} = \gamma a/t_1 \tag{7.82}$$

7.5.2.3 Recent progress

A research program has recently been developed at the University of Cambridge [23] in order to determine a new calculation method for plate elements under compression. This approach was also presented to the ECCS committee TC 2-2 during revision of the European recommendations.

A basic behavioral distinction has been made between web elements supported on both longitudinal edges, and outstand elements supported on only one edge with the other edge free.

Aluminum alloy structures

Fig. 7.117

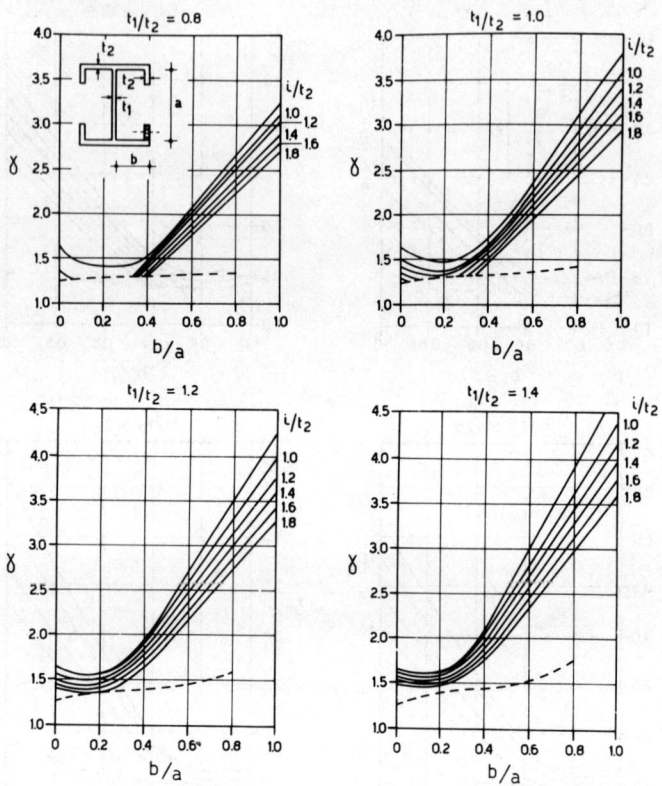

Fig. 7.118

Fig. 7.119 Calculations for web elements

	Computation data		Use data	
Curve	n	Out-of-flatness	Type of plate	$f_t/f_{0.2}$
1	15	$0.001b$	nonwelded	<1.2
2	8	$0.001b$	nonwelded	>1.2
3	15	$0.005b$	welded	<1.2
4	8	$0.005b$	welded	>1.2

The calculations for web elements are based on the definition of four theoretical strength curves, all corresponding to the simply supported free-pull-in edge condition (see the nondimensional representation of Fig. 7.118 and the data of Fig. 7.119). They cover both welded (3 and 4) and nonwelded (1 and 2) plates made of heat-treated ($n = 15$, $f_t/f_{0.2} < 1.2$) and non-heat-treated ($n = 8$, $f_t/f_{0.2} > 1.2$) alloys.

In the calculations for the welded cases a more severe out-of-flatness ($0.005b$ instead of $0.001b$) has been considered, together with a distribution of residual stresses. For curve 4 it has been assumed that the residual compressive stress is that needed to equilibrate a tensile zone of width $4t$ and stress $f_{0.2}$ at either edge. For curve 3 the assumed residual compressive stress is half of this value.

In the case of sections with outstand elements there is good evidence that the basic strength curve lies above that for the web. This is also shown by the numerous NACA test results (1950) (see ref. [23]) on short aluminum columns of I, Z and channel section. When these are plotted in Fig. 7.120 by assuming σ_{cr} for the section as a whole, it is seen that they

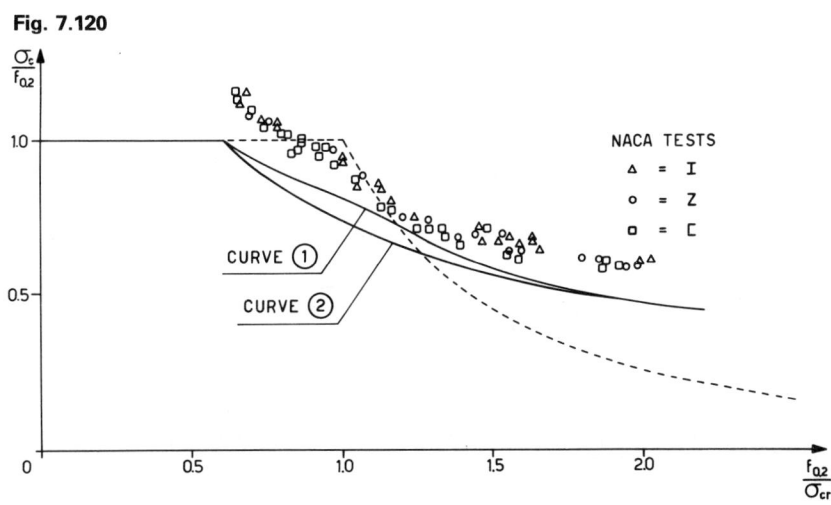

Fig. 7.120

Aluminum alloy structures

all lie above the theoretical curves for nonwelded webs (curves 1 and 2) because of the presence of the outstands.

Particular attention must be given to beams with slender outstands, which tend to fail more violently than webs due to shedding of load from the free edges. In this case (range $f_{0.2}/\sigma_{cr} > 1.5$) the use of σ_c seems to be unsafe, and it has been suggested that the compressive stress in outstands is limited to σ_{cr} (dotted curve of Fig. 7.120) rather than σ_c.

7.5.2.4 New British Standard method

On the basis of the research at Cambridge the British Standard for the new edition of the revised code CP 118 will incorporate a simplification of the theoretical results (see Section 7.5.2.3). It introduces three design curves A, B, C which cover both web and outstand elements in columns and beams for all aluminum alloys, both nonwelded and welded (Fig. 7.121).

The highest curve A has been introduced for outstands in columns. Curve B represents the average of curves 1 and 2 of Fig. 7.118. Curve C is virtually equivalent to curve 4. They correspond to the following cases:

Curve A	nonwelded outstands
Curve B	nonwelded webs, welded outstand
Curve C	welded webs

These curves may be represented by the empirical equation

$$\sigma_c = 1.1 f_{0.2}\left(\frac{1}{\beta} - k_1 \frac{1}{\beta^2} + k_2 \frac{1}{\beta^3}\right) \tag{7.83}$$

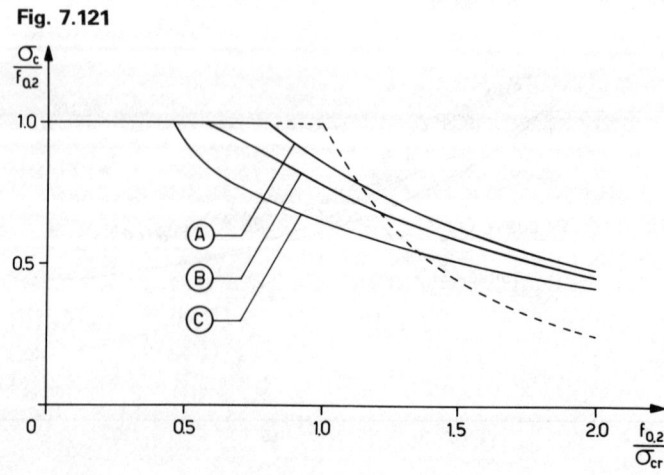

Fig. 7.121

where

	A	B	C
k_1	0.22	0.31	0.53
k_2	0	0.02	0.12

The β values are taken as follows:

web elements $\quad \beta = 0.52\sqrt{(f_{0.2}/E)}(b/t)$

outstands $\quad \beta = 1.59\sqrt{(f_{0.2}/E)}(b/t)$

The actual code presentation is given by plotting $\sigma_c/f_{0.2}$ versus b/t (Fig. 7.122) for different values of $f_{0.2}$ (in N mm^{-2}). The practical use of the design curves of Fig. 7.122 is the following:

(a) Nonwelded web elements (derived from curve B)
(b) Welded web elements (derived from curve C)
(c) Nonwelded outstands in columns (derived from curve A)
(d) Welded outstands in columns (derived from curve B)
(e) Nonwelded outstands in beams (derived from curve A corrected with σ_{cr})
(f) Welded outstands in beams (derived from curve B corrected with σ_{cr}).

The check of a column having a section comprised of web and outstand elements is indertaken by evaluating the limiting stress of the section as the sum of those of each element of area A_i:

$$\sigma_c = \frac{\sum_i [\sigma_c/f_{0.2}]_i A_i}{\sum_i A_i} f_{0.2} \tag{7.84}$$

The term $[\sigma_c/f_{0.2}]_i A_i$ represents the effective area of each element, reduced because of the local buckling phenomenon. The ratio $\sigma_c/f_{0.2}$ is given for each element from the curves (a), (b), (c) and (d) of Fig. 7.122. This value for σ_c (Eq. 7.84) is assumed as the elastic limit fictitiously reduced ($f_{0.2,\text{red}} = \sigma_c$) for entering the column curve of overall buckling. For sections containing reinforced webs or lipped outstands (Fig. 7.123), two determinations of σ_c should be made:

Treating each web or outstand with its reinforcement as a single reinforced element (cases (a) and (c))
Treating every flat element as an individual web or outstand (cases (b) and (d))

The lower value of σ_c must be taken for the buckling check.

The problem of the local buckling check in beams seems to be rather more complex, owing to the uncertainty whether the elastic moment M_e or the fully plastic moment M_{pl} should be assumed as the ultimate

Aluminum alloy structures

Fig. 7.122

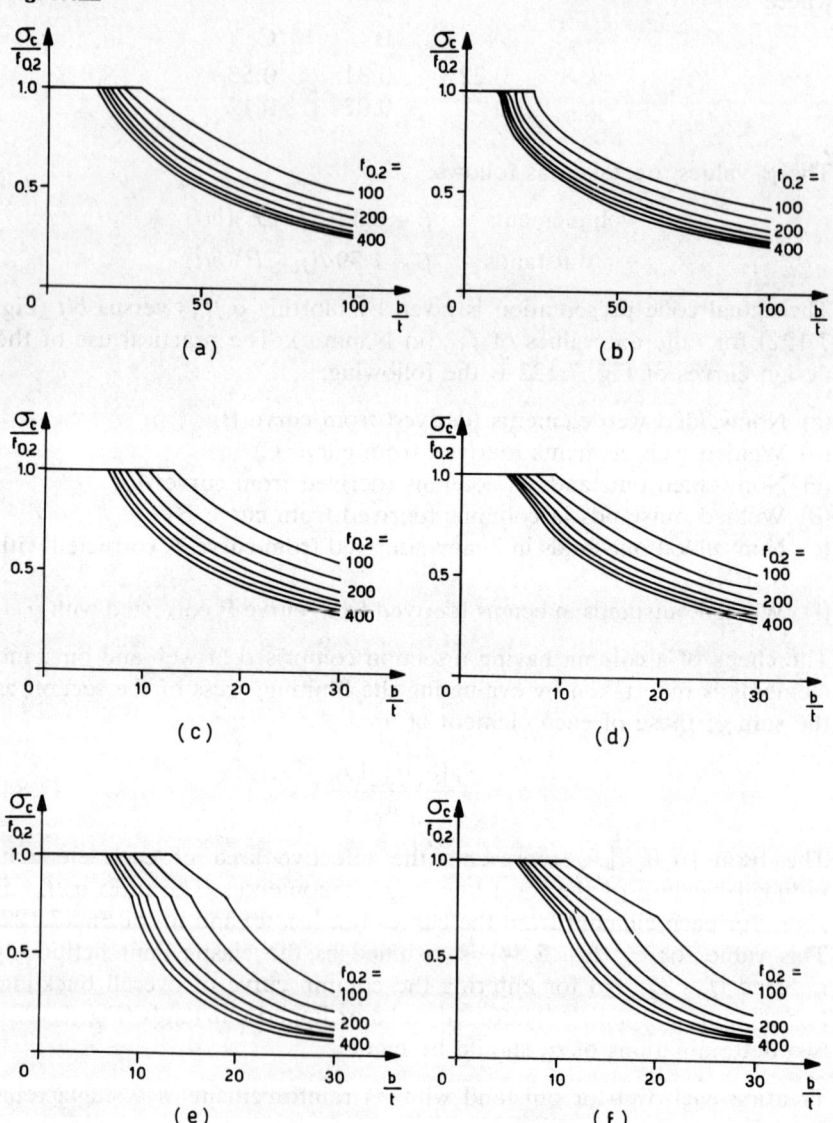

moment M_u of the cross-section:

$$M_u = \begin{cases} M_e = W f_{0.2} & (7.85) \\ M_{pl} = Z f_{0.2} & (7.86) \end{cases}$$

For calculating elastic and plastic section moduli, W and Z, the effective section is considered. It is derived from the actual section by reducing the

Fig. 7.123

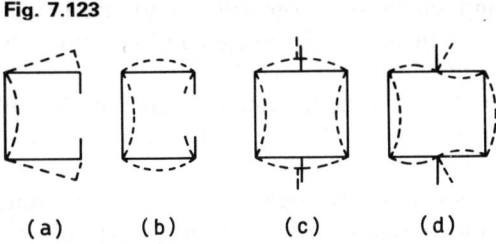

(a)   (b)   (c)   (d)

compressive parts of the flanges and web by the reduction factor $\sigma_c/f_{0.2}$. This factor is given by curves (a) or (b), (e) or (f) of Fig. 7.122.

Even though it can be argued that the use of the plastic moment (Eq. 7.86) may sometimes be unsafe, it has been justified [23], at least in the case of sections in which the compression flange consists of elements parallel to the neutral axis. In fact, the general method itself is already conservative because it neglects the hardening effect of the alloy, it assumes simply supported edge conditions for the plate elements, and it substitutes the values of σ_c with σ_{cr} for slender outstands.

7.5.3 *Stiffened Plates*

7.5.3.1 ECCS method

Despite the numerous studies and research programs on stiffened plates in metal constructions, there are as yet no results sufficiently reliable to be used in a general calculation method. This is the reason why the European Convention for Constructional Steelwork, while waiting for further developments of the problem, decided to base its methodology upon the classical theory of elasticity in the elastic range. Some coefficients are used in order to take into account the different typologies and the subsequent post-critical behavior.

The plate is assumed to have no initial geometrical imperfections, with stiffeners symmetrical with respect to the midplane of the plate. Plane stress conditions are assumed. For a given value of the load defined as the critical load, bifurcation of equilibrium occurs and the structure buckles outside its plane. Reaching this load does not represent collapse of the structure, which can experience a post-elastic range to different degrees depending upon the loading and the boundary conditions. However, elastic theory cannot take into account this behavior. The coefficients provided by ECCS have to be applied to the critical load and they are reduction coefficients if the post-critical range of the structure is negligible, or amplifying coefficients if the structure can experience a plastic range.

Aluminum alloy structures

The degrading effect of geometrical and mechanical imperfections cannot be assessed in this methodology, and therefore the restrictions on manufacturing tolerances must be respected.

It should be noted that the method explained in the following has been used in the first edition (1978) of the ECCS recommendations for steel structures. Since – as previously pointed out – this method is based upon elastic theory, it seems logical to adopt it in aluminum structures provided that the manufacturing tolerances typical of light alloys are respected.

7.5.3.2 Design of web plates

The ideal critical stresses for a rectangular plate of dimensions a, b loaded by normal stresses σ_x, σ_y along two opposite sides or by tangential stresses τ, each acting separately, are given by

$$\sigma_{cr,x,0} = k_{\sigma,x}\sigma_{cr,0}$$
$$\sigma_{cr,y,0} = k_{\sigma,y}\sigma_{cr,0} \qquad (7.87)$$
$$\sigma_{cr,0} = k_{\tau}\sigma_{cr,0}$$

$\sigma_{cr,0}$ represents the reference stress, which corresponds to the Euler critical load of an infinite plate of width b and thickness t with simple support conditions at the edges. It is given by

$$\sigma_{cr,0} = \frac{\pi^2 E}{12(1-\nu^2)}(t/b)^2 \qquad (7.88)$$

and in the case of aluminum alloys it becomes:

$$\sigma_{cr,0} = \frac{61\,940}{(b/t)^2}\,(\text{N mm}^{-2}) = \frac{6320}{(b/t)^2}\,(\text{kg mm}^{-2}) \qquad (7.89)$$

Values of $k_{\sigma,x}$, $k_{\sigma,y}$, called buckling coefficients, depend upon the stress distribution, the side ratio ($\alpha = a/b$), the boundary conditions and the stiffeners' properties.

In the case of panels without internal stiffeners and with simple support conditions, the values of the buckling coefficients can be calculated from Fig. 7.124 as a function of the ratio α and with respect to the four loading conditions which cause normal stresses along two opposite sides (cases 1, 2, 3, 4) and one corresponding to tangential stresses only (case 5). The ψ coefficient is representative of the σ normal stress field.

For the other support conditions of the panel, k values are given in Fig. 7.125 for different α values and three loading conditions.

In the case of panels with internal stiffeners, the value of k also depends upon flexural and extensional rigidities defined by Eqs 7.111 and 7.112,

Stability of structural elements

Fig. 7.124

	LOADING CONDITION IN THE PLATE		
1	Compression stresses linearly varying $0 \leq \psi \leq 1$	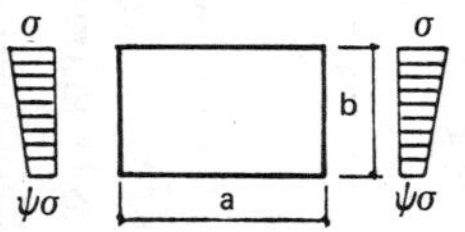 $\alpha \geq 1$	$k_\sigma = \dfrac{8.4}{1.1+\psi}$
		$\alpha < 1$	$k_\sigma = \left(\alpha + \dfrac{1}{\alpha}\right)^2 \dfrac{2.1}{1.1+\psi}$
2	Compression greater than tension $-1 < \psi < 0$		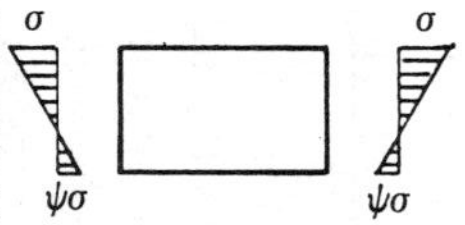 $k_\sigma = (1+\psi)k_{\sigma,1} - \psi k_{\sigma,3}$ $+ 10\,\psi(1+\psi)$ with: $k_{\sigma,1}$ from case 1 for $\psi = 0$ $k_{\sigma,3}$ from case 3
3	Maximum values of compression and tension are equal $\psi = -1$	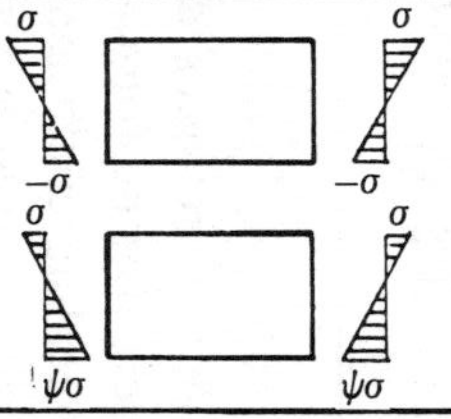 $\alpha \geq \dfrac{2}{3}$	$k_\sigma = 23.9$
		$\alpha < \dfrac{2}{3}$	$k_\sigma = 15.87 + 8.6\alpha^2 + \dfrac{1.87}{\alpha^2}$
4	Tension greater than compression $\psi < -1$		
5	Shear stress uniformly distributed	$\alpha \geq 1$	$k_\tau = 5.34 + \dfrac{4.00}{\alpha^2}$
		$\alpha < 1$	$k_\tau = 4.00 + \dfrac{5.34}{\alpha^2}$

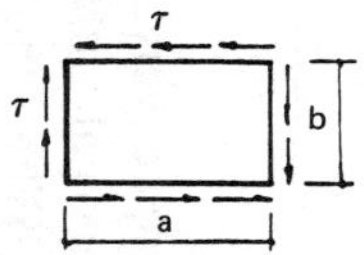

Cases 1–4: Compression and tension stresses

Aluminum alloy structures

Fig. 7.125

Buckling coefficients k_σ / Boundary conditions					
Loading conditions	$\alpha \geq 1.0$	$\alpha \geq 0.8$	$\alpha \geq 0.7$	$\alpha \geq 1.6$	$\alpha \geq 1.5$
uniform compression	4.00	5.40	6.97	1.28	0.43
triangular (0 to σ)	7.81	12.16	13.56	6.26	1.71
bending (±σ)	7.81	9.89	13.56	1.64	0.57

respectively. The expressions of the warping coefficient of panels with internal stiffeners can be found in the specific literature.

When the buckling coefficients are known, Eqs (7.87) allow the ideal values of the critical stresses acting separately to be calculated. When the stresses σ_x, σ_y and τ act together, the combination which causes buckling is characterized by the three values of the ideal critical stresses

$$\sigma_{cr,x} \qquad \sigma_{cr,y} \qquad \tau_{cr} \tag{7.90}$$

which define in each case a point of the stability domain of the panel. This locus can be defined in an approximate way by interaction formulas which can solve, on the conservative side, the most simple and frequent cases. In the case of panels without stiffeners, for example, with the same support conditions at the edges and subjected to a monoaxial state of stress, Massonnet proposes:

$$0.25(1+\psi)\frac{\sigma_{cr,x}}{\sigma_{cr,x,0}} + \sqrt{\left[0.25(3-\psi)\left\{\frac{\sigma_{cr,x}}{\sigma_{cr,x,0}}\right\}^2 + \left\{\frac{\tau_{cr}}{\tau_{cr,0}}\right\}^2\right]} = 1 \tag{7.91}$$

In the case of $\psi = 1$ (uniform compression) the linear interaction formula can be used:

$$\frac{\sigma_{cr,x}}{\sigma_{cr,x,0}} + \frac{\tau_{cr}}{\tau_{cr,x,0}} = 1 \tag{7.92}$$

The three values of Eq. 7.90 can be combined with the Von Mises criterion, in order to get the ideal stress equivalent to the biaxial stress state:

$$\sigma_{cr,id} = \sqrt{(\sigma_{cr,x}^2 + \sigma_{cr,y}^2 - \sigma_{cr,x}\sigma_{cr,y} + 3\tau_{cr}^2)} \qquad (7.93)$$

or in the monoaxial case:

$$\sigma_{cr,id} = \sqrt{(\sigma_{cr,x}^2 + 3\tau_{cr}^2)} \qquad (7.94)$$

In this case if σ_x and τ are the stresses before buckling, the reference stress

$$\sigma_{id} = \sqrt{(\sigma_x^2 + 3\tau^2)} \qquad (7.95)$$

has to be multiplied by a coefficient ν in order to obtain critical conditions:

$$\sigma_{cr,id} = \nu \sigma_{id} \qquad (7.96)$$

If it is assumed that the loads increase in a proportional way, then:

$$\frac{\sigma_x}{\sigma_{cr,x}} = \frac{\tau}{\tau_{cr}} = \frac{1}{\nu} \qquad (7.97)$$

If the two parts of Eq. 7.91 are multiplied by $1/\nu$ from Eq. 7.97, and the expression for ν obtained is substituted in Eq. 7.96, the following results:

$$\sigma_{cr,id} = \frac{\sqrt{(\sigma_x^2 + 3\tau^2)}}{0.25(1-\psi)(\sigma_x/\sigma_{cr,x,0}) + \sqrt{[0.25(3-\psi)(\sigma_x/\sigma_{cr,x,0})^2 + (\tau/\tau_{cr,0})^2]}} \qquad (7.98)$$

which is valid for the monoaxial state of stress. It gives:

$$\begin{aligned} \text{for } \tau = 0 & \quad \sigma_{cr,id} = \sigma_{cr,x,0} \\ \text{for } \sigma_x = 0 & \quad \sigma_{cr,id} = \sqrt{(3)}\tau_{cr,0} \end{aligned} \qquad (7.99)$$

In the case of a biaxial state of stress characterized by σ_x, σ_y and τ, an analogous procedure leads to the following formulas:

$$\sigma_{cr,id} = \frac{\sqrt{(\sigma_x^2 + \sigma_y^2 - \sigma_x\sigma_y + 3\tau^2)}}{0.25(4-\rho)(\sigma_y/\sigma_{cr,y,0}) + \sqrt{[0.25\rho(\sigma_y/\sigma_{cr,y,0})^2 + (\sigma_x/\sigma_{cr,x,0})^2]}} \qquad (7.100)$$

for $\rho \leq 1$, or

$$\sigma_{cr,id} = \frac{\sqrt{(\sigma_x^2 + \sigma_y^2 - \sigma_x\sigma_y + 3\tau^2)}}{0.25(2+\rho)(\sigma_y/\sigma_{cr,y,0}) + \sqrt{[0.25(2-\rho)(\sigma_y/\sigma_{cr,y,0})^2 + (\tau/\tau_{cr,0})^2]}} \qquad (7.101)$$

for $\rho > 1$, where

$$\rho = \frac{\tau k_{\sigma,x}}{\sigma_x k_\tau} \qquad (7.102)$$

Aluminum alloy structures

and is a function of the stress state and of the buckling coefficients of the panel.

In the case of $\tau = 0$ ($\rho = 0$), the stability domain expressed by Eq. 7.100 becomes a line:

$$\frac{\sigma_{cr,x}}{\sigma_{cr,x,0}} + \frac{\sigma_{cr,y}}{\sigma_{cr,y,0}} = 1 \qquad (7.103)$$

whereas for $\sigma_x = 0$ Eq. 7.101 represents a parabola.

When $\sigma_{cr,id}$ is higher than the elastic limit of the material, which can be conventionally assumed equal to

$$f_e = 0.8 f_{0.2} \qquad (7.104)$$

$\sigma_{cr,id}$ has to be replaced by a reference stress

$$\sigma_{cr,red} = \eta \sigma_{cr,id} \qquad (7.105)$$

reduced with the appropriate corrective coefficient $\eta < 1$, given by the ratio

$$\eta = \sqrt{\frac{E_t}{E}} \qquad (7.106)$$

to take into account the inelastic range. This reduction factor η is a function of the stress state it is referred to. In fact, if the Ramberg–Osgood law is assumed to be valid, the tangent modulus can be obtained in the form

$$E_t = \frac{d\sigma}{d\varepsilon} = E \left[\frac{1}{1 + (0.002nE/f_{0.2}^n)(\sigma_{cr,red})^{n-1}} \right] \qquad (7.107)$$

calculated for the stress state

$$\sigma = \sigma_{cr,red}$$

If Eq. 7.107 is substituted in Eq. 7.106 and then in Eq. 7.105, the following relation is obtained:

$$\sigma_{cr,red}^2 + \frac{0.002nE}{f_{0.2}^n} \sigma_{cr,red}^{n+1} = \sigma_{cr,id}^2 \qquad (7.108)$$

which allows $\sigma_{cr,red}$ to be calculated.

Having calculated the reference ideal stress of the elastic theory $\sigma_{cr,id}$ or its reduced value $\sigma_{cr,red}$ (when $\sigma_{cr,id} > 0.8 f_{0.2}$), the ECCS method suggests that this stress is corrected with an appropriate coefficient α_c, which has to be smaller than α_c^*. In order to check buckling of the panel it has to be ensured that

$$\sigma_{id} < \alpha_c^* \eta \sigma_{cr,id} = \alpha_c^* \sigma_{cr,red} \qquad (7.109)$$

since, in the more general case,

$$\sigma_{id} = \sqrt{(\sigma_x^2 + \sigma_y^2 - \sigma_x\sigma_y + 3\tau^2)} \tag{7.110}$$

The criteria to be used to define the corrective coefficient α_c^* are given in Section 7.5.3.3.

7.5.3.3 Design of stiffeners

As was anticipated in Section 7.5.3.1, the possibility of changing the critical load through corrective coefficients, within the limits imposed by α_c^*, is strictly related to the post-critical capacity of the structure and to the extensional and flexural rigidities.

The *relative flexural rigidity* is given by the ratio between the flexural rigidity of the stiffener and that of the panel:

$$\gamma = \frac{EI_s}{bD} = \frac{12(1-\nu^2)I_s}{bt^3} \tag{7.111}$$

or for aluminum alloys ($\nu = 0.3$):

$$\gamma = 10.92 I_s/bt^3$$

where I_s is the moment of inertia of the member whose cross-section is composed of the stiffener together with the effective width b_{eff} of the panel.

The *relative extensional rigidity* is given by the ratio between the area of the stiffener A_s and the cross-sectional area of the panel:

$$\delta = \frac{A_s}{bt} \tag{7.112}$$

The experimental behavior of metallic plates showed that the stiffener very often buckles at rigidity values which are higher than those indicated by the linear elastic theory as rigid values. This is mainly due to manufacturing imperfections, to the asymmetry of the stiffeners and to the stress state which is not the perfectly plane stress condition.

If the behavior of a stiffened panel is analyzed by elastic theory, a limiting value of its relative flexural rigidity γ^* is obtained. In fact the curve relating the critical local buckling stress σ_{cr} to the rigidity γ (Fig. 7.126) shows that σ_{cr} increases up to a value σ_{cr}^*, which corresponds to the value of rigidity, and then remains constant.

γ^* is called the *optimal relative rigidity*. Depending on its value, stiffeners can be classified into two categories (Fig. 7.127):

flexible if $\gamma < \gamma^*$
rigid if $\gamma > \gamma^*$

Fig. 7.126

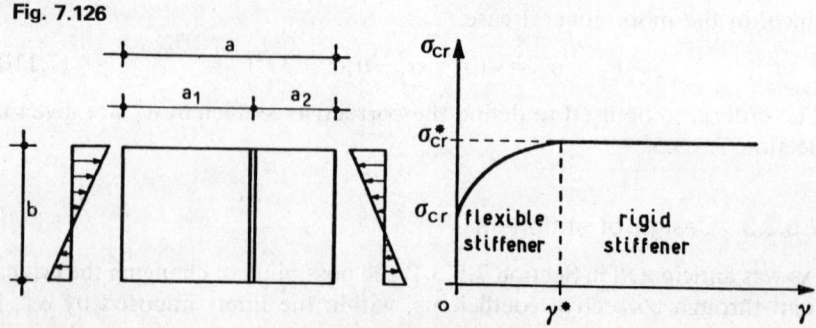

In the first case (Fig. 7.127a) local buckling occurs in the global stiffened panel, even though the critical stress is higher than without a stiffener ($\gamma = 0$).

In the second case (Fig. 7.127b) local buckling occurs separately in the two panels separated by the stiffener, with the stiffener not deforming. In this case it is not worth while increasing the rigidity in order to increase the critical stress of the overall system, because this corresponds to the smallest value from the two subpanels.

γ^* values, for the most common cases and for the different loading conditions, are given in tables as functions of the geometric properties of the panel ($\alpha = a/b$) and of the stiffeners ($\delta = A_s/bt$, $\eta = \gamma_t/\gamma_L$). The specific literature can be consulted in order to find these tables [65].

Compared with the prediction of linear theory Massonnet's test results adversely showed that stiffeners designed on the basis of the theoretical value of the optimal rigidity do not in reality deform until the critical load is reached.

On the basis of these experimental results, ECCS considered two different procedures, the first taking account of post-critical strength and the second neglecting this capacity. If the advantages of the post-critical capacity of the panels are to be utilized, the stiffeners should not deform

Fig. 7.127

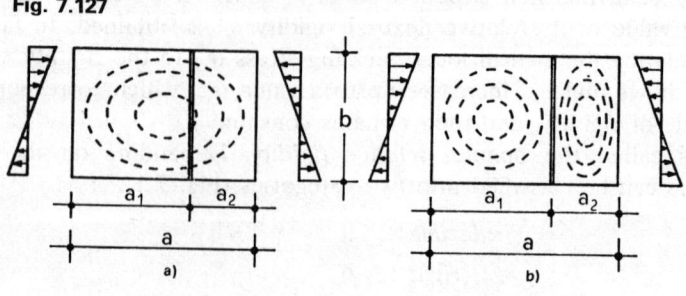

until collapse of the structure. In this case stiffeners have to be dimensioned with an effective relative flexural rigidity γ^{**} which is ξ times larger than that given by elastic theory:

$$\gamma^{**} = \xi \gamma^* \tag{7.113}$$

The values of the amplifying coefficient ξ were evaluated on the basis of the given results to take into account geometrical and mechanical imperfections. Values varied from 3 to 7 depending upon the position of the stiffeners and the loading condition. Average ξ values suggested by the ECCS recommendations for steel structures are:

$\xi = 4$ for open section stiffeners
$\xi = 2.5$ for box section stiffeners

If the stiffeners are dimensioned with:

$$\gamma > \gamma^{**} \tag{7.114}$$

then the frame of the panel will stay rigid, allowing the formation of diagonal stress bands which characterize the post-critical behavior.

If instead

$$\gamma < \gamma^{**} \tag{7.115}$$

then the stiffener is considered flexible and, to evaluate the local buckling coefficients, the value of γ of all stiffeners has to be divided by ξ. In this way the stiffener is not fully effective and only the $(1/\xi)$th part can be considered effective.

If this procedure is followed, the limiting values of the correction coefficient α_c^*, proposed by the ECCS recommendations, are obtained. They are shown in Fig. 7.128 in the column 'first method'. The parameter ψ, which defines the variation of the stress diagram, cannot be smaller than -1.

In the case of a stress state characterized by σ_x, σ_y and τ, it can be

Fig. 7.128 Values of correction coefficient

Correction factor	Simple stress states	Imposed values	
		First method	Second method
$\alpha_{c,y}^*$	transversal compression (σ_y)	0.83	0.83
$\alpha_{c,x}^*$	normal stress (σ_x)	$1.05 - 0.11(1+\psi)$	$0.95 - 0.06(1+\psi)$
$\alpha_{c,x,y}^*$	shear stress (τ)	1.05	0.95

Aluminum alloy structures

assumed that:

$$\alpha_c^* = \sqrt{\frac{(\alpha_{c,x}^*\sigma_x/\sigma_{cr,x,0})^2 + (\alpha_{c,y}^*\sigma_y/\sigma_{cr,y,0})^2 + (\alpha_{c,xy}^*\tau/\tau_{cr,0})^2}{(\sigma_x/\sigma_{cr,x,0})^2 + (\sigma_y/\sigma_{cr,y,0})^2 + (\tau/\tau_{cr,0})^2}} \qquad (7.116)$$

and the optimal rigidity γ to be used in Eq. 7.114 through the interaction formula:

$$\gamma^* = \frac{1}{\alpha_c^*}\frac{\gamma_x^*}{\nu_{c,x}} + \frac{\gamma_y^*}{\nu_{c,y}} + \frac{\gamma_{xy}^*}{\nu_{c,xy}} \qquad (7.117)$$

can be evaluated.

α_c^* in Eq. 7.117 is derived from Eq. 7.116. γ_x^*, γ_y^* and γ_{xy}^* values represent the optimal rigidity values of the stiffeners using the hypothesis of panels separately loaded by σ_x, σ_y, τ_{xy}. $\nu_{c,x}$, $\nu_{c,y}$ and $\nu_{c,xy}$ coefficients represent the ratios between the values of the reduced critical stress $\sigma_{cr,red}$, calculated with the hypothesis of σ_x, σ_y and τ_{xy} acting separately, and the corresponding real values of σ_x, σ_y and τ_{xy} which act on the structure.

As an alternative to these methods, if the designer does not wish to utilize the post-critical capacity of the structure, linear theory can be used. The values of γ have to be adopted to calculate the local buckling coefficients. However, in this case the limiting values of the corrective coefficient have to be smaller. The second procedure suggested by the ECCS recommendations provides these values, which are given in the last column of Fig. 7.128. In the case of a combined state of stress, Eq. 7.116 can be used.

In any case the corrective coefficients sometimes have to be decreased even more when the stability of the panel, either stiffened or unstiffened, is governed by the 'column effect', which emphasizes the influence of the imperfections and reduces the post-critical range. This behavior is typical of longitudinally stiffened panels with mainly longitudinal compression. In this case the Euler critical stress of the stiffened plate, imagined as a column of length a which is simply supported at its ends and whose cross-section is given by the plate and the stiffeners, is larger than one-half of the ideal local buckling stress (Eq. 7.87). This means:

$$1 > \frac{\sigma_{cr}}{\sigma_{cr,x,0}} > 0.5 \qquad (7.118)$$

independent of the compression stresses. Having

$$\sigma_{cr} = \frac{\pi^2 EI}{a^2 A} \qquad (7.119)$$

$$\sigma_{cr,x,0} = k_{\sigma,x}\frac{D}{b^2 t} \qquad (7.120)$$

where

$$\begin{cases} I = \dfrac{bt^3}{12} + \sum I_{s,L} \\ A = bt + \sum A_s \\ D = \dfrac{Et^3}{12(1-\nu^2)} \end{cases} \qquad (7.121)$$

with the positions

$$\alpha = a/b; \qquad \gamma_L = \frac{EI_{s,L}}{bD}; \qquad \delta_L = \frac{A_{s,L}}{bt} \qquad (7.122)$$

the following results:

$$\frac{\sigma_{cr}}{\sigma_{cr,x,0}} = \frac{(1-\nu^2) + \sum \gamma_L}{\alpha^2 k_{\sigma,x}(1 - \sum \delta_L)} \qquad (7.123)$$

where $\sum \gamma_L$ and $\sum \delta_L$ respectively represent the relative flexural and extensional rigidities of all the longitudinal stiffeners.

In the range defined by Eq. 7.118 the corrective coefficient has the following expression:

$$\alpha^*_{c,red} = \alpha^*_c \frac{1}{1 + \{(\sigma_{cr,red}/\sigma_c) - 1\}\{2(\sigma_{cr}/\sigma_{cr,x,0}) - 1\}} \qquad (7.124)$$

where $\sigma_{cr,red}$ is the reference stress (Eq. 7.105) and σ_c is the highest stress that the panel can withstand when it behaves like a column compressed parallel to the stiffeners.

As was previously pointed out, the α^*_c correction values to be used when verifying the panel (Eq. 7.109) are closely related to the dimensions of the stiffeners. The ECCS recommendations provide some indications for these dimensions.

In the case of transverse web stiffeners the condition of rigid behavior $\gamma_t \geq \sum \gamma^*_t$, referred to the inertia of the stiffener, gives

$$I_{s,t} \geq \xi \left[2.5 \left(\frac{h_w}{a}\right)^2 - 2 \right] at_w^3 \qquad (7.125)$$

provided $I_{s,t} > 0.5 \xi a t_w^3$, where h_w, t_w are the depth and the thickness of the web, respectively, and a is the distance between transverse stiffeners.

Another approach to the checking of transverse stiffeners is to assume that the diagonal stress fields occur in the panels. In this case, transverse stiffeners can be checked as compressed bars subjected to an axial force equal to 0.3 times the shear V. In this case the minimum area of the

Aluminum alloy structures

stiffener is given by:

$$A_s = \chi \frac{0.3V}{f_y} - 12 t_w^2 \tag{7.126}$$

where

$$\chi = \begin{cases} 1 & \text{in the case of stiffeners symmetrical with respect to the plane of the plate} \\ 2 & \text{in the case of a stiffener on one side} \end{cases}$$

provided its slenderness is smaller than 25.

For longitudinal stiffeners the condition $\gamma_L \geq \xi \gamma_L^*$ gives an optimal value of the ratio $I_{s,L}/\xi$ corresponding to the same critical ideal stresses in the single panel and in the whole stiffened plate.

7.5.3.4 Fabrication processes and tolerances

The ECCS recommendations pay particular attention to the manufacturing aspects in order to limit them to acceptable bounds (geometrical and mechanical imperfections, manufacturing tolerances, incorrect design of connections). If these bounds are exceeded the linear theory with correction factors is inapplicable.

In order to rely on post-critical capacities, all the stiffeners must be adequately connected to the panels. This can be ensured by extruding the profile together with the stiffeners or by welding the stiffeners to the panels with continuous welds.

When stiffeners cross each other, continuity conditions have to be guaranteed. Longitudinal stiffeners must pass through the transverse stiffeners or be welded to these stiffeners in the case when shrinkage is not of concern. Transverse stiffeners have to be connected to the compression or to the tension flanges if concentrated loads are present.

Besides the general criteria of the details, the ECCS recommendations provide precise manufacturing tolerances which also guarantee the applicability of linear theory. They provide the limiting values of the geometrical imperfections by means of the initial midspan deflection v_0 of the single parts which form the stiffened plate. The values of v_0 are usually limited to 1/500 of a characteristic dimension of the stiffened panel, with a maximum of 4 mm in the panel and 8 mm in the stiffeners.

References

1. Mazzolani, F. M., Proposition de travail pour établir les courbes européennes de flambement des profiles en aluminium (Working proposal to establish the

Stability of structural elements

European buckling curves for alu-alloy profiles), *Annual Meeting of ECCS Chairmen*, Puteaux, December 1971.
2. Bernard, A., Frey, F., Janss, J. and Massonnet, Ch., *Recherches sur le comportement au flambement de barres en aluminium* (Research on the buckling behavior of aluminum columns), Report CIDA, Liège-Paris, May 1971; IABSE Mem., Vol. 33-I, Zurich, 1973.
3. Mazzolani, F. M., La caratterizzazione della legge $\sigma-\varepsilon$ e l'instabilità delle colonne di alluminio (Characterization of the $\sigma-\varepsilon$ law and buckling of aluminum columns), *Costruzioni Metalliche*, 1972, No. 3.
4. Bernard, A., *Etude sur le flambement de barres industrielles en aluminium* (Study on buckling of aluminum industrial bars), ECCS Committee 16, Doc. 1.1-73-3, February 1973.
5. Flahaux, M. and Frey, F., Etude d'une serie de problèmes posés par le flambement centré plan des colonnes industrielles en alliages d'aluminium (Study of a series of problems for plane buckling of aluminum alloy industrial columns), Travail de Fin d'Etudes, University of Liège, 1974.
6. Mazzolani, F. M., *Shape effect on buckling of alu-alloy columns*, ECCS Committee 16, Doc. 16-75-2, 1975.
7. Valtinat, G. and Muller, R., *Alu-alloy welded column buckling research program: numerical computations*, ECCS Committee 16, Doc. 16-76-3, 1976.
8. Valtinat, G. and Muller, R., Ultimate load of beam-columns in aluminum alloy with longitudinal and transversal welds, *Proceedings of the 2nd International Colloquium on Stability of Steel Structures*, Liège, April 1977.
9. Mazzolani, F. M. and Frey, F., Buckling behavior of aluminum alloy extruded members, *Proceedings of the Second International Colloquium on Stability of Steel Structures*, Liège, April 1977.
10. Frey, F., *Alu-alloy welded column buckling research program: test results*, ECCS Committee 16, Doc. 16-77-3, 1977.
11. Frey, F., *Buckling, lateral buckling and eccentric buckling of alu-alloy columns, beams and beam-columns*, ECCS Committee 16, Doc. 16-77-1, March 1977.
12. Faella, C. and Mazzolani, F. M., *Buckling behavior of aluminum alloy welded columns*, ECCS Committee 16, Doc. 16-78-2, June 1978.
13. Frey, F. and Rondal, J., *Aluminum alloy buckling curves a, b, c: table and equations*, ECCS Committee 16, Doc. 16-78-1, March 1978.
14. Faella, C. and Mazzolani, F. M., European buckling curves for aluminum alloy extruded members, *Alluminio*, 1980, No. 10.
15. Faella, C. and Mazzolani, F. M., European buckling curves for aluminum alloy welded members, *Alluminio*, 1980, No. 11.
16. Mazzolani, F. M., Welded construction in aluminum. European Recommendations: welded members, *Proc. I.I.W. Colloquium on Aluminum*, Porto, September 1981.
17. Faella, C. and Mazzolani, F. M., Simulazione del comportamento di aste industriali inelastiche sotto carico assiale (Simulation of the behavior of inelastic industrial bars under axial load), *Costruzioni Metalliche*, 1974, No. 4.
18. Faella, C. and Mazzolani, F. M., Sul comportamento post-critico di aste sotto

carichi generici (On the post-buckling behavior of bars under generical loads), *2nd National Congress AIMETA*, Naples, October 1974.
19. Mazzolani, F. M., *Inelastic Buckling of Metal Bars*, Stavebnicky Casopis, September 1975.
20. Faella, C., Influenza delle imperfezioni geometriche sul comportamento instabile delle aste compresse in alluminio (Influence of geometrical imperfections on the buckling behavior of aluminum compression bars), *La Ricerca*, 1976, May–August.
21. Mazzolani, F. M., The influence of mechanical imperfections on the structural behavior of welded aluminum alloy members, *2nd Int. Conference on Aluminum Weldments*, Munich, May 1982.
22. Gilson, S. and Cescotto, S., *Experimental research on the buckling of alu-alloy columns with asymmetrical cross-section*, ECCS Committee T2, Document, 1982.
23. Dwight, J. B. and Mofflin, D. D., *Local buckling of aluminum: preliminary proposals*, ECCS Committee T2, Document, 1982.
24. Valtinat, G. and Dangelmaier, P., Column in aluminum with longitudinal and transversal welds, *2nd Int. Conference on Aluminum Weldments*, Munich, May 1982.
25. Beer, H. and Schulz, G., Die Fraglast des Planning Mittig gedruckten Stabs mit Imperfektionen, *VDI-Zeitschrift*, **III,** Nos. 21, 23, 24, 1969.
26. Frey, F., Calcul de flambement des barres industrielles (Buckling computation of industrial bars), *Bulletin Technique de la Suisse Romande*, May 1971, No. 11.
27. Mazzolani, F. M., Buckling curves of hot-rolled steel shapes with structural imperfections, *IABSE Proc., Int. Coll. on Column Strength*, Paris, November 1972.
28. Mazzolani, F. M., Influenza delle imperfezioni strutturali sulla stabilità delle colonne in acciaio (The influence of structural imperfections on the buckling of steel columns), *Costruzioni Metalliche*, 1973, No. 6.
29. Arnault, P. and Sfintesco, D., *Recherches sur le flambement des profilés en alliages légers*, CTICM et Centre Technique de l'Aluminium, Paris, October 1967.
30. Djalaly, H. and Sfintesco, D., Recherches sur le flambement des barres en aluminium, *IABSE Proc., Int. Coll. on Column Strength*, Paris, November 1972.
31. Djalaly, H., Molina, C. and Michaut, C., *Etude du flambement de flexion des profilès en alliages d'aluminium*, Centre Technique de l'Aluminium, Report No. 1389, Paris, 1975.
32. Klöppel, K. and Barsch, W., Ein Beitrag zur Bemessung von Sruckstaben aus Aluminium, *Aluminium*, 1971, No. 47, pp. 146–53.
33. Sowa, W., Untersuchungen zur knickung von Druckstaben aus Material mit beliebiger Werkstoffkennilinie (Study on the buckling of compression specimen for materials of different stress–strain law), *Aluminium*, 1971, No. 47, pp. 140–5.
34. Klöppel, K. and Barsch, W., Versuche zum Kapitel "Stabilitatsfalle" der Neufassung von DIN 4113, *Aluminium*, 1973, No. 10, pp. 690–9.

35. Templin, R. L., Sturm, R. G., Hartman, E. C. and Holt, M., *Column strength of various aluminum alloys*, ARL Technical Paper No. 1, ALCOA, 1938.
36. Osgood, W. R. and Holt, M., *The column strength of two extruded aluminum alloy H-sections*, NACA Report 656, 1939.
37. Holt, M., Tests on built-up columns of structural aluminum alloys, *Trans. ASCE*, **105**, 1940, pp. 196–219.
38. Sutter, K., (a) Die theoretischen Knickdiagramme bei Aluminium-legierungen; (b) Die exzentrische Biegeknickung und die praktischen Knickdiagramme bei Aluminiumlegierungen; (c) Das ortliche Knicken von ebenen Aluminiumprofilteilen, *Technische Rundsschau, Bern*. (a) Nos. 20, 24, 1959; (b) No. 51, 1960; (c) Nos. 3, 26, 30, 1961.
39. Brungraber, R. J. and Clark, J. W., Strength of welded aluminum columns, *Trans. ASCE*, **127**, 1962, Part II.
40. Clark, J. W. and Rolf, R. L., Design of aluminum tubular members, *J. Struct. Div., ASCE*, **ST6**, December 1964, pp. 259–89.
41. Baehre, R., *Bemessungsgrundlagen fur Aluminiumkonstruktionen nach der schwedischen Norm*, Zentralinstitut fur Schweisstechnik, DDR, 1966.
42. Baehre, R., *Theoretische Untersuchungen zum Tragverhalten von Druckstaben aus Elastoplastischem Material*, Arne Johnson Tech. paper, Nos. 21, 22, Stockholm, 1968.
43. Clark, J. W., *Formulas for design of welded columns*, ALCOA Report No. 12-71-23, 1971.
44. Marincek, M., Some remarks regarding buckling curves, *IABSE Proc. Int. Coll. on Column Strength*, Paris, November 1972.
45. Steinhardt, O., Drehknicken von Aluminium Staben im elastischen und plastischen Schlankheitsbereich, *IVBH-LISSABON Symposium*, September 1973.
46. Clark, J. W., Aluminum structures. In *Structural Engineering Handbook*, Eds. Gaylor and Gaylor, McGraw-Hill, New York.
47. Chapuis, J. and Galambos, T., Restrained crooked aluminum columns, *J. Struct. Div., ASCE*, **ST3**, March 1982, pp. 511–24.
48. Chapuis, J. and Galambos, T., Reliability of aluminum beam-columns, *J. Struct. Div., ASCE*, **ST4**, April 1982, pp. 709–27.
49. Como, M. and Mazzolani, F. M., Ricerche teorico–sperimentali sullo svergolamento nel piano e fuori del piano di profilati in presenza di tensioni residue, (Theoretical and experimental research on the plane and out of plane buckling of profiles with residual stresses), *Costruzioni Metalliche*, 1969, No. 3.
50. Mazzolani, F. M., Plane, torsional and lateral buckling of I-metal shapes taking into account residual stresses, *Proc. 4th Sci. Technol. Conf. Metal Construct.*, Warsaw, June 1970.
51. Como, M. and Mazzolani, F. M., Influence des contraintes rémanentes sur les problèmes fondamentaux de l'instabilité de forme (Influence of residual stresses on basic buckling problems), *Costruzioni Metalliche*, 1971, No. 1.
52. Faella, C. and Mazzolani, F. M., Instabilità flesso–torsionale inelastica di travi metalliche sotto carichi trasversali (Inelastic lateral–torsional buckling of metal beams under transversal loads), *Costruzioni Metalliche*, 1972, No. 6.

Aluminum alloy structures

53. ECCS Committee 8, *Stability* (in cooperation with IABSE, SSRC and CRCJ). *Manual on the stability of steel structures*, Introductory Report to the 2nd International Colloquium on Stability, Liège, 1977.
54. Dumont, C. and Hill, H. N., *The lateral instability of deep rectangular beams*, NACA Tech. Note 601, 1937.
55. Dumont, C. and Hill, H. N., *The lateral stability of equal-flanged alu-alloy I-beams in pure bending*, NACA Tech. Note 770, 1940.
56. Hill, H. N., The lateral instability of unsymmetrical I beams, *J. Aero. Sci.*, **9,** 1942, p. 175.
57. Clark, J. W. and Jombock, J. R., Lateral buckling of I-beams subjected to unequal end moments, *J. Eng. Mech. Div.*, *ASCE*, **EM3,** 1957.
58. Clark, J. W. and Rolf, R. L., Buckling of aluminum columns, plates and beams, *J. Struct. Div.*, *ASCE*, **ST3,** 1966, p. 17.
59. Hill, H. N. and Clark, J. W., Lateral buckling of eccentrically loaded I-section columns, *ASCE Trans.*, **116,** 1951, p. 1179.
60. Hill, H. N. and Clark, J. W., Lateral buckling of eccentrically loaded I-section and H-section columns, *ASCE Proc. First U.S. Nat. Congr. Appl. Mech.*, 1951, p. 407.
61. Clark, J. W., Eccentrically loaded aluminum columns, *ASCE Trans.*, **120,** 1955, p. 1116.
62. Hill, H. N., Hartmann, E. C. and Clark, J. W., Design of aluminum alloy beam-columns, *ASCE Trans.*, **121,** 1956, p. 1.
63. Marsh, C., *Strength of Aluminum*, First edn, Alcan, 1983.
64. ECCS, *European recommendations for aluminum alloy structures*, First edn, 1978.
65. Ballio, G. and Mazzolani, F. M., *Theory and Design of Steel Structures*, Chapman and Hall, 1983.
66. Lay, M. G., Flange local buckling in wide-flange shapes, *J. Struct. Div.*, *ASCE*, **91,** December 1965.
67. Valtinat, G., *Outstandings of compressed aluminum profiles*, ECCS Committee 16 Report, November 1976.
68. Mazzolani, F. M. and Frey, F., ECCS stability code for aluminum-alloy members: present state and work in progress, *Third International Colloquium on Stability of Metal Structures*, Preliminary Report, Paris, November 1983.

Index

Allowable stress 130, 138, 150, 189, 193, 203

Bar
  ideal 284, 302
  industrial 64, 283, 302, 372
Bauschinger effect 64, 111, 287, 293
Bolt
  aluminum alloy 179, 185
  high strength steel 179, 197, 200
Bolted connections 177, 186, 210
Brittle fracture 28
Buckling
  flexural–torsional 301, 369, 372, 378, 381, 383, 387
  local 214, 218, 377, 396, 402, 411, 420
  plane 301, 381, 383, 387
  twist 301
Buckling test 291, 297, 299

Camber 112
Casting
  alloys 7, 9, 10, 14
  process 6, 38
Characteristic value 140–1
Chromium 7
Cold forming 38, 41
Column
  built-up 362
  extruded 302, 339
  welded 302, 341, 354, 358
Connection
  bolted 177, 186, 210
  friction 186, 197
  riveted 176, 186, 202
  welded 107, 109, 164, 210
Copper 7, 21, 29
Corrosion 29, 31
Crippling 214

Deflection (mid-span) 66, 205, 224, 227, 230
Deformation
  axial 205
  bending 205
  initial 287
  residual 224, 227, 230
  shear 205, 208
  thermal 33
  uniform 274

Density 26
Design load combination 142, 145
Draw plate 40
Ductility 210, 213, 246, 264, 268, 274, 277

Eccentricity 67, 298, 302, 314, 324
Effective length 302, 373
Elastic limit
  distribution 107, 109, 302, 314
  reduction 155
  stress 27, 32, 48, 58, 88, 294
Elongation
  at rupture 17, 27, 58
  ultimate 155
  uniform 237
Enhancement coefficient 147
Extrusion process 5, 38

Fabrication 4
Flexural–torsional buckling 301, 369, 372, 378, 381, 383, 387
Force
  shrinkage 102, 107
  tendon 102, 106, 109
Friction connection 187, 197

Geometrical imperfection 64, 112, 229, 291, 302, 307, 321, 372, 414, 421, 424

Heat affected zone 8, 91, 153, 162, 380

Imperfection
  geometrical 64, 112, 291, 302, 307, 321, 372, 414, 421, 424
  mechanical 64, 245, 293, 302, 314, 316, 342, 353, 372, 414, 421, 424
Iron 7, 29

Limit state
  deformation 215
  plastic adaptation 219
  plastic collapse 220
  service 129, 144, 186, 206
  ultimate 129, 144, 186, 209, 216
Local buckling 214, 218, 377

Magnesium 7, 21, 23, 24, 25, 29
Manganese 22, 29

Index

Mechanical imperfection 64, 245, 293, 302, 314, 316, 342, 353, 372, 412, 414, 424
Modulus
  elasticity (Young's) 26, 32, 44, 54, 61, 88, 291, 293, 300
  hardening 116
  secant 116
  tangent 44, 50, 303
Moment
  elastic 213, 220, 247, 411
  plastic 250, 252, 264, 382, 411
  ultimate 213, 250, 252, 381, 387

Nickel 7, 29

Out-of-straightness 298, 302, 307, 314, 321, 325, 342

Plane buckling 301, 381, 383, 387
Plastic adaptation 219, 225, 229
Plastic hinge 263, 268, 274, 277
Plate 395, 413
Plate girder 206
Probability 140
Protection 30

Reduced strength zone 90, 107, 161, 245, 298, 316, 342, 354, 361, 363
Residual strain 61
Residual stress
  distribution 107, 109, 291, 317, 342, 409
  effect 245
  measurement 69, 297
  model 83
Rivet 176, 185, 203
Riveted connections 176, 186, 202

Safety factor 129, 138–9, 191, 221, 223
Section
  critical 196
  extruded 38
  throat 170
  welded 40
Section modulus
  elastic 219, 412
  plastic 220, 247, 412
  reduced 162, 380
Service limit state 129, 144, 186, 206
Shape factor 232, 247, 252

Shrinkage 102, 107
Simulation 283, 302, 307, 314, 316, 322, 327, 341
Silicon 7, 22, 24, 29
Stiffeners 215
Straightening 6, 64, 112
Strain-hardening 112, 114
  parameter 43, 51, 55
Strength
  design 141, 163, 165, 167, 169, 172, 188, 193
  slipping 202
  ultimate 27, 32, 58, 139, 155, 188, 236
Stress
  allowable 130, 138
  bearing 194, 204
  characteristic 294, 300
  collapse 303
  critical 414
  elastic limit 27, 32, 48, 58, 88, 139, 162
  failure 302
  ideal 149, 417
  proportionality 26, 44, 48
  residual 64, 68, 76, 83, 107, 109, 291, 294, 302, 314, 409
  yield 17, 27, 188
Stub column test 291, 297, 303, 325, 327

Tendon force 102, 106, 109
Thermal diffusion 71, 76
Thermal expansion 26
Tightening 178
Titanium 7
Tolerances 414, 424
Tractioning 6, 112
Trusses 207
Twist buckling 301

Ultimate limit state 129, 144, 186, 209, 216

Weld
  butt 106, 161, 164, 169
  corner 106
  fillet 107, 161, 169
Welded connections 107, 109, 164, 210
Welding 38, 75, 153
Wrought alloys 8, 9

Zinc 7, 25, 29
Zirconium 7